Dyslexia and Development

Dyslexia and Development

Neurobiological Aspects of Extra-Ordinary Brains

Edited by

Albert M. Galaburda

Harvard University Press
Cambridge, Massachusetts
London, England 1993

Printed in the United States of America
10 9 8 7 6 5 4 3 2 1

This book is printed on acid-free paper, and its binding materials have been chosen for strength and durability.

Library of Congress Cataloging in Publication Data

Dyslexia and development : neurobiological aspects of extra-ordinary brains / edited by Albert M. Galaburda.
p. cm.
Includes bibliographical references and index.
ISBN 0-674-21940-6 (alk. paper)
1. Dyslexia—Pathogenesis—Congresses. 2. Dyslexia—Pathophysiology—Congresses. I. Galaburda, Albert M., 1948–
[DNLM: 1. Brain—physiopathology. 2. Dyslexia—physiopathology.
WL 340.6 D998]
RC564.A56 1992
616.85'53—dc20
DNLM/DLC
for Library of Congress
92-49952
CIP

To Emily Landau

Contents

Foreword

The problem of developmental dyslexia came to light as interest arose in child neuropsychiatry, because of its intrinsic characteristics as well as its relationship to problems of right/left orientation, learning disorders, and personality deficits. In Barcelona, Eulalia Torras dedicated a monograph based on her doctoral thesis to the problem, and in 1964 a distinguished student of aphasiology, MacDonald Critchley, published a monograph on developmental dyslexia, which had significant influence. Later, with the contributions of Hier and colleagues and Galaburda and Kemper, the neurobiological substrate became the chief interest after the description of cytoarchitectonic abnormalities of the brain in dyslexic children, including a deviation from the standard pattern of brain asymmetry. Even now, though, it is not known whether abnormalities are present in all dyslexics, and it remains difficult to relate innate factors to educational and developmental problems of personality development.

On the basis of work by Patricia Goldman, Norman Geschwind suggested the possibility that, together with deficits related to the affected area, certain brain lesions could result in "superior functions" in the unaffected hemisphere. He commented that "such a mechanism may explain the superior right hemisphere capacities exhibited by dyslexic children." With regard to subjects with hyperlexia, who despite their apparent precociousness in reading aloud show mediocre reading comprehension, Geschwind was of the opinion that "one could postulate that the area implicated in the learning of graphemic-phonemic conversions was appropriately developed, but its connection to other brain regions was deficient."

Another issue of interest regarding developmental reading disorders has to do with their eventual similarity, or lack thereof, to acquired reading disorders in the adult. Elizabeth Warrington has referred to acquired disorders as "dyslexias," but they are more commonly called "alexias" in the United States and in the European continent. In 1895, Déjerine and Vialet made a fundamental contribution to our understanding of acquired reading disorders by singling out the condition called "pure word blindness," or agnosic alexia without agraphia. In 1958, Alajouanine, Lhermitte, and de Ribaucourt-Ducarne insisted on the differentiation between agnosic and aphasic alexias. At that stage, four types of alexias were recognized: (1) pure, or agnosic, alexia without agraphia; (2) aphasic alexia; (3) alexia with agraphia, which was also recognized by Déjerine and which corresponds to the limiting cases (complete dissociation) of Wernicke's aphasia type III of Roch-Lecours and Lhermitte; and (4) spatial alexia, linked to the syndrome of left hemineglect, which results from right-hemisphere damage.

This classification of acquired alexias in the adult, accepted over the past twenty years, has recently been replaced by one based on a deeper understanding of the reading problems themselves. The previous classification was based more on the context in which the deficit occurred than on the intrinsic characteristics of the deficit. The difference between the two will be more deeply understood if one takes into consideration the strategies used by patients to carry out "residual" reading activities and the characteristics of the paralexias—semantic, morphologic, derivative, and others—produced. That which the patient does badly—the manner in which errors are committed and the "paralexic substitutions"—has more significance to the new classification than that which he or she *cannot* do. An example is the new view of aphasia, in which the presence of stereotypies, agrammatisms, and paraphasias is more illuminating than the initial mutism that follows some strokes.

The present classification of acquired alexias begins with the analysis of paralexias as carried out in 1973 by John Marshall and Freda Newcombe. In an illustrative schematic diagram, Marshall examines three positive mental routes to reading: (1) the phonologic route, (2) the direct (holistic, gestalt) route, and (3) the lexical route, in which morphologic decomposition relies on the visual-gestalt or holistic form of the word, which is segmented into its morphometric constituents (basic form, prefixes, suffixes).

Recently, Jordi Peña from our group has proposed the following functional classification of alexias.

PERIPHERAL ALEXIAS

Neglect alexia

Attentional alexia

Spelling dyslexia (Kinsbourne and Warrington). This is characterized by the ability of the patient to spell out words, letter by letter. Each letter must be named (often aloud) before the word can be identified. In this case, what is characteristic is not the type of error but the manner in which words are identified. Indeed, the name *spelling dyslexia* appears paradoxical or erroneous, in that it is precisely by spelling that reading is accomplished.

CENTRAL ALEXIAS

Surface dyslexia, characterized by paralexias derived from the inappropriate application of the rules for graphemic-phonemic conversion (Deloche and collaborators). In this syndrome there is an alteration in the direct route from the written word to the extraction of semantic meaning. There may also be a problem in orthographic knowledge. The capacity that remains reflects the use of a route mediated by graphemic–phonemic (word)–lexical (semantic) correspondences.

Phonological dyslexia (Marie-France Beauvois and J. Derousné). This syndrome is the opposite of the former, in that there is the inability to go from spelling to sound because of a failure to apply the rules of graphemic-phonemic correspondence. Instead, the direct route is used. High-frequency words are recognized more easily, whereas the ability to read nonsense words is markedly affected. Paralexic errors consist of substitutions by visually related words.

Semantic access alexia (Warrington and Shallice). The patient has difficulty in accessing meaning from the written word, as compared with the spoken word.

Deep alexia, in which there is an alteration in graphemic-phonemic conversion. The patient is unable to read nonsense words but in this case will produce semantic paralexias (e.g., "table" is read as "chair"). Reading concrete words is superior to reading abstract words, and there is marked diminution in the ability to read verbs and particles.

The anatomic localization of these syndromes remains tentative. In general, the topography of the dyslexic lesions is viewed in the context of the topography of classical aphasic syndromes. It is also possible to find similar clinical syndromes with different lesion localizations. In principle, surface dyslexia would be found in the context of Wernicke's aphasia or during some stage of the evolution of the syndrome of pure alexia. Deep dyslexia can be seen in cases of Broca's aphasia, while phonological dyslexia would be seen in association with Broca's aphasia, conduction aphasia, or Wernicke's aphasia. In such instances, specific symptom dissociations and/or neighborhood signs would help pinpoint the exact localization. Additional anatomic verification of the deficits in autopsy specimens, as well as in life through computerized tomography scans and magnetic resonance imaging, will lead to a better understanding of the anatomical bases for the various forms of alexia.

On the other hand, there is great interest in the relationship between acquired alexias and developmental reading disorders. John Marshall has pointed out that "there has been some progress in the phenomenological description of both acquired and developmental dyslexias, within a common frame." More recently, the same author debated with John Morton concerning the question of whether developmental dyslexia reflects arrest or metamorphosis of normal reading acquisition. In other words, does the anomalous reading represent reading at an earlier stage or reading that is not ever seen in normal development? In his commentary on the debate, Galaburda wrote that in order "to defend one position or the other, it is more appropriate, it seems to me, to compare the reading characteristics of adults with acquired dyslexia and those of developmental dyslexia to those of normal children learning to read, because it is not at all clear that acquired or developmental dyslexics make solely errors of earlier stages of reading acquisition."

The work reported in this book reflects the variety of questions and approaches I have tried to present in these modest introductory comments.

Lluis Barraquer-Bordas

Preface

Plasticity is a property of the brain during development, and development lasts a lifetime. An important function of this plasticity is the adaptation of the brain to environmental demands. The brain's environment may be defined in broad terms: a gene product can be the environment of another gene; an incoming axon can cause a membrane to express a receptor on its surface; a signal from a distant endocrine gland, a hormone, can trigger a glial cell into undergoing mitosis—each of these being examples of the internal environment of brain structures. There is, moreover, the ordinary understanding of environmental influences: light coming in through the open eyes, which causes activity-dependent synaptic reorganization; or a baby's first exposure to its mother's language, which selects that language among all possible human languages to become the child's native tongue.

In these and many other ways, the environment affects the true expression of the genetic potential of the brain. Environment can be, and often is, abnormal. Brain injury is a common occurrence, and it may take place at any stage in life—before the cradle until the grave. The effect of brain injury is at least two-fold: it destroys components of the brain, the cells, connections, blood vessels (and the functions they support); and it changes the environment of the remaining "intact" brain, which leads to additional changes in brain structure and in behavior. The changes that take place in the brain as a response to injury are probably more important than the actual loss of brain tissue, particularly when the latter is relatively minor.

All types of brain changes made as a response to injury are not possible at every stage of development. It is not known in great detail how much change is possible at any given stage, but it is likely that major reorganization of the brain's cellular architecture and connectivity becomes impossible soon after birth in the human. On the other hand, changes in local connections and in the synaptic architecture may continue much longer, perhaps throughout life. Even subtler changes, detectable only in the molecular characteristics of neurons, probably are caused by ordinary environmental changes from day to day.

It is reasonably likely that the same mechanisms of plasticity triggered by injury to the brain are also active during normal environmental fluctuation, although probably to a different extent and with different consequences. In other words, the ability of the brain to react to environmental changes was probably acquired over the course of evolution for the purpose of adaptation to gentle changes, but the violent changes associated with brain injury activate the same mechanisms of plasticity. Although the mechanisms for brain plasticity in response to ordinary environmental stimuli may be adaptive, the same mechanisms when initiated by brain injury need not always produce positive behavioral results. On the one hand—as is the case of the bone scar that is stronger than the surrounding undamaged bone—plastic repair of the brain might conceivably lead to functional strengths. On the other hand, repair after brain injury at any age could lead to anomalous circuitries that could predispose an individual to epilepsy, motor and sensory disturbances, or cognitive deficits. It is even possible to conceive that the same repair could produce strengths in some areas as well as weaknesses in others.

Evidence is presented in this volume that early brain injury and the response to injury may lead to brain malformations. Brain malformations are often associated with functional deficits. This is particularly clear for the major malformations, those that are incompatible with survival or that lead to mental retardation, disorders of motor and sensory behavior, or epilepsy. The effects of minor malformations are less well understood. Oftentimes small brain defects are missed in routine autopsies, and when they have been noticed, they have been either ignored or thought to be trivial. But the need to consider them as potential culprits in developmental neurological disorders is illustrated by the fact that they have been found in individuals with epilepsy, learning disorders, fetal alcohol syndrome, and autism. Furthermore, the definition of a "minor" abnormality is based mainly on the appearance of brain matter after application of classical neuropathological techniques. These techniques cannot address questions about the disorganization of connections—local or distant—nor about significant distortion of the specific neuronal makeup of affected brain regions and their neighbors. Both types of change could indeed be accompanied by substantial and measurable disorders of behavior. In addition, animal models have been recently developed that imitate naturally occurring minor malformations in the human brain, and these have begun to show that "small" anatomic abnormalities may not be so minor after all.

A major impediment to the establishment of functional relevance for the minor malformations lies in the fact that their proposed effects may be subtle enough to require specialized testing during life. The early-developmental nature of the brain injury under question makes it likely that there will be compensation for the cognitive loss and that deficits will be seen only when specific cognitive strategies are assessed. In most cases it is not likely that a routine battery of tests will uncover meaningful information. The proper testing, however, is seldom available in medical records. Subtle learning or emotional disturbances of childhood are likely to be all but forgotten in the examination of adults, or they may be buried under the more dramatic events that led to death.

But this unfortunate state of affairs is likely to change, thanks partly to increased awareness among families and health care professionals, better assessment tools, including cognitive and morphologic assessments administered during life, and growing interest in the research community. It may be added that advances in the fields of neuroscience and cognitive science in the past fifteen years can now be gainfully applied to the generation of coherent theoretical perspectives and empirical approaches to the study of living humans and animal and computational models.

In 1988, I organized a conference in Florence to gather together a group of researchers and thinkers whose work I thought would be relevant to a research program aimed toward an understanding of learning disorders, especially developmental dyslexia. Topics discussed were language development, reading acquisition, literacy, machine models, inter-individual differences, developmental neuropathology, neurobiology of learning and memory, and more—indeed, a broad-brush approach. As predicted, many of the contributors were shy about establishing specific links between their fields of study and developmental dyslexia. Their timidity is understandable: developmental dyslexia is only informally defined in cognitive terms; there may be several unrelated types; the condition is likely to arise from a complex set of interactions between biology and culture, about which we know little; the biologic substrates of the disorder are only preliminarily described, leaving plenty of room for caution; and, last but not least of this list of problems, a well-developed cognitive neuroscience of normal reading and even language is not as yet available. This conference and related efforts to establish a cognitive neuroscience of learning disorders, however tentative, have already paid off, and they promise to continue to give results. There is growing interest among cognitivists and computational scientists in disorders of cognition, including reading—the Society for Neuroscience has an entry in its "key words" list for *dyslexia*, and the concept itself appears to have acquired tangible scientific as well as educational validity.

This book came about as a result of a second meeting, which took place in Barcelona. Each chapter emphasizes one aspect of research on the neurobiology of developmental dyslexia. My goal was to gather together a selective body of research on normal and abnormal brain development and on child neurology that would focus on the discovery that abnormalities of brain development, albeit possibly minor in comparison to others, may play a significant role in the pathogenesis of learning disorders, particularly developmental dyslexia.

The major themes in ongoing research bearing on the neurobiology of dyslexia are represented in this volume: plasticity during brain development leading to variation; the effect of sex hormones on brain differentiation; developmental neuropathology, including human descriptive neuropathology and animal models; cerebral lateralization and asymmetry; vision research; genetic factors in behavioral disorders; neuropsychology of language development; and the imaging of functioning, living brains. It is anticipated that in the next decade, "The Decade of the Brain," we will see a convergence of research findings from each of these fields and arrive at a greater understanding of the development of the brain for cognitive and emotional behaviors.

The conference in Barcelona came about as a result of the warm generosity of Emily Landau and the personal involvement and support of Caryl Frankenberger of the Fisher-Landau Foundation of New York.

The hard work and dedication of Loraine Karol, administrative assistant to our Neurological Unit at the Beth Israel Hospital, guaranteed a flawlessly organized conference, comfortable travel, and pleasant stay for all speakers and guests in Barcelona. William Baker, executive director of the National Dyslexia Research Foundation, was responsible for establishing links with dyslexia researchers in North America and Europe, for organizing the dyslexia research community in Spain around the conference, and for many of the welcome amenities. Montserrat Estilles and her able assistants were our hosts and hostesses in Barcelona. Ruth Davis, U.S. consul in Barcelona, was supportive and helped us establish contacts with Barcelona officials. Judy Sharp in Madrid carried on as hostess when some of the group continued on to the capital city.

Speakers and guests alike made the conference a success beyond all our expectations. We were fortunate to have among us, in perhaps her last public appearance, Isabelle Liberman, whose contribution to the field of dyslexia is legendary.

My colleagues Gordon Sherman and Glenn Rosen offered helpful advice during the editing stages of the book, and Eileen Moran's and Meggin Sullivan's help were invaluable during the preparation of the typescript. Finally, Angela von der Lippe, Linda Howe, Pat Pershing, and Kate Schmit of Harvard University Press demonstrated great patience and encouragement when dealing with me on the long multiauthored manuscript.

To each I offer my grateful appreciation.

Albert M. Galaburda

1

Regressive Events in Early Cortical Maturation: Their Significance for the Outcome of Early Brain Damage

Barbara L. Finlay
Brad Miller

Two features of the neocortex must be taken into account if we are to understand the mechanisms of its normal and abnormal development. First, the entire neocortex has a common organizational structure: the tangential organization of layers with their characteristic cell types, inputs and outputs, and the radial organization of the cortical column. Second, the cortex is locally differentiated with identifiable cytoarchitectonic areas that differ in their total cell complement, the proportions and sizes of cells in various layers, and specific types of input and output. We need to understand the general mechanisms that produce overall cortical structure and the means by which local differentiation develops.

The basic sequence of generative events that make the cortex is well known. Cortical neurons are generated from columnar epithelial cells residing in the ventricular zone (Sidman, Miale, and Feder, 1959). After their last cell division, neuronal cells migrate out to form the cortical plate, with each wave of cells coming to reside nearer the pial surface than the previous generation; thus the cortex is said to develop in an "inside-out" sequence (Rakic, 1974; Angevine and Sidman, 1961, 1962; see McConnell, 1988, for review). Cortical cells migrate along radial glial fibers, remaining in very close apposition to the fibers until they reach their destination (Rakic, 1971, 1972, 1974). The fundamental uniform scaffolding of the cortex, its laminar and columnar organization, is laid down in this way.

After the cortical plate is generated and in place, regressive events are significant in the continuing differentiation of the cortex. Here we will consider the regressive changes of cell death and axon retraction. A large proportion of the original cell complement that forms the cortical plate and its subplate normally dies during development. The callosal and subcortical axons of pyramidal cells typically extend their processes to a large number of locations they will not innervate by adulthood. Axon retraction and cell death serve in the production of local laminar and cytoarchitectonic differences and in the establishment of local, specialized patterns of connectivity. As is discussed elsewhere in this volume (Chapter 5), regressive events have also been considered as one mechanism involved in the establishment of cerebral lateralization. Regressive events are so timed in development as to be directly affected by complications of premature birth and other types of perinatal trauma.

In this chapter, we will review a number of experiments, principally from our own laboratory, in which we perturbed some feature of normal cortical development. These perturbations were all intended as probes for the types of processes that operate in normal cortical maturation, but they can also serve as models for likely reorganizations consequent to particular types of brain damage. We will be interested in the control of neuron number in the developing cortex, particularly as this factor interacts with developmental cell death, in early axon exuberance and retraction, and in the relationship between these two factors.

We will review briefly normal cell death in the cortex and then describe our attempts to modify cell loss in the cortex with afferent and target manipulations. Next we will describe the changes in axon-terminal distribution that parallel the findings of alterations in cell number. Finally, we will examine the alterations in axonal distribution consequent to early hypoxia-ischemia and the role that regressive events might play in these changes.

Control of Cortical Cell Number

What is the normal pattern of cell death in the neocortex?

Cell loss in the neocortex is of very significant magnitude. Absolute cell counts indicate cell losses in layers II–IV of up to 50 percent (Heumann and Leuba, 1983). Several years ago, we examined cell death in five areas of hamster neocortex: areas 17, 18b, 29b, 29d, and 27, chosen because they show a great deal of variation in final thickness and density (Finlay and Slattery, 1983). Nearly all cell death occurs in the last-generated, upper layers (II–IV), and particularly in II–III. Very little degeneration was seen in layers V–VI (also noted by Heumann and Leuba, 1983). Furthermore, the amount of degeneration seen during development in these upper layers predicted the eventual cortical cell number in each of the five areas (Figure 1.1). These results suggest that differential cell death is a factor that gives rise in part to the differences in cortical cell number between various cortical areas. We have termed this effect "sculpting." However, even in area 17 of the hamster, the area that has the greatest final number of neurons per column, there is still some cell loss, which suggests that cell death must have some other function.

Parenthetically, we note here that this discussion concerns the incidence of cell death in the cortical plate itself. There is another population of cells in the developing cortex that is also marked by transience and has attracted much recent interest. Luskin and Shatz (1985) have recently called attention to a population of cells that lie underneath the cortical plate, the subplate cells, which are generated quite early in development (Marín-Padilla, 1978; Kostovic and Rakic, 1980). These cells appear to form transient connections with developing thalamocortical afferents and may play a central role in cortical specification. There are considerable differences among species in the proportion of these cells that die, which may provide a key to their developmental function (Valverde and Facal-Valverde, 1987; Woo, Beale, and Finlay, 1991). The relationship of the mechanisms that control cell death in the subplate and in the cortical plate will clearly be of great interest.

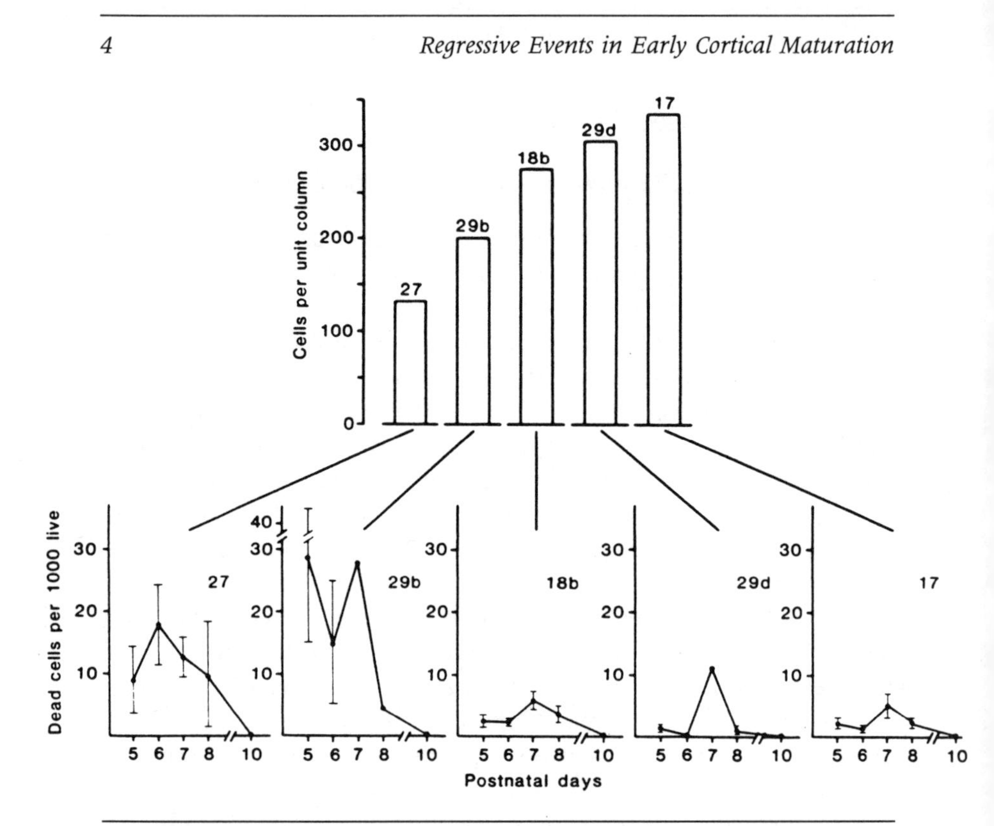

Figure 1.1 Pyknotic cell indices of various cortical areas in the hamster posterior neocortex in early development, compared with the number of cells per unit column in the same areas. Shown here are the number of neurons in the upper cortical laminae in a column (200 by 30 μm with variable height) for five cortical areas and the corresponding cell death indices for postnatal days 5–10. The upper cortical laminae include layers II, III, and IV for areas 17, 18b, and 29d, and layers II and III for agranular 29b and 27. A high incidence of early cell death in the upper laminae is associated with reduced cell numbers in those laminae at maturity. (Reprinted with permission from Finlay and Slattery, 1983.)

Efferent and afferent control of cortical cell number

A demonstrated function of cell death in other neural systems is the matching of afferent and target populations (Hamburger and Levi-Montalcini, 1949; Hollyday and Hamburger, 1976; Pilar, Landmesser, and Burstein, 1980; O'Leary and Cowan, 1984; but see Lamb, 1984). In a series of studies, we have investigated the dependence of cortical neurons on various subcortical and cortical targets. We have examined both reactive cell death and the eventual number of neurons in probes of cortical columns, both overall and by layer.

The experimental animal we have used, the hamster, allows us to make transections and ablations largely before innervation has occurred. All the ablations to be described were done on the day of birth. At this time, in hamsters, the last cortical neurons have been generated, but migration to the cortical plate will continue over the next four or five postnatal days (Shimada and Langman, 1970). Thalamocortical fibers can be found in residence under the cortical plate but have not yet innervated layer IV (Naegele, Jhaveri, and Schneider, 1988). Innervation of subcortical targets by cortical neurons has just begun, and although the corpus callosum has formed, the majority of the neurons that will send axons to form it have not yet reached their mature positions in the cortex. Cell death in the cortical plate begins around postnatal day 6 or 7, depending on area, and continues over the next several days (Figure 1.1). At postnatal day 15, when the hamster opens its eyes, the cortex appears relatively mature.

The neocortex shows evidence of a strong trophic dependence on one of its afferent sources, the thalamus. We have investigated the effect of unilateral thalamic lesions in neonatal hamsters on cortical cell numbers in the cortical projection areas of the ablated thalamic nuclei (Windrem and Finlay, 1991). By adulthood, a thalamic lesion results in a cortex of strikingly different architectonic appearance: it is thinner overall, with increased cell density, and there is a notable absence of the small stellate cells of layer IV. On postnatal day 7 (seven days after the lesion), just after the cessation of migration to the cortex and before normally occurring cell death in the cortical plate, no significant differences were found in cell number per cortical column or in the laminar composition of the cortex. Thus, the thalamus does not influence the last phases of cortical cell generation or migration to the cortex. During the period of normal cell death, pyknotic cell indices were elevated in the granular and supragranular layers and in layer VI. By adulthood, there was a significant reduction in the total number of neurons per unit cortical column. This difference was due to the absence or substantial reduction of an identifiable layer IV and variable loss of layer VI. Numbers of cells in layers II, III, and V, whose cells are neither afferent nor target for the thalamus, were unaffected.

Rakic (1988, 1990), for a related study in monkeys, reduced the number of thalamic afferents to visual cortex by removing both eyes on embryonic day 60. This reduces the number of cells in the lateral geniculate nucleus (LGN) by at least 50 percent and results in the appearance of a new area ("X"), as defined by cytoarchitectonic features, and a reduction in area 17. Rakic has interpreted this set of new cytoarchitectonic features to be the result of the interaction of the set of cortical neurons normally destined to be area 17 with a new set of afferents (or possibly, a lack of afferents). This study differs from ours in that we have found a relatively uniform alteration in the radial extent of the cortex, while Rakic finds a decrease in the tangential extent of area 17. In both cases, however, novel cytoarchitectonic organization is found. Given the many differences between these studies (type and timing of lesion, completeness of lesion, and species), and without knowledge of the manner in which the remaining lateral geniculate innervates the cortex in the monkey, it is difficult to give an appropriately detailed reconciliation of the difference in results. For this discussion, however, the important point of both studies is that the thalamus has a profound effect on cortical organization and on cortical cell number.

Additional evidence for a special role of thalamic afferents in producing local cytoarchitectonic specificity comes from transplant studies. Preliminary evidence (Schlaggar and O'Leary, 1989) has shown that when embryonic visual cortex is heterotopically transplanted into parietal cortex of neonatal host rats, the visual cortex displays "barrel-like" features typical of parietal cortex. This suggests that thalamic afferents specify certain cytoarchitectonic characteristics, in this case features of layer IV, in the developing cortex (see also O'Leary, 1989).

By contrast, removal of cortical and subcortical efferent targets of the cortex appears to have little effect on cortical cell number. Ramirez and Kalil (1985) have demonstrated that neonatal lesions of the pyramidal tract in the hamster have no effect on the final number of pyramidal cells in motor cortex, although they did observe that the somata of target-deprived cells were smaller in size than normal. We have explored target dependence of visual cortical cells in the hamster (Pallas, Gilmour, and Finlay, 1988). Neonatal unilateral deletions were made in the superior colliculus, a major target of layer V pyramidal cells in the primary visual cortex, to assess the effects of target loss on cortical cell death. The incidence of pyknotic cells during the major period of cell death (postnatal days 5–10) and in adults was monitored, as well as the total number of cortical cells in layer V and in all cortical layers combined. The target deletions were found to have no effect either on the incidence of cell death during development or on final cell number, either in layer V alone or in all cortical layers combined (Figure 1.2).

A second study in our laboratory found no effect of loss of callosal targets or afferents on the survival of cells in several cortical areas, including the area 17–18 border (Windrem, Jan de Beur, and Finlay, 1988). Callosal projections were prevented by the insertion of a foil barrier along the cortical midline in neonatal hamsters. This target loss had no effect on cell survival (number of neurons per unit column) either in the callosally projecting layers (II–III and V) or in the entire cortical column (Figure 1.3).

If we assume that all neurons do in fact require some trophic support from their efferent and afferent connections to survive, it is not difficult to come up with alternate sources of trophic support for these two populations of pyramidal cells. In the case of thalamic lesions, we have directly explored the potential for reorganization of axonal connectivity from the remaining thalamus and the corpus callosum, which we will discuss below. There is ample evidence from a number of laboratories, however, that the axons of pyramidal cells may branch widely and contact a number of terminal fields, both transiently in development and permanently in adulthood (Bates and Killackey, 1984; Caminiti and Innocenti, 1981; Innocenti, 1981; Innocenti, Berbel, and Clarke, 1988; Innocenti and Caminiti, 1980; Innocenti and Clarke, 1984a, 1984b; Innocenti, Clarke, and Kraftsik, 1986; Innocenti and Frost, 1979; Ivy, Akers, and Killackey, 1979; Ivy and Killackey, 1981, 1982; Killackey and Chalupa, 1986; O'Leary, 1989; O'Leary and Stanfield, 1985, 1986; O'Leary, Stanfield, and Cowan, 1981; Stanfield, O'Leary, and Fricks, 1982; Stanfield and O'Leary, 1985; Tolbert, 1987; Tolbert and Der, 1987; Wise and Jones, 1978). Presumably, the preservation of any one of these branches might be enough to sustain a cell.

In summary, we have found the cortex to be highly reactive to removal of its thalamic input and much less reactive (as determined by cell number) to ablations of the afferents and targets of the pyramidal cells of layers II–III and V. We view it as quite likely that the normal cell death seen in the cortex is related to the normal

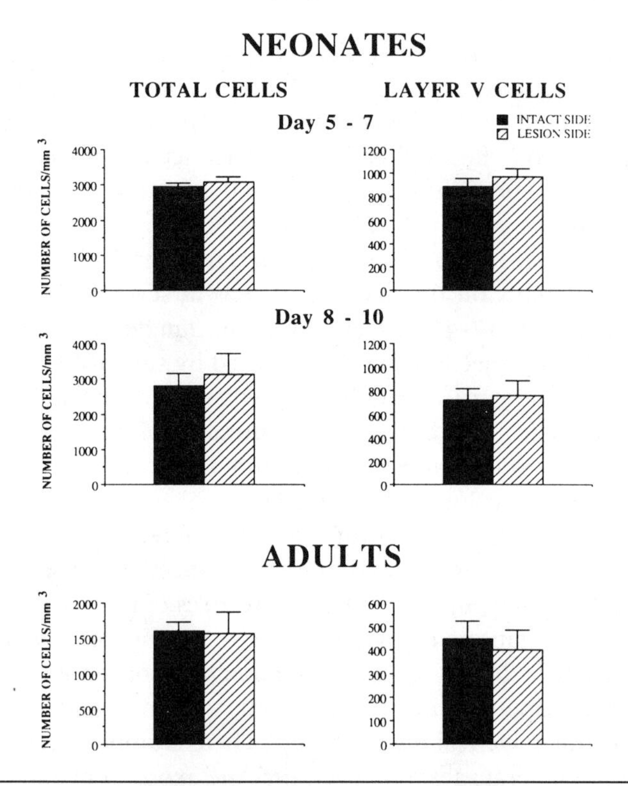

Figure 1.2 No changes were found in cell number in area 17 overall and in layer V after neonatal ablations of the superior colliculus. In all panels, the intact side of the cortex is represented by black bars, the lesion side by hatched bars. Error bars represent the standard error of the mean. *Top left:* total cells per unit column in the cortex for hamster pups at 5–7 days of age ($N = 6$). *Top right:* layer V cells per unit column for hamster pups at 5–7 days of age. *Middle left:* total cells per unit column in the cortex for hamster pups at 8–10 days of age ($N = 4$). *Middle right:* layer V cells per unit column for hamster pups at 8–10 days of age. *Bottom left:* total cells per unit column in the cortex for adult hamsters ($N = 3$). *Bottom right:* layer V cells per unit column for adult hamsters. (Reprinted with permission from Pallas, Gilmour, and Finlay, 1988.)

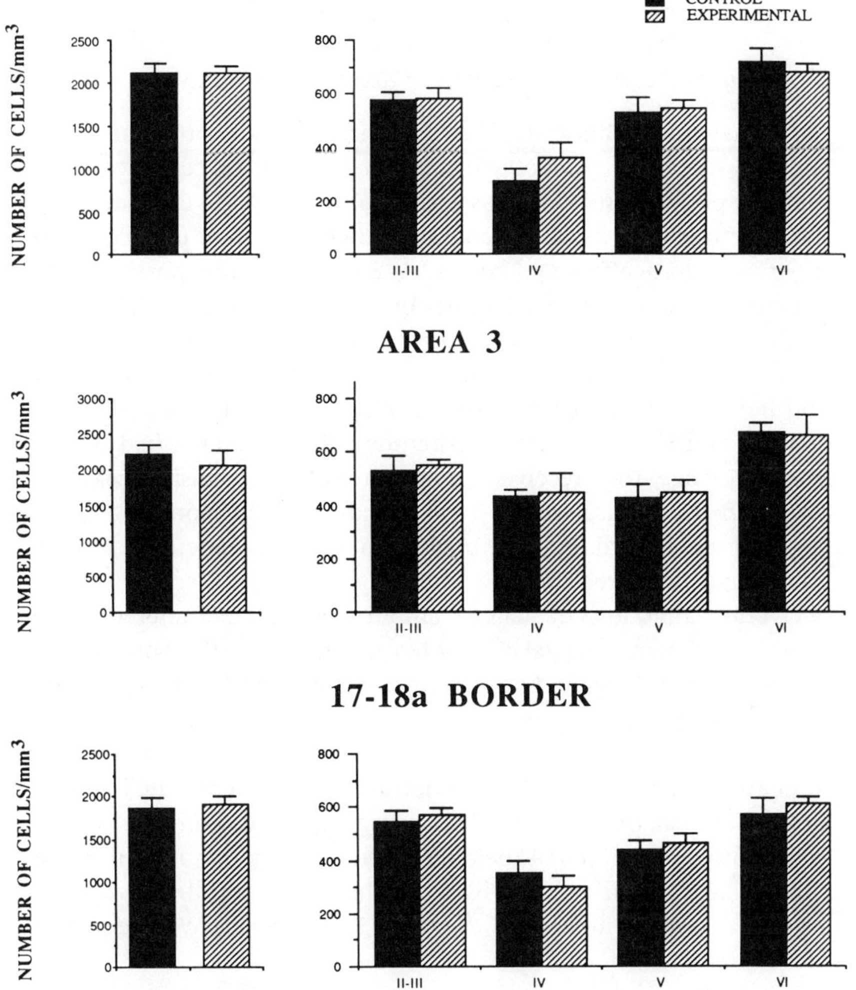

Figure 1.3 Cell number per unit column after corpus callosum transection, shown for the whole cortex *(left)* or by layer *(right)* in standardized cortical columns from three cortical regions: area 6 and the area 17–18a border, which have heavy callosal input, and area 3, which is largely acallosal. (Reprinted with permission from Windrem, Jan de Beur, and Finlay, 1988.)

matching of variable thalamic afferents to the number of cells in layer IV in the developing cortex. In the following section we will relate the observed alteration or lack of alteration in cell number to axonal reorganization consequent to the same manipulations.

Thalamic and callosal projections to deafferented cortex

Early thalamic ablation deafferents a large area of cortex, and presumably alters the competitive state for terminal space between remaining populations as well as cortical physiology. Do the remaining thalamic nuclei in this case reorganize their pattern of projections to innervate the deafferented cortex? We examined this possibility by ablating the primary visual (LGd) or somatosensory (VB) thalamic nucleus on the day of birth with an electrolytic lesion. At this time thalamic fibers are lying in the subplate under their cortical target and have not yet penetrated the developing cortical plate (Naegele, Jhaveri, and Schneider, 1988). After the animals had reached at least 30 days of age, they received an injection of horseradish peroxidase (HRP) in the cortical region corresponding to the normal afferent target of the ablated thalamic nucleus to assess any alterations in the connectivity between these two areas.

Overall, after loss of a large amount of thalamic afferents, the remaining thalamic nuclei did not reroute or expand their projections to innervate the deafferented cortex. An example of one of these animals is shown in Figure 1.4. Seven of nine experimental animals had no unusual thalamic projections to the denervated cortex. Two animals did have anomalous projections to the denervated cortex. In one animal this amounted to just a few cells from the medial geniculate and paratenial thalamic nuclei, and in the other it was a substantial proportion of cells from the anterodorsal and anteroventral thalamic nuclei (Miller et al., 1987; Miller, Windrem, and Finlay, 1991).

We also looked for changes in the strength of projections from remaining nuclei that normally project to the deafferented region. A given cortical field receives thalamic input from multiple thalamic nuclei (Herkenham, 1986), and one compensation for removing a source of primary thalamic input is that minor thalamic projections will increase. Only one of the nine animals had an increase in afferents from a normally projecting thalamic nucleus, and it was one of the two animals that demonstrated anomalous thalamic projections (Miller, Windrem, and Finlay, 1991). The mediodorsal and reuniens thalamic nuclei in one animal had a greater-than-expected number of projections to the visual cortex. Statistically, there was no evidence for any kind of numerical compensation for the deafferentation: overall, the number of cells labeled by the HRP injection in the damaged thalami was halved. Most of this projection comes from residual parts of the appropriate but partially damaged thalamic nuclei. About 10 percent of the normal labeled projections comes from diffusely projecting nuclei, which were often undamaged in the experimental animals. If only these nuclei, which are in the best position to be able to take advantage of increased terminal area, are considered, this population increases in number by about 30 percent. This is numerically trivial, however: it is only an increase from 10 percent to 13 percent in the contributions from these nuclei if they are considered with respect to the original complement.

The reason that thalamic projections are unlikely to reroute to a denervated cortical region may lie in the developmental program they follow. Studies of the development of the thalamocortical system have shown a high degree of early specificity in connectivity (Dawson and Killackey, 1985; Parnavelas and Chatzissavidou, 1981; Crandall and Caviness, 1984), though some evidence for minor transient projections exists (Bruce and Stein, 1988; Laemle and Sharma, 1986). If thalamic projections normally do not make transient exuberant projections to widespread cortical regions, the opportunity for reorganization of thalamocortical projections may be rather limited since it is often these exuberant projections which can be stabilized.

The relative absence of transient exuberant projections seen in the developing thalamocortical system contrasts quite sharply with developing cortical efferent systems. Neocortical efferent systems (subcortical, cortical, and callosal) show these transient exuberant projections to the vicinity of potential targets. The discrete pattern of connections seen in mature animals arises through development as the exuberant projections are retracted. For example, pyramidal cells in the developing visual cortex send axons down the spinal cord which are later retracted (Bates and Killackey, 1984; Stanfield, O'Leary, and Fricks, 1982). Corticocortical cells send a transient collateral projection to the contralateral as well as the ipsilateral

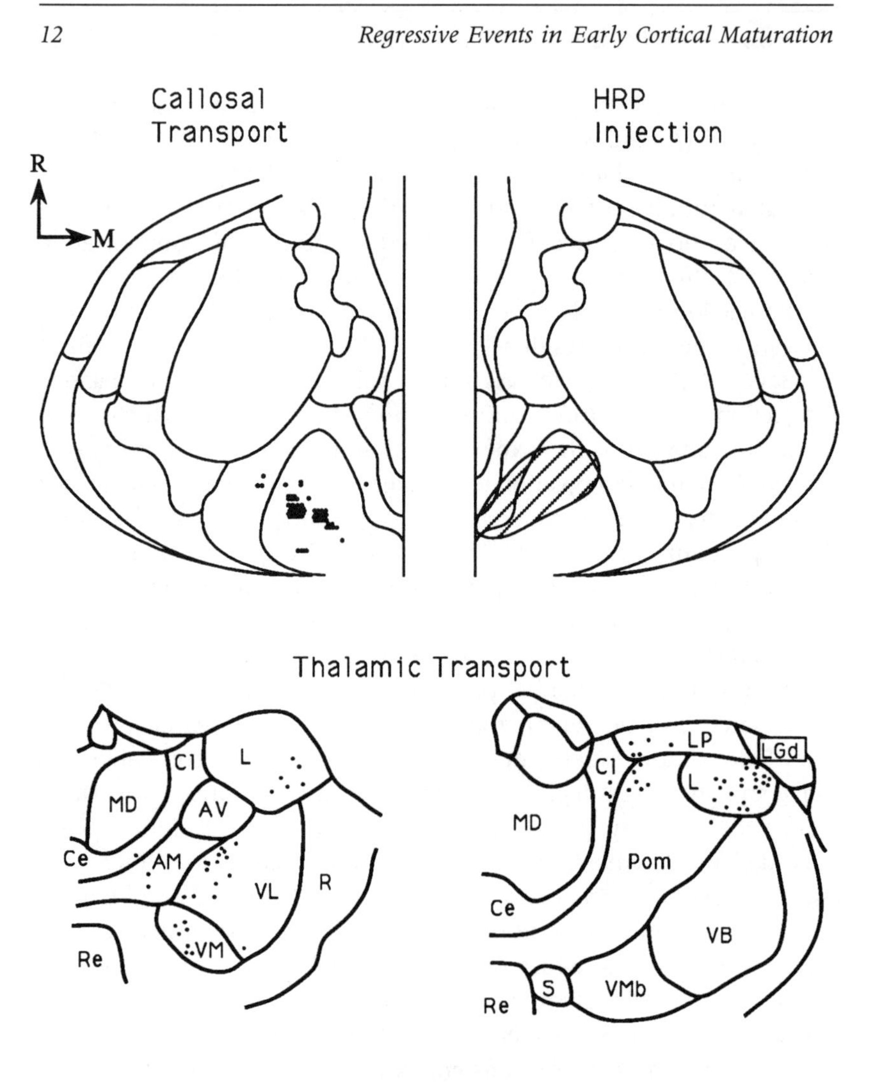

Figure 1.4 An example of thalamocortical and callosal projections to visual cortex in a normal hamster and in a hamster with neonatal damage to the visual thalamic nuclei. Upper drawings show an unrolled dorsal view of cortex, with the HRP injection site *(striped region)* in visual (posterior) cortex. The callosal projection is a tight, homotopic group of cells in the normal animal; callosal projections from the animal with neonatal thalamic damage are diffusely spread and originate from different cortical regions. The tha-

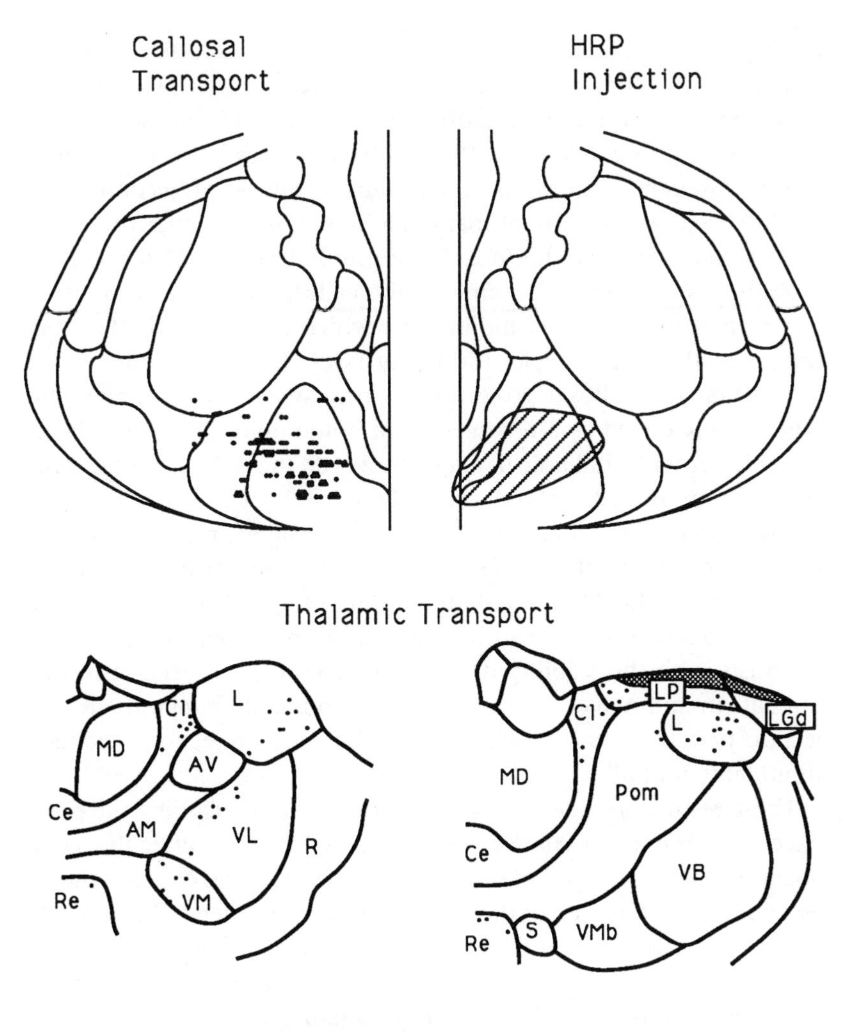

lamic projections to the HRP injection site are represented in two coronal drawings. Each drawing represents the sum of three alternately adjacent sections. The thalamic damage *(stippled region)* in the animal on the right extends posteriorly and has reduced the size of the visual nuclei (LGd and LP) by about 50 percent. The distribution of thalamic projections in the damaged animal does not differ from the normal distribution. (In all drawings each dot represents a single retrogradely labeled cell.)

hemisphere (Innocenti, Clarke, and Kraftsik, 1986; Ivy and Killackey, 1982). The origin of cells in the developing corpus callosum is much wider than in the mature callosum (Ivy and Killackey, 1981; Olavarria and Van Sluyters, 1985). Many studies have shown that these exuberant projections (for instance, callosal projections) can be stabilized by a variety of manipulations (Caminiti and Innocenti, 1981; Cusick and Lund, 1982; Innocenti and Frost, 1979).

With this in mind we examined the callosal projections to visual cortex in normal animals and in those which received early thalamic lesions to the visual nuclei (primarily LGd). In normal animals we found that the callosal projections to the small HRP injection site were characterized by a relatively tight homotopic cluster of cells. In the animals with thalamic damage, however, the tight clustering was no longer preserved. Callosal cells originated from a relatively diffuse homotopic region. In addition, there were callosal projections from auditory cortex to the deafferented visual cortex (Howard, Miller, and Finlay, 1989; Miller, Windrem, and Finlay, 1991). An example of this altered projection in one animal, also shown in Figure 1.4, contrasts with the stability of the thalamocortical projection in this animal. In summary, the callosal projections did change their projection pattern substantially, whereas thalamic projections changed minimally, if at all.

These results provide a reasonable context for explaining the presence and absence of changes in cortical cell number after the same ablations. After removal of thalamocortical input, the cortex remains essentially deafferented of that input, and neurons depending on that trophic support would die. Pyramidal cells have the opportunity and capacity to reorganize their projections and locate alternate targets for trophic support. In the case of brain damage, these observations make some interesting predictions: damage to the thalamus or layer IV of the cortex, or interruption of their connectivity, will result in the loss of related cells; in contrast, damage to pyramidal cells or to their targets or interruption of their connectivity will result in a reorganization of connections, not loss of cells. The absence of layer IV cells in Down syndrome (Ross et al., 1984) might be an example of this type of thalamocortical reactivity to developmental disruptions. Alterations in lateralization after unilateral brain damage could be an example of changes in the projection patterns of pyramidal cell axons (see, for example, Chapter 5).

Axon Retraction and Diffuse Brain Damage

Hypoxia-ischemia and callosal connectivity

In collaboration with the laboratory of Britton Chance, University of Pennsylvania, we have recently examined how a more "natural" pathology, hypoxia-ischemia, might affect regressive events in cortical development. Hypoxia-ischemia in infants is known to produce a variety of anatomical and behavioral pathologies; it also produces changes in neurotransmitter concentrations and protein synthesis, which may contribute to the cellular pathology (Hagberg et al., 1987; Johnston and Silverstein, 1987; Dwyer et al., 1987). We have examined the effects of early hypoxia-ischemia on gross neurological development and on the development of visual cortical projections in the corpus callosum in cats.

The organization and development of callosal connections between the visual cortices in cats have been well studied (Innocenti, 1981; Innocenti, Clarke, and Kraftsik, 1986; Payne et al., 1988). In normal adult cats the visual-callosal connections are located near a region around the border between the primary and secondary visual cortex (areas 17 and 18, respectively). Medial 17 and much of lateral 18 are acallosal. The callosal projections in newborn kittens are very different; the entire region of areas 17 and 18 send projections to the contralateral visual cortex. The discrete adult pattern emerges by the end of the third postnatal month as exuberant projections are selectively eliminated (Innocenti, 1981, 1986).

Changes in the number and distribution of callosal cells in developing cats have been seen in various experience-dependent alterations, such as dark-rearing and monocular or binocular deprivation (Innocenti and Caminiti, 1980; Innocenti and Frost, 1979, 1980; Frost and Moy, 1989), and after neonatal cortical damage (Caminiti and Innocenti, 1981). Callosal connectivity thus seemed a possible candidate for modification in the case of diffuse brain damage.

One-week-old kittens were made hypoxic-ischemic by bilateral ligation of the common carotid artery. The metabolic state of the neocortex was monitored by phosphorus nuclear magnetic resonance (NMR) after the ligation and at biweekly intervals. The cats were also given neurological examinations every week for two months and then biweekly up to three months of age. The neurological examinations consisted of tests of locomotor, superficial, and deep reflex and sensory nerve function and visual cliff and visual placing tasks.

The neurological development of the hypoxic-ischemic kittens was not different from that of normal animals, except for one test. Three of five hypoxic-ischemic animals failed to perform the visual cliff task. Two of these animals were found by magnetic resonance imaging (MRI); to be grossly hydrocephalic: although this suggests that venous return as well as arterial supply might have been compromised by the ligation, the arterial and venous supplies are quite distinct in cats, and some other type of explanation will be required. The hydrocephalus was, however, entirely unrelated to the callosal alterations described below. No gross brain pathology was noted in the case of the third animal that failed the visual cliff task, or in the other two hypoxic-ischemic cats. The body of the corpus callosum was of normal size or thinned in these animals.

After the cats reached the age of at least 90 days, multiple HRP injections were made into the visual cortex of one hemisphere during aseptic surgery. The tangential extent and number of retrogradely labeled callosal cells was measured from the area 17–18 border. The hypoxic-ischemic animals showed a 50 percent increase in the number of callosally projecting cells (Figure 1.5). There was a greater increase in area 17 than in 18 (Finlay et al., 1990). Most of the additional cells are preserved near the border, but the proportion of additional cells varies with distance from the border. The width of the callosal zone (the tangential extent of callosal cells) was not different from the width in normal animals. Thus, the gross topography of axonal connections was not affected by hypoxia-ischemia, but the number of cells contributing to the projection was. Since the corpus callosum was not larger overall, but nevertheless appeared to be accommodating more axons, some alteration in myelination or axon caliber must also have occurred.

The additional cells seen in the hypoxic-ischemic animals are quite likely to be remnants of the exuberant neonatal projections (Innocenti, 1981, 1986; Payne et al., 1988). It is not known what metabolic changes occur in response to hypoxia-ischemia that preserve these exuberant projections. It is possible that changes in neurotransmitter level and protein synthesis (Hagberg et al., 1987; Johnston and Silverstein, 1987; Dwyer et al., 1987) could alter events leading

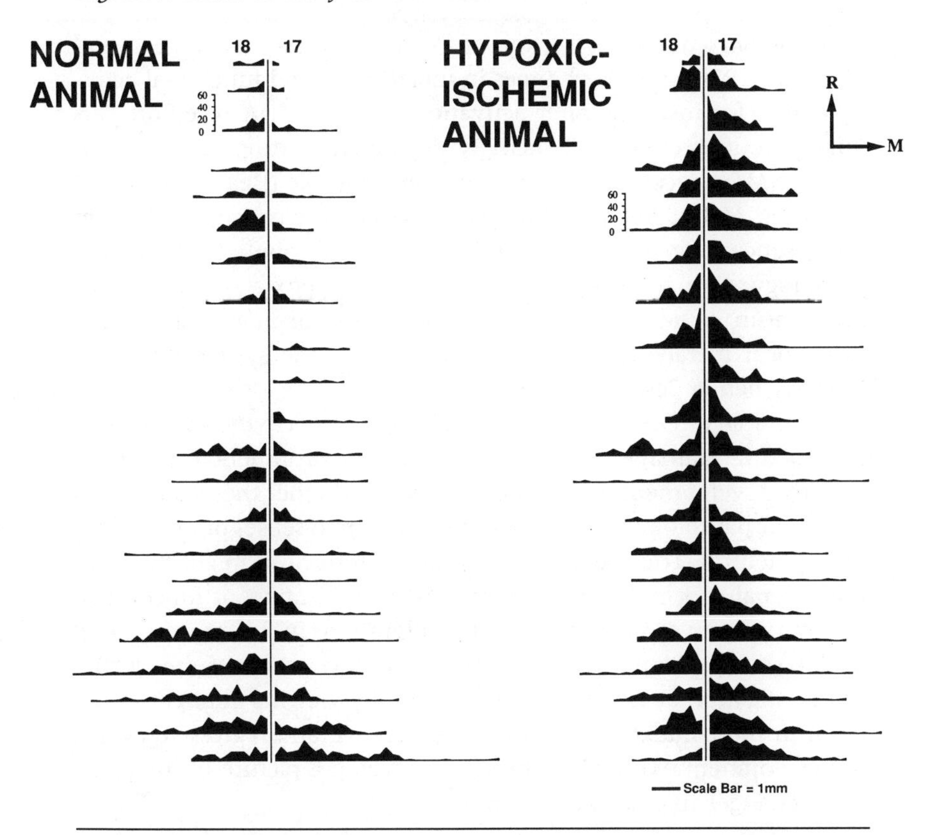

Figure 1.5 An example of the density and extent of callosally projecting cells in areas 17 and 18 in a normal cat and in a cat subjected to early hypoxia-ischemia. The drawing represents 21 equally spaced (every 640 μm) coronal sections taken from the visual cortex for each animal. The number of cells was counted in 200-μm units from the area 17–18 border. The representation is centered on the cytoarchitectonic border of areas 17 and 18, with distance from the border on the horizontal axis and cell number on the vertical axis. The hypoxic-ischemic animal has many more callosally projecting cells than the normal animal, but the tangential extent of the callosal zone did not differ from normal. Data are missing in area 18 in those regions where the callosal label in area 18 was continuous with adjacent area 19; cells in these sections were not counted.

to the stabilization or retraction of axonal processes (Changeux and Danchin, 1976). Subtle changes in metabolism and functional activity could, for instance, alter activation of the NMDA receptor. This might produce changes in activity-dependent sorting, a process in which NMDA receptors play a very important role (Bear et al., 1987; Gu et al., 1989). Such alterations in the anatomical connections in the absence of gross cortical pathology may contribute to behavioral pathology seen in humans after perinatal hypoxia-ischemia. This mechanism could presumably operate at any degree of specificity, from local metabolic accidents or infarcts to the system-wide hypoxia-ischemia described here.

In summary, removal of transient axonal connectivity is vulnerable to structural damage, metabolic insult, or environmental alteration in early development. We do not yet know whether these alterations represent pathology, are compensatory in some sense, or are behaviorally neutral. The speculation that axon retraction might figure in the normal process of establishing the lateralization of function in the human nervous system, and that alterations in this process could lead to such problems as developmental dyslexia, seems quite worthy of investigation at this point. In any case by viewing changes in the brain in the context of the normal generative and regressive events of development, we are beginning to develop a picture of the paths these changes might take after injury.

Summary

Experimental manipulations of the developing cortex have provided a window on the mechanisms that establish the normal cytoarchitecture and connectivity of the cortex. They also show potential routes for brain reorganization after early damage. Over the past several years, we have investigated the consequences for cortical organization of early ablation or alteration of cortical afferents and efferents, with particular attention to how early ablations interact with naturally occurring axon retraction and cell death.

The cortex acquires its full complement of cells over an extended period of migration (Rakic, 1974). Within the cortical plate, normal cell death begins after migration has ended, coincident with the early phases of establishment of cortical efferent and afferent connectivity. Cell death is confined principally to the last-generated granular and supragranular layers, and its magnitude is quite variable by cortical area (Finlay and Slattery, 1983). Early thalamic lesions alter this pattern: early cell degeneration increases in the prospective granular and supragranular layers and causes a marked loss of cells in layer IV. By contrast, the proportion of pyramidal cells of layers V and II–III that die in early development is resistant to afferent and target manipulation. Ablation of the superior colliculus has no effect on the number of cells in layer V of the neocortex; transection of the corpus callosum has no effect on the disposition or survival of cells in the normally callosal zones of the cortex (Finlay and Pallas, 1989; Pallas, Gilmour, and Finlay, 1988; Windrem, Jan de Beur, and Finlay, 1988).

Relative stability of the number of cortical neurons mirrors the potential for exuberance and plasticity of the appropriate axonal populations. Numerous laboratories have demonstrated early exuberance and stabilization of subcortical, intracortical, and callosal axons of layers V, III, and II. By contrast, thalamocortical axons are rather more restricted in their initial innervation of the cortex (Crandall and Caviness, 1984). Moreover, after partial thalamic damage, the remaining thalamocortical axons do not substantially reorganize to innervate denervated cortex, whereas callosal axons in the same animals do reorganize. This persisting denervation is a likely cause of the loss of cells in layer IV after thalamic ablation. Considering the significance of these observations for early acquired brain damage, it is clearly predicted that damage to the thalamus and layer IV of the cortex would typically result in the loss of the relevant neuronal population, but damage to supra- and infragranular layers or connections would result in reorganization.

Finally, we have investigated axon retraction or rearrangement following developmental insult similar to an acquired pathology in humans, early hypoxia-ischemia. We have pursued this research to explore how "natural" pathologies may intersect regressive events of development. In collaboration with the laboratory of Britton Chance, we have investigated the behavior, gross cortical pathology, and visual callosal connectivity in cats after early bilateral ligation of the carotid artery. At adulthood, there is a substantial increase in the number of neurons that contribute to the visual callosal projection in the hypoxic-ischemic animals, as measured by retrograde transport of HRP. This rearrangement can occur in the absence of gross cortical pathology and may account for behavioral disorders in the absence of obvious lesions.

Acknowledgments

This chapter reviews work supported by NIH grants RO1 NS19245 and KO4 NS00783 to B. Finlay and NS22881 to B. Chance. Thanks to Amy Baernstein, Larry Chou, and Lael Hinds for their helpful criticisms of early drafts of this manuscript.

2

Androgens and Brain Development: Possible Contributions to Developmental Dyslexia

Darcy B. Kelley

Certain tissues, including the central nervous system, differentiate according to sexually specific programs. These programs are controlled by the secretion of steroid hormones from developing and adult gonads. A key hormone in sexual differentiation is the androgen testosterone, which directs the masculinization of target tissues during a restricted developmental period. The ability of testosterone to induce sexual differentiation relies on the presence of functional receptor proteins within the target tissues. These proteins, when complexed to the hormone, act as nuclear transcription factors and can alter gene expression.

A striking feature of androgen-induced masculinization is its pleiotropic nature: a wide variety of coordinated molecular and cellular events are set into motion by hormone action. At the cellular level we can consider that androgen acts on developing tissues as a morphogen—a substance that controls tissue development so as to produce a differentiated, functioning organ from undifferentiated precursors. In fact, the androgen receptor protein is very closely related to the receptor for the only identified vertebrate morphogen, retinoic acid.

Development of the nervous system is strongly influenced by the hormonal milieu. In particular, exposure to androgenic steroids has been shown to affect a number of cellular control points in development, including the commitment, proliferation, migration, differentiation, and survival of cells. Androgen effects can be direct (action on a cell that contains receptor) or indirect (action on synaptic or endocrine partners of receptor containing cells). A variety of neuronal cell types—motor neurons, hypothalamic neurons, cortical neurons—express steroid receptors during development. This latter finding, especially pronounced in primates, has raised the possibility that cortical development is influenced by steroids. Particularly intriguing is the possibility that developmental disorders of cognitive function that are more pronounced in males than in females (dyslexia is one) might be due to abnormalities in androgen-induced neural differentiation.

Males are more vulnerable than females to a variety of developmental insults. Some of these developmental problems are associated with abnormalities in neurological and/or psychological function that result, for example, in dyslexia, stuttering, and autism. Multiple forces conspire to produce sex differences in behavior, and it can be difficult to track down the developmental origins of these differences—especially in our own species, for the constant reinforcement of cultural stereotypes is capable of negating even the most powerful biological constraints. This point is made most poignantly by success in raising baby boys as girls following accidental damage to the genitalia at circumcision (reviewed in Money and Erhardt, 1972). Nonetheless, certain sexually differentiated behaviors (such as the higher incidence of rough-and-tumble play in boys) are so prevalent across human cultures (and so frequently observed in other species) that an underlying biological determinant for the behavior is widely accepted. That biological determinant is the secretion of sex hormones—the androgenic steroids—during prenatal and early postnatal development.

This paper has two aims. The first is to review data on the mechanisms through which androgen secretion can affect neurological development. The use of experimentally powerful animal model systems has produced striking advances in this area (reviewed in Kelley, 1986, 1988). The second aim is to ascertain whether these experimental studies can generate useful insights into the origins of developmental neurological abnormalities in humans, particularly dyslexia. This second aim has as an inspiration a global hypothesis that was advanced by Norman Geschwind and his colleagues in the

1980s (Geschwind and Behan, 1982; Geschwind and Galaburda, 1985) and that has been followed since Geschwind's death by his colleague Albert Galaburda. Geschwind noted an association between left-handedness, dyslexia, and certain autoimmune disorders and suggested that all were attributable to androgen action in the fetus. Subsequent work has focused on associating androgen action with the neuroanatomical basis of cerebral asymmetries believed to contribute to language functions (Galaburda et al., 1986, 1987; Rosen et al., 1990a).

Androgens and the Development of the Nervous System

Steroid hormones

The steroid hormones, which include the androgens, estrogens, and progestins, can act directly on cells that contain receptor proteins. Organs that contain such cells are called steroid target tissues and include structures associated with reproduction (uterus, prostate), metabolism and secretion (liver, kidney), and behavior (brain and skeletal muscles). Steroid action on these tissues often involves the activation or repression of specific genes within target cells. The ability of steroid hormones to alter gene transcription depends on the ability of the receptor, when complexed with hormone, to act as a DNA-binding protein (Yamamoto, 1985). Steroid hormones can also influence target organs indirectly, through the interaction of target cells and nontarget cells. Distinguishing between the direct and indirect actions of steroid hormones can be difficult because of the extreme dependence of some developing cells on interactions with others. Neural development is particularly dependent on interactions with synaptic partners.

Location of steroid hormone receptors

Where are steroid target cells located in the central nervous system? Beginning in the 1960s, this topic was extensively studied in vertebrates (reviewed in Morrell et al., 1975a; see also Simerly et al., 1990); studies have recently been extended to invertebrates as well (Fahrbach and Truman, 1989). A number of different experimental methods have been used to address this question. Steroid targets will accumulate and retain radioactive hormones because these bind to intracellular receptors and are trapped for a time in the cell. The presence of such steroid-binding cells can then be detected biochemically for quantitative studies, and autoradiography may be used to visualize individual hormone target cells. More recently, specific antibodies to different receptor proteins have become available, as have cDNA probes that detect mRNAs of different receptors.

Results of these experimental approaches are in good general agreement. The vertebrate brain is a steroid target organ in that it contains neurons expressing specific steroid hormone receptors. The distribution of steroid target cells is different for different hormones. For example, androgens are accumulated by specific groups of motor neurons while estrogens, by and large, are not (Kelley et al., 1975; Morrell et al., 1975b; Sar and Stumpf, 1977). Estrogen- and androgen-concentrating cells have been identified in the developing cerebral cortex of mammals, including primates (MacLusky et al., 1979; Michael et al., 1986; Clark et al., 1989). Brain nuclei in the hypothalamic-preoptic-limbic continuum that surrounds the third ventricle contain a large number of hormone target cells, particularly those sensitive to estrogen (Pfaff and Keiner, 1973; Watson and Adkins-Regan, 1989; Simerly et al., 1990). The location of hormone target cells within the hypothalamic-limbic nuclei is well conserved phylogenetically (Morrell et al., 1975a; Kelley and Pfaff, 1978). Locations of other cells that express hormone receptors are more variable across species. For example, which cranial-nerve motor nuclei express androgen receptors differs markedly in primates, rodents, birds, and frogs (Kelley et al., 1975; Arnold et al., 1976; Sar and Stumpf, 1977; Kelley, 1978; Watson and Adkins-Regan, 1989).

Intersection with behavioral effectors

A different way to look at the neural targets of developmentally important hormones is to ask how they intersect with neurons that are thought to participate in specific behavioral functions. This sort of analysis yields the surprising result that the presence of a specific steroid hormone receptor protein characterizes sets of neurons that participate in specific behavioral patterns. The result is especially compelling because these behavioral patterns are often sexually dimorphic, occurring more often or in a different form in one sex than in the other (Kelley, 1988). My colleagues and I have mapped a number of the brain nuclei that participate in the production of courtship song by male African clawed frogs, *Xenopus laevis* (Kelley et al., 1975; Kelley, 1980, 1981; Wetzel et al., 1985). If this map is compared with the locations of cells that concentrate androgens (hormones that control song behavior; Wetzel and Kelley, 1983), it is evident that the presence of the receptor protein functions as a "stain" for brain regions that participate in the production of the behavior (Figure 2.1). This coincidence is not confined to frogs, as it is also evident in brain regions that produce courtship song in birds (Arnold et al., 1976); nor is it confined to courtship, as the neural system producing copulatory reflexes in rats is similarly sensitive to androgens (Breedlove and Arnold, 1980). The coincidence is also not specific to androgens, as brain regions participating in estrogen- or progesterone-controlled behaviors also have large numbers of cells containing the appropriate steroid hormone receptor in a wide variety of species, including primates (Pfaff and Keiner, 1973; Michael et al., 1986; Watson and Adkins-Regan, 1989; Simerly et al., 1990).

In addition to neural targets, steroid hormones also directly affect muscles. Every skeletal muscle contains a small amount of receptor for androgens; this phenomenon is believed to contribute to the "anabolic" effects of these steroids and the overall greater mass of muscles in males than in females. In addition, certain muscles have markedly elevated levels of androgen receptor proteins. These muscles include some associated with reproductive functions, such as the levator ani and bulbocavernosus muscles that control penile reflexes in the rat (Jung and Baulieu, 1972), and with courtship, such as the muscles of the vocal organs (Lieberburg and Nottebohm, 1979; Segil et al., 1987). It is noteworthy that these muscles are often sexually dimorphic and usually much larger in the male than in the female. Brain nuclei that participate in sexually dimorphic behaviors are also quite different in size between the sexes. The most striking example here is the telencephalic nuclei that control song in oscine birds (Nottebohm and Arnold, 1976). These nuclei are so much larger in males than in females that it is possible to determine the sex of a bird simply by looking, with the naked eye, at a brain section through the song-control areas.

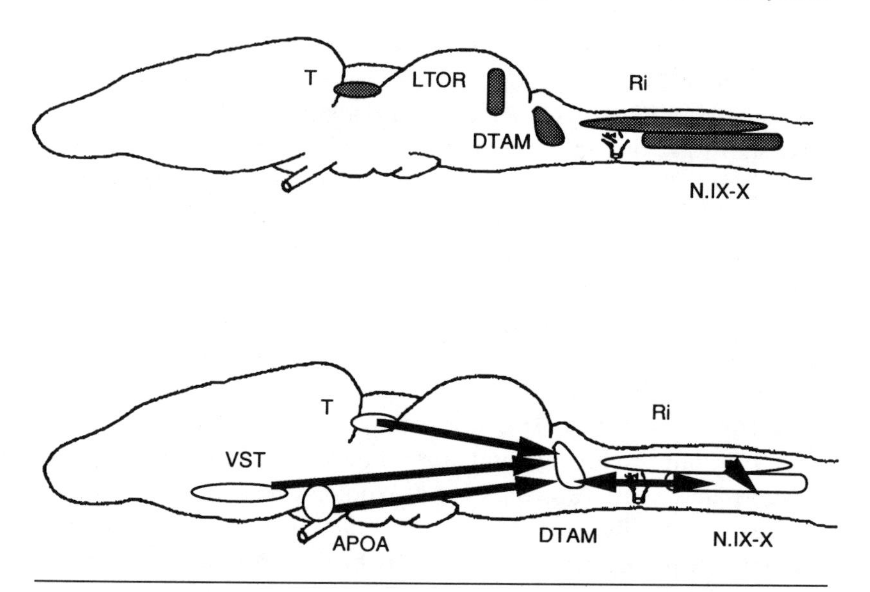

Figure 2.1 Schematic side views of the brain of *Xenopus laevis.* The locations of brain nuclei targeted by androgens (those that concentrate dihydrotestosterone) are shown in the upper panel; connections between brain nuclei that participate in the production of vocal behaviors are shown in the lower panel (based on Kelley, 1981; Wetzel et al., 1985). The anterior preoptic area *(APOA)* also contains steroid target cells; cells in this nucleus are labeled after administration of radioactive testosterone or estradiol (Kelley et al., 1975; Morrell et al., 1975b) but not dihydrotestosterone (Kelley, 1981), suggesting the presence of estrogen receptors and intracellular aromatases, enzymes that convert testosterone to estradiol. Aromatization to estrogen probably also accounts for some aspects of androgen action in the cortex and other regions of mammalian brains (Michael et al., 1986). The ventral striatum *(VST)* does not appear to contain aromatizing activity in *Xenopus laevis* but it does have estrogen receptors. Other abbreviations: *DTAM,* pretrigeminal nucleus of the dorsal tegmental area of the medulla (participates in production of the vocal pattern in frogs and toads); *LTOR,* laminar nucleus of the torus semicircularis (an auditory nucleus); *N. IX-X,* cranial nerve motor nucleus of the IX-X nerve (contains the vocal motor neurons); *Ri,* nucleus reticularis inferior (part of the vocal pattern generator); *T,* thalamus (auditory, somatosensory, and lateral-line sensory nuclei are hormone targets).

Sexual dimorphism in effectors

A central goal of this field has been to account for sex differences in behavior in terms of sex differences in the underlying neural and muscular effectors (Kelley, 1986, 1988). Sexually dimorphic features in the brain include the number of neurons in a particular brain nucleus, the extent of the dendritic arbors (and number of afferent synaptic contacts), and the actual synaptic connectivities among brain nuclei (Greenough et al., 1977; Gurney, 1981; DeVoogd and Nottebohm, 1981; Wetzel et al., 1985; Konishi and Akutagawa, 1985). Sex differences in behavioral effector muscles include the number of muscle fibers, the size and efficacy of neuromuscular synapses, and the physiological properties that contribute to the speed and force of contraction (Sassoon and Kelley, 1986; Marin et al., 1990; Bleisch et al., 1989; Tobias and Kelley, 1987, 1988; Sassoon et al., 1987).

Sex differences in behavioral repertoires are largely controlled by hormonal secretions early in development (the "organizational" effects of steroid hormones) and in adulthood (the "activational" effects of hormones). Sex differences in the central nervous system (CNS) and in behavioral effector muscles also display two phases of hormone sensitivity: early in development and later, after puberty or the first breeding season. It is thus essential to determine how the secretion of sex hormones produces sex differences in neural and muscular effectors. This effort has yielded a number of insights into the cell biology of hormone action on the developing nervous system, some of which are reviewed below (see also Kelley, 1986, 1988).

Of particular interest to the problem of the neurological basis of dyslexia are studies of how cell numbers are controlled by hormone action. Language-related skills in humans are strongly associated with functional and anatomical asymmetry of the cerebral cortex. The best-studied example is the planum temporale, a region of cerebral cortex that can be markedly different in size in the left and right cerebral hemispheres (Geschwind and Levitsky, 1968). If cell number plays a role in the development of cerebral asymmetries (see below), androgenic hormones might produce disorders in language function by interfering with the development of asymmetries. It is then of interest to ask whether androgen can influence cell numbers in developing systems. In what follows I review recent work on this topic with the goal of generating a roster of possible cellular mechanisms from animal experimental work that may include those used in the assembly of language-related areas of the human nervous system.

Control of cell numbers

The way in which neurons in a particular brain area are identified greatly influences counts of neuronal numbers. The vertebrate brain can be subdivided into groups of similar neurons separated geographically from other groups of functionally distinct cells. Functional groups of neurons are often bordered by axon bundles that contain their afferents, efferents, or unrelated passing fiber tracts. These groups of neurons, visualized with various cytological stains, are called "nuclei," and counts of neurons usually refer to numbers in a particular CNS nucleus. Some nuclei can be further subdivided by the appearance of subclusters of cells. When distinctions based on subdivision are applied to a large neural expanse, such as the cortex, one is said to distinguish subregions by their "cytoarchitectonic" appearance. Counts of neuron numbers in a particular region can be difficult if cytoarchitectonic boundaries are not clear; this is often the case in immature brains. Migration of developing neurons into or away from particular cytoarchitectonic zones can also influence neuronal counts.

The functional subdivisions of the nervous system do not always correspond to cytoarchitectonic boundaries. Sets of neurons with common anatomical connections may lie on both sides of a fiber bundle used for nuclear demarcation. Similarly, a brain nucleus may be characterized by a population of large cells distinguished by intense staining for Nissl substance (cytoplasmic RNA). If cells at the margin of the nucleus differ in size or in staining properties from interior cells, the cytoarchitectonic boundaries of the nucleus will appear to shift and the neuron count will be affected. An example is the telencephalic song-control nucleus HVc in male canaries, which, by cytoarchitectonic criteria, changes in volume with the seasons (Nottebohm, 1981). If another method, such as directing antibodies to the estrogen receptor or back-labeling from HVc synaptic targets, is instead used to outline the nucleus (Gahr, 1990), the volume of HVc does not vary seasonally. Attempts to apply connectivity or receptor-expression criteria to counts of neuron number in developing systems are especially difficult because young cells may not yet have connected with synaptic targets or be expressing hormone receptors. One can resort to labeling the cells with tritiated thymidine if their "birthdays" are sufficiently different from those of neighboring neurons. Most of the studies described below have used multiple criteria to identify functional groups of neurons. It might be argued that these sorts of functional criteria, especially those derived from neuroanatomical tracing methods, cannot be applied to the human brain. However, it is now possible to use carbocyanine dyes (DiI and others) in fixed material to trace the connections of neural groups (Godemet et al., 1987). Such an approach would be a valuable adjunct to the present, widely used cytoarchitectonic criteria.

Given that one has identified a functional CNS nucleus, what processes control the number of neurons in that nucleus? The majority of cells in the CNS arise from proliferative zones at the neural epithelium. Cells then differentiate into neurons or glia, migrate to specific destinations, and establish synaptic connections. Some neurons die during formation of the nervous system; the process is called "ontogenetic cell death" (Oppenheim, 1985). Control of cell number could thus be attributable to the number of neuroepithelial precursor cells, the number of cell divisions, the number of cells committing to neural or glial phenotypes, and/or cell death.

Which of these cellular processes is responsible for sex differences in neuron number? Currently, the best-established mechanism is ontogenetic cell death. A good example is the spinal nucleus of the bulbocavernosus (SNB), a group of motor neurons that innervate the penile muscles in the rat (Breedlove and Arnold, 1980, 1981; Breedlove et al., 1983; Nordeen et al., 1985; Breedlove, 1986). The number of neurons is the same in males and females at birth, but females subsequently lose about 80 percent of these cells during the postnatal period. Cell loss is accompanied by morphological evidence of neuronal degeneration, suggesting that the underlying process is ontogenetic cell death. Ontogenetic cell death in females can be prevented by administration of androgen. Similarly, differences in cell loss are believed to contribute to sex differences in the number of vocal motor neurons in frogs (Kelley and Dennison, 1990). Ontogenetic cell death also participates in control of the number of other sorts of neurons, including telencephalic neurons that participate in the production of bird song (Nordeen et al., 1987).

Does differential cell proliferation in males and females contribute to differences in neuron numbers? This question has been examined in regard to the motor neurons of rats and frogs (Breedlove et al., 1983; Gorlick and Kelley, 1987). In both cases, cell proliferation (as judged from the incorporation of tritiated thymidine) is complete before sex differences in cell number are present and before those cells express androgen receptors (Breedlove, 1986; Gorlick and Kelley, 1986). For telencephalic cells (of some interest here because their development could be analogous to that of cortical neurons), the answer is less clear-cut. The telencephalic vocal control nuclei of male and female song birds are sexually dimorphic in neuron number (Gurney, 1981). New neurons are continuously generated in the telencephalon, even in adult life (Goldman and Nottebohm, 1983.). During development, cell proliferation occurs at a time when sex differences in area HVc (an important participant in bird song) are established and could contribute to those sex differences (Nordeen and Nordeen, 1988). In some song birds, nucleus HVc is functionally lateralized in that lesions in the left HVc of adults will severely disrupt song while lesions of right HVc produce considerably milder effects (Nottebohm et al., 1976). Because of the functional lateralization of song control, any hormonal influences on the proliferation of cells destined for song-control areas could have bearing on the development of cerebral asymmetries in humans, even though bird brains do not evince any apparent morphological asymmetries in these nuclei.

Another idea worth discussing at this point is the possibility that hormones alter the lineage of cells in the CNS, biasing commitment to a neuronal rather than a glial fate. If supplies of the neuronal and glial precursors are limited, such a switch in cell fate could profoundly affect cell numbers. In the original report of adult neurogenesis in bird brains, androgen was shown to increase the number of tritiated-thymidine-labeled glial cells (Goldman and Nottebohm, 1983). The possible effects of androgen on these two cell types in developing song birds have not been described. If neuronal precursors remain poised at the lineage fork (neuron or glia?) for some time, the hormonal milieu could affect the lineage commitment of these cells.

Mechanisms of androgen action in control of cell number

Current evidence thus suggests that males have more neurons than females in certain sexually dimorphic brain nuclei because androgen secretion during development prevents ontogenetic cell death. The mechanism by which androgen acts to prevent ontogenetic cell death is under investigation in the sexually dimorphic motor neurons of the bulbocavernosus–levator ani (BC–LA) complex in rats and the vocal motor neurons in frogs. In the rat, recent studies suggest that the effects of androgen are mediated through action on the synaptic targets of the motor neurons: the BC–LA muscles (Breedlove, 1986; Fishman et al., 1990). In adult rats, both motor neurons and muscles express androgen receptor protein, and thus both could be directly affected by hormone secretion (Breedlove and Arnold, 1980; Jung and Baulieu, 1972). During the postnatal period, however, only the motor neurons express detectable levels of androgen binding, suggesting that androgen may be acting at the level of the muscle (Fishman et al., 1990). Further evidence comes from studies of chimeric mice (*TFM* × *sxr*) in which motor neurons that do not express the receptor can be rescued from ontogenetic cell death if muscles do express the receptor (Breedlove, 1986).

What does androgen do to the BC–LA muscle that affects its ability to support motor neurons? For many years it was claimed that the levator ani is present in both sexes at birth but involutes in females unless the muscle is supplied with exogenous androgen (Venable, 1966). The loss of BC–LA motor neurons could thus be viewed simply as a secondary result of the loss of muscle targets (Arnold, 1984). Recently, however, it has become clear that the BC–LA muscle persists, albeit in diminished form, in females (Tobin and Joubert, 1988). Since proliferation of muscle precursor cells (myoblasts) can be induced in adult females by supplying androgen (Joubert and Tobin, 1989), androgen may act not by preventing myofibers from dying but instead by promoting proliferation of myoblasts. The way in which androgen rescues motor neurons could thus be to control a required cell-to-cell interaction with myoblasts. Some evidence for this mechanism comes from studies of vocal muscles in frogs.

Genesis of a sexually dimorphic system for vocal control

The African clawed frog, *Xenopus laevis,* uses sexually dimorphic vocalizations during courtship (reviewed in Kelley and Tobias, 1989; Kelley and Gorlick, 1990). The male song, the mate call, is a repetitive fast and slow trill; males can sing continuously for long periods of time. Trills are made up of clicks produced in the larynx, the vocal organ, when contractions of paired bipennate muscles cause cartilaginous discs to pop apart. Males have about twice as many muscle fibers and laryngeal motor neurons as do females (Kelley and Dennison, 1990; Marin et al., 1990).

The greater number of muscle fibers in the male is due to androgen-induced myogenesis during post-metamorphic development (Sassoon et al., 1986; Sassoon and Kelley, 1986; Marin et al., 1990). Androgen stimulates myoblast proliferation; myoblasts then fuse with each other to form new muscle fibers. Levels of androgen receptor are extremely high during this period of muscle-fiber addition in males (Kelley et al., 1989).

We have studied the relation between muscle development and the innervation of the larynx by axons of motor neurons. Interestingly, sex differences in innervation (numbers of axons that enter the larynx) arise during tadpole development (Kelley and Dennison, 1990) whereas sex differences in muscle fiber number are not evident until considerably later, during post-metamorphic life (Sassoon and Kelley, 1986). This finding raises the possibility that the degree of innervation contributes to muscle fiber number, perhaps through an interaction with myoblasts. We do not know when sex differences in myoblast number arise. One possibility is that males have more muscle precursor cells even as tadpoles and that it is the interaction of axons with myoblasts that influences motor neuron. Alternatively, the motor neuron may be the prime target of androgen action in the tadpole, and sex differences in innervation may contribute to sexually dimorphic myogenesis.

This question can be addressed by determining when androgen receptors arise in neurons and muscle and how receptor levels are regulated. Androgen receptor levels are the same in male and female laryngeal muscle at metamorphosis; thereafter receptor levels start to decline in females (Kelley et al., 1989). We do not know when muscle first expresses androgen receptor but suspect that it is before metamorphosis, because androgen can induce increased laryngeal proliferation in pre-metamorphic tadpoles (R. Spiera, D. Sassoon, and D. Kelley, unpublished observations). We have not yet detected any sex difference in the number or intensity of androgen-accumulating neurons (Kelley et al., 1975; Morrell et al., 1975b; Kelley, 1981). Using steroid autoradiography, we can detect androgen receptors at tadpole stage 64 but not at tadpole stage 60 (Gorlick and Kelley, 1986). This negative result may have been due to competing androgen secretion at stage 60, and we are thus reinvestigating this issue using *in situ* hydridization with a cDNA probe for the *X. laevis* androgen receptor. Sex differences in axon number in the laryngeal nerve are present at tadpole stage 59 (Kelley and Dennison, 1990) and may be due to pre-metamorphic sex differences in androgen secretion.

Androgens and the Development of Language-Related Areas of the Human Brain

The Geschwind hypothesis

Norman Geschwind put forth a set of ideas (later elaborated with Albert Galaburda: Geschwind and Behan, 1982; Geschwind and Galaburda, 1985) on the relation between left-handedness, language disorders, and autoimmune function. These ideas were provoked by clinical observations of the association between dyslexia and autoimmune diseases. Some immune disorders are more prevalent in males (particularly before puberty) than in females. Dyslexia is also more prevalent in males and can be associated with left-handedness. The latter association raises the possibility that anomalous cerebral dominance contributes to dyslexia. Geschwind's idea was that the association between left-handedness, dyslexia, and autoimmune disease is due to testosterone secretion during development. Testosterone secretion is greater in developing males than in females. What differences in the human brain are associated with dyslexia and how might testosterone act to produce or amplify these differences?

Asymmetry of the planum temporale

For the most part, the left hemisphere is dominant for language-related functions; this left-hemispheric dominance is associated with a gross morphological asymmetry in the planum temporale, a cortical area that is the extension, on the upper surface of the temporal lobe, of Wernicke's speech area (Geschwind and Galaburda, 1985). The asymmetry in size of the planum temporale can be extreme; in some individuals this region on the left is ten times the size of that on the right (Geschwind and Galaburda, 1985). Within the planum temporale is found cytoarchitectonic area tpt. Area tpt also displays a size asymmetry closely correlated with the overall asymmetry in the planum temporale (Galaburda et al., 1978).

Some insight into possible cellular mechanisms for generating size differences in the planum temporale (PT) comes from studies of the total (left plus right) volume of the PT in brains of different degrees of symmetry. ("Symmetric" and "asymmetric brains" are used here as shorthand terms for brains in which the size of the left and right PT are either similar or different in size; I do not mean to imply that such brains are necessarily asymmetric in a global sense.) Asymmetry is associated with an overall smaller size of the PT; the total volume of the PT is larger in symmetrical brains than in asymmetrical ones (Galaburda et al., 1987). This correlation suggests that the number, size, or spacing of neurons in asymmetric brains should be less than in symmetric brains. Further, for asymmetric brains, these cellular characteristics should be relatively less expressed in the smaller than in the larger PT. Differences in cell-packing density apparently do not account for size asymmetries, at least in some lateralized cortical regions of rats (Galaburda et al., 1986). Dendritic differences have been argued against on a *priori* grounds since they would distort the cytoarchitectonic features of these regions and such distortions are not seen (Galaburda et al., 1986). Such dendritic differences do occur, however, in male and female bird brains in certain song-control nuclei without marked cytoarchitectonic distortion (DeVogdt and Nottebohm, 1981). A study of dendritic architecture in symmetric and asymmetric brains should prove helpful in settling this question. By exclusion, then, the most likely basis for size differences in these regions of the human brain are differences in neuron numbers. I note that such differences are inferred and have not been directly demonstrated.

Control of cell number in the planum temporale

What could account for presumed differences in the number of neurons in the left and right PT? A tritiated-thymidine study in rat cortex revealed no difference between left and right sides in the ratio of labeled to unlabeled cells, although there was an asymmetry in the number of labeled cells at the shortest intervals after thymidine administration (Rosen et al., 1990b). Such a result is consistent with the idea that the left and right cortex do not differ with respect to neuronal proliferation but instead differ in rates of ontogenetic cell death. Moving now to the human brain, if we assume that the relation between left-handedness and dyslexia implies more symmetry in the case of developmental language disorders, we might expect that dyslexics would experience less ontogenetic cell death in the left planum temporale (and perhaps less in the right PT as well, given the greater overall volume of the PT in symmetric brains). This progression is quite obviously an oversimplification, given that the majority of left-handers are not dyslexic. The brains of individuals with severe language-related disorders do, however, show deviations from the expected pattern of cerebral asymmetry as well as a variety of abnormalities in cortical organization.

Androgen and symmetry, a hypothesis

To continue in this speculative vein, we may ask how testosterone secretion could affect the development of the cerebral cortex and thus contribute to the greater incidence of language-related disorders in males. First, it is worth noting that during development the mammalian cortex is a steroid target organ. In the cortex of both rodents and primates, there are at birth large numbers of hormone-accumulating cells, but these cells decrease in number as the nervous system matures. The cingulate cortex, which participates in vocal control in nonhuman mammals, is especially rich in steroid-accumulating cells (Clark et al., 1988). Many target cells accumulate estrogen, not androgen. Estrogen, however, is often the active form of androgen in cells that contain the appropriate enzymes, the aromatases; it is possible, therefore, that cortical cells could be affected by testicular androgen secretion (Michael et al., 1986). Thus, mammalian cortex has the potential to be directly affected by androgen secretion during development.

Given cortical sensitivity, how might androgen affect anomalous dominance? Androgen could decrease cerebral symmetry by decreasing ontogenetic cell death, particularly by affecting the degree of ontogenetic cell death in the right hemisphere. A hypothetical scheme for this effect is as follows. Let us assume that the development of one hemisphere is normally slower than that of the other. Exposure to testicular androgen secretion would occur at different stages in the development of cortical neurons, during the critical period for rescue from ontogenetic cell death for one hemisphere and before that critical period in the other hemisphere. One hemisphere would thus have more cells than the other because its neuron numbers were established during the period of androgen secretion. Now let us suppose that the overlap between the critical period and the androgen-secretion period is not complete; development of the slower hemisphere falls outside the critical period entirely while growth in the faster hemisphere is partially coincident with androgen secretion. Both hemispheres would experience ontogenetic cell death but the effect would be attenuated in the faster-developing hemisphere. Increased length of exposure or sensitivity to testosterone would decrease ontogenetic cell death in both hemispheres, potentially wiping out the asymmetry in cell numbers (illustrated in Figure 2.2). Such a mechanism could account for the observation that the

Figure 2.2 A hypothetical scheme for the effects of androgen on asymmetry of the planum temporale (PT) of the human cerebral cortex. The underlying hypothesis is that the PT usually undergoes asymmetric ontogenetic cell death; androgen can reduce cell death (increasing the overall size of the PT) and asymmetry (exacerbating developmental language disorders). In the upper panel, the number of the neurons in the PT of the right and left hemispheres is illustrated during cortical development. I assume that the left hemisphere leads the right in development and that cell death in the left hemisphere's PT overlaps partially with the period of androgen secretion. The right hemisphere shows a greater decline in cell number than the left hemisphere because cell death occurs outside of the period of androgen secretion or sensitivity. In the lower panel, the number of neurons in the PT is illustrated for the case where androgen secretion or sensitivity is prolonged. The period of ontogenetic cell death now overlaps with the period of androgen secretion for both hemispheres, and cell number in both hemispheres is maintained with a resultant decrease in PT size asymmetry. This reduction in asymmetry will tend to decrease extremes of laterality in hand-

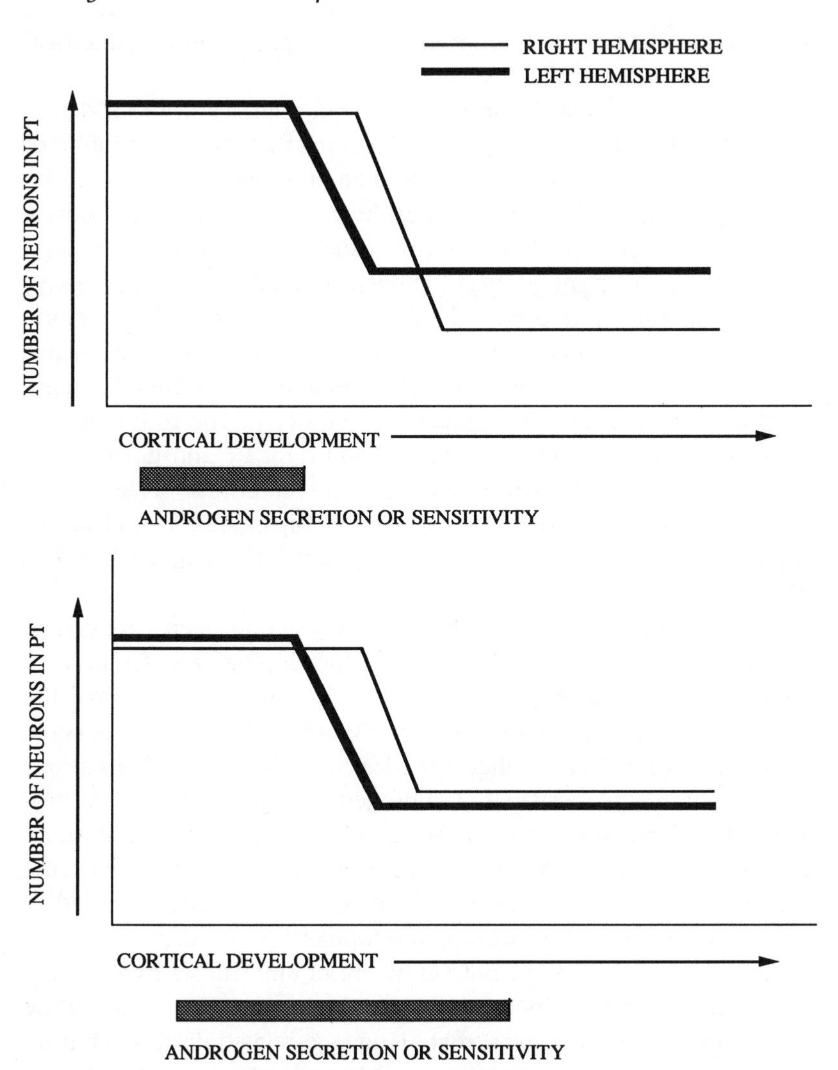

edness. This scheme can be applied to PT development in both sexes. If we assume that both sexes are exposed to androgen secretion but that secretion or sensitivity is greater in males, then development of the male PT will more closely resemble that illustrated in the lower panel while PT development in females will be closer to that illustrated in the upper panel. Males will show a greater incidence of anomalous cerebral dominance, such as more ambidexterity or left-handedness. Developmental abnormalities in cortical formation, leading to disorders such as dyslexia, will be exacerbated in males because of their potential for reduced asymmetry.

total volume of the PT is larger in symmetrical than in asymmetrical brains (Galaburda et al., 1987).

How does this scheme account for abnormalities in brain development that contribute to dyslexia? I assume that there is an inherent tendency for one hemisphere to develop at a different rate than the other. For example, the right hemisphere may develop more slowly than the left for right-handers (the usual case) and the left more slowly than the right for left-handers. In either case, the slowly-developing hemisphere would miss being exposed to testosterone during the critical period for prevention of ontogenetic cell death while the faster-developing hemisphere would benefit from hormone exposure by reduction in ontogenetic cell death. The results would be size asymmetries between the left and right PT and normal language function. If, however, testosterone secretion or sensitivity to the hormone is prolonged, the degree of asymmetry would be reduced, contributing to developmental language disorders (Figure 2.2).

Does this scheme apply to females, whose androgen exposure is generally less than that of males? In developing rats, females do experience considerable exposure to androgen during prenatal life (though less than males do: Weisz and Ward, 1980), and it is possible that androgen secretion affects cortical development in both sexes. Because of the greater levels of androgen in males, biasing of cortical development toward symmetry would still be expected to occur more frequently in males than in females. I would expect that the occurrence of language disorders might be greater in females exposed to excessive androgen titers during development (as in congenital adrenal hyperplasia) but I am not yet aware of any studies that address this issue. Note that in this scheme abnormalities in testosterone sensitivity or exposure exacerbate language-related disorders but do not cause them. The underlying neurological defect is assumed to be disordered development of a specific cortical area, including abnormalities in connectivity.

The significance of cell numbers

Attention to the phenomenology of cerebral asymmetries and to underlying cellular mechanisms obscures, in my opinion, an important issue: what can we learn from differences in cell number? This issue surfaces repeatedly in studies of sexual differentiation of the nervous system. It is widely assumed that because adult males have more neurons than females in some brain areas, those additional neurons must contribute to male-specific brain functions. Sex differences in cell number are, however, established during development and it is possible that the functional significance of these differences is instead ontogenetic.

I have argued (Kelley, 1988) that the control of cell number may be key for establishing qualitative sex differences in synaptic connectivity. Cell number would affect connectivity by altering the competitive interactions among cells that refine connections in the developing nervous system (Figure 2.3). Thus having more or fewer cells (within limits) would not imply that the nervous system could perform a certain function more or less well but would instead be an adult reflection of the developmental events that establish different neural pathways. Intracortical connectivity has been investigated in rodent brains of differing degrees of asymmetry by examining terminations of the corpus callosum (Rosen et al., 1990a). An unexpected finding was that the reduction in interhemispheric termination within the asymmetric brains was more exaggerated than one would expect from simple differences in volume of cortical area. This finding suggests that intrahemispheric connections in asymmetric brains may be maintained at the expense of interhemispheric connections. One bias in the maintenance of connections is the size of synaptic targets. By increasing symmetry in cell numbers as described above, androgen could bias one set of synaptic connections at the expense of another. In this scheme androgen acts on the synaptic targets of cortical neurons and not necessarily on the targets themselves. Good candidates for these targets are thalamic neurons and other cortical neurons. Androgens could also act to increase afferent input to cortical neurons, a circumstance which is known to decrease ontogenetic cell death (Oppenheim, 1985). The synaptic connectivity of the adult planum temporale is not well understood and its development is largely unknown. Again we may hope that the use of DiI in fixed tissue will provide much needed information.

Geschwind's Contributions

As scientists we know that our work has a highly personal and idiosyncratic side, especially in the choice of questions whose resolution seems to us burning issues. What I admired most about Norm

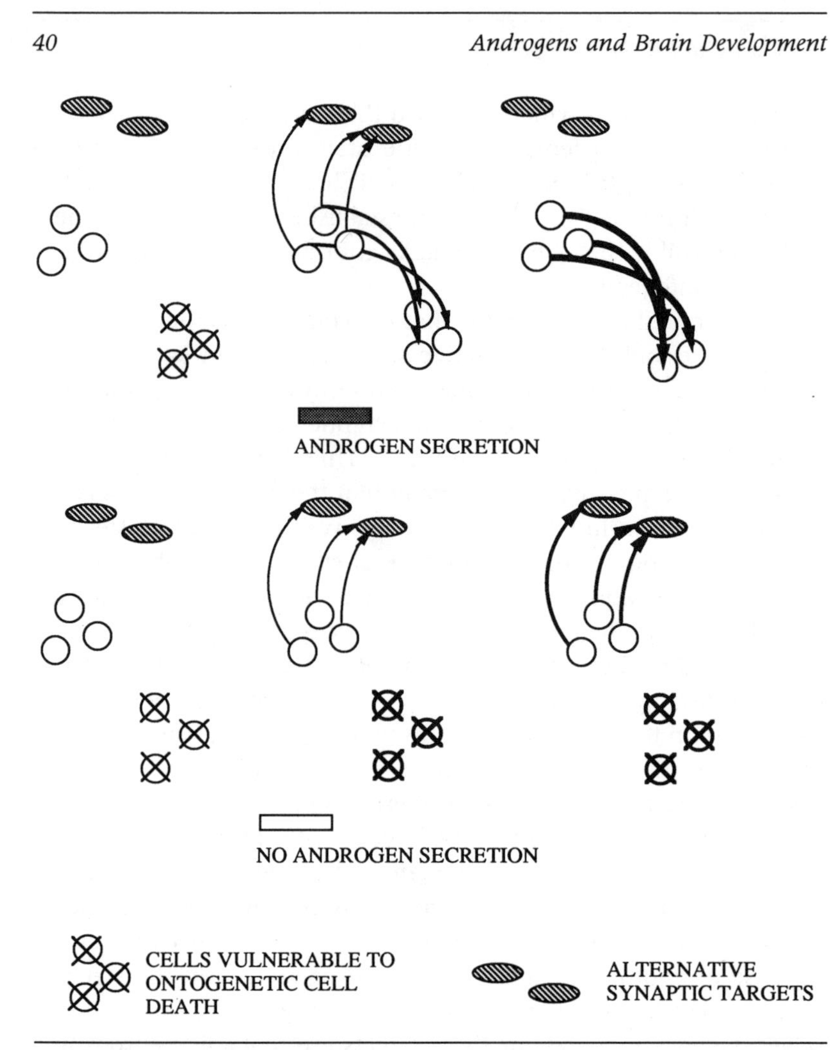

Figure 2.3 How control of cell number during neural development could produce qualitative changes in synaptic connectivity. This scheme assumes, again, that androgen can reduce cell death in target neurons *(circles)*. Certain of these neurons have two possible synaptic partners, a group of cells vulnerable to ontogenetic cell death *(circles with crosses)* and a less attractive alternative group *(shaded ovals)*. In the middle panel at top, hormone target cells extend processes toward both sets of potential synaptic partners. If one set of cells is dead *(lower panel)* because of lack of androgen during a sensitive period, the connection with the alternative synaptic partners will be maintained and strengthened. If androgen targets survive *(upper panel, right)* because of exposure to androgen during the sensitive period, the synaptic connections to the less attractive synaptic partner will be withdrawn at the expense of connection to androgen target neurons.

Geschwind, clearly the inspiration for many of the ideas outlined above, was that he could take a very difficult and complicated problem and attack it at many levels of analysis. Complexity was a challenge to Geschwind; he was never intimidated into reducing a problem to the most experimentally manageable size. This kind of approach carries the risk of frequent error. Geschwind's wrong ideas were not only inspirational; they were frequently more interesting than the correct notions of others. To his writings can be traced large portions of the research agenda in the field of language and human brain development. Geschwind would have been delighted to read about the new cellular findings, even those that contradicted mechanisms that he had suggested, and he would have been, as always, leading the field in invention and imagination.

Acknowledgments

Research in the author's laboratory is supported by NIH grants NS19949 and NS23684. I gratefully acknowledge the contributions of past and present laboratory colleagues, especially Leslie Fischer, Dennis Gorlick, Melanie Marin, David Sassoon, Martha Tobias, and Daniel Wetzel, to research described here.

3

Peptidergic Neurons in the Hypothalamus: A Model for Morphological and Functional Remodeling

Damaso Crespo

In a pioneering work, Ernst Scharrer (1928) proposed that the histologic features of the supraoptic (SON) and paraventricular (PVN) nuclei of the hypothalamus were the morphological expressions of the secretory activity of their constituent neurons. Later (1940), he and his wife, Berta, summarized their preliminary studies postulating that the nuclei had a "glandular function" similar to those described for other secretory cells. This bold hypothesis was based on the peculiar anatomy of the nuclei and the presence of colloid-like granules in the cytoplasm of their neurons. Bargmann and Scharrer (1951) introduced the term *neurosecretion* in reference to the capacity of these cells to release their products into the neurohypophysis. These preliminary works were carried out using the Gomori technique to show the reactivity of the disulfide bridges of the molecules synthesized in these cells. The researchers were able to define a system of large neurons located in the SON and PVN whose axons form the hypothalamo-neurohypophyseal tract which terminates in the posterior lobe of the hypophysis. These neurons constitute the magnocellular neurosecretory system (MNS) of the hypothalamus (for a historical review, see Silverman and Zimmerman, 1983).

In recent years our knowledge of the MNS in general and SON neurons in particular has increased with the application of immunocytochemical methods. Antibodies have been used against neurohormones, their associated neurophysins (NPs), and the neurotransmitters that operate in the supraoptic nucleus. Radioimmunoassay (RIA) has allowed hormone levels in blood plasma to be determined. Most recently, the neurohormone genes have been cloned and characterized, and the techniques of *in situ* hybridization have been employed to study neurohypophyseal hormone mRNAs.

Dierickx, Vandesande, and Goessens (1978) first demonstrated that neurons of the SON synthesize either oxytocin (OT) or vasopressin (VP). At the central nervous system (CNS) level these neuropeptides are involved in a variety of functions, including cardiovascular regulation, memory, behavior, and nociperceptive and body-temperature regulation (reviewed by Sofroniew and Weindl, 1981). In addition, Ikeda et al. (1989) have demonstrated recently that VP has a neurotrophic action on motoneurons of the rat embryo spinal cord in culture. Both VP and OT neurohormones have a depolarizing action on motoneurons but OT has no neurotrophic action. In the peripheral organs, these neuropeptides are linked to functions such as water balance, milk ejection, and parturition (Aravich and Sladek, 1987). As a result of these biological properties, neurosecretory neurons (NSNs) are a useful model for analyzing the cell biology and function of peptide-producing neurons. They share the properties of secretory cells, and their soma and dendritic membranes receive afferent excitatory and inhibitory inputs, which cause neuronal membrane potentials. These potentials are conducted through the fibers of the supraoptico-neurohypophyseal tract until they reach axon terminals in the neurohypophysis, producing a release of the neurohormones by exocytosis (Brownstein, Russell, and Gainer, 1980). In this regard, a number of physiological and experimental conditions can enhance hormone demands. The release of neurohormones determines cellular changes and leads to modifications in their structural relationships with one another and with SON glial cells as well as to rearrangements in the local synaptic circuitry.

Probably the most striking feature of these changes is the fact that they are reversible. When the physiological or induced stress ceases, the NSNs are rearranged, after a variable recovery period, and attain their normal functional state. This plasticity has made these cells an experimental model for analyzing the behavior of the neuronal protein synthesis machinery and the reorganization of the synaptic circuitry in the adult CNS. In the present chapter I will discuss the biology of SON neurons as well as the influence of several stresses on their structure and relationships.

Neurohormone Gene Expression

Oxytocin and vasopressin are synthesized in NSNs in the same way as other secreted proteins are produced in glandular cells (Figure 3.1). In the cell nucleus, gene transcription and splicing of the newly formed messenger RNA (mRNA) produce a readable mRNA. After

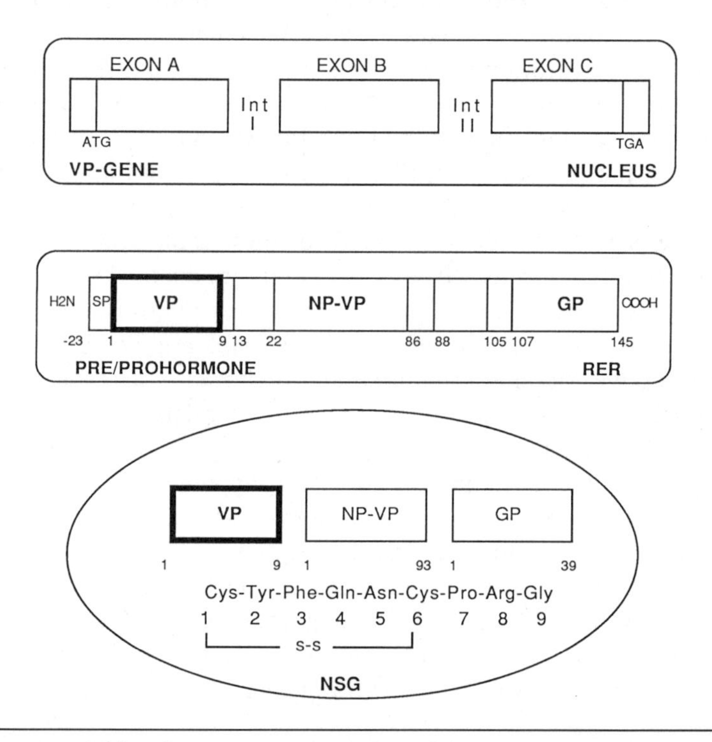

Figure 3.1 Highly schematic representation of the vasopressin (VP) gene and the steps in the production of the final active neurohormone. The VP gene is composed of three exons (A, B, and C) linked by two introns (I and II). It presents a glycoprotein (GP) sequence that does not occur in the oxytocin (OT) gene. OT differs from VP in the third and eighth amino acids (isoleucine and leucine for OT and phenylalanine and arginine for VP). (Data taken from Ivell and Richter, 1984, and Gainer, Alstein, and Hara, 1988.)

the mRNA is translated on ribosomes of the rough endoplasmic reticulum (RER) of the perikarya, the synthesis of a large peptide chain called preprohormone is initiated. This long precursor carries at its N-terminal a "signal peptide" that is cleaved when it enters into the RER cisternae lumen, forming the prohormone. Each prohormone consists of the nine neurohormone amino acids (Aa), the NP (93 Aa), and an "end" Aa (histidine for OT and arginine for VP prohormones, respectively). The arginine "end" of VP is joined to a glycopeptide of 39 Aa, which does not occur in the case of OT. This glycosylation starts at the RER and ends when the prohormone reaches the Golgi apparatus. Prohormones arriving at the Golgi bodies are condensed and packed into neurosecretory granules (NSGs), which suffer several proteolytic cleavages (post-translational processes) as they travel through the axons to the neurohypophysis. This process leads to the formation of mature neuropeptide hormones, which are sliced from the NP- and VP-joined glycopeptide. Neurosecretory granules together with their final biological products are stored in the axon-terminal swellings in the neural lobe to be released to the perivascular space in response to organic demands. The biological functions of NPs and the glycopeptide have not been clarified to date.

Organization of the Supraoptic Nucleus

In contrast to the neighboring hypothalamic areas, the SON appears as a compact mass of large neurons attached to the lateral borders of the optic chiasm and optic tracts (Figure 3.2). Neurosecretory neurons of the SON can be located among the axons of the optic chiasm and medially to the optic tracts, but these two locations represent a very small proportion of the neuronal population of this nucleus. SON neurons are termed "magnocellular neurons" because they have very large somas (160 square micrometers in cross-sectional area). When stained with basic dyes, NSNs have a well-defined Nissl substance, typically located in a marginal position, and a light perinuclear region surrounding a large, light nucleus with predominantly dispersed chromatin, which allows their nucleoli to be distinguished easily. Scattered among these neurons there are blood vessels and glial cells, mainly astrocytes.

Quantitative studies have shown that there are about five thousand

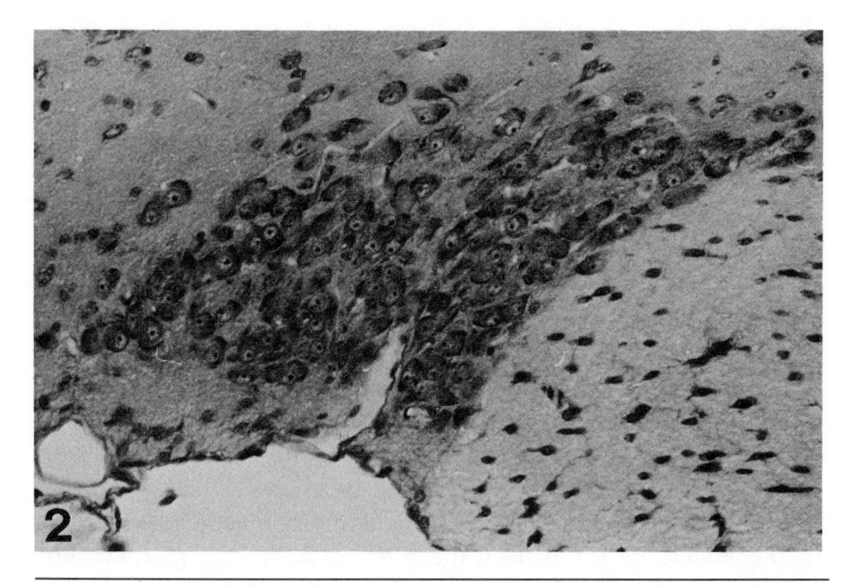

Figure 3.2 The SON appears as a compact mass of large neurons attached to the lateral borders of the optic chiasm. SON neurons have very large somas with Nissl substance located in a marginal position. The cell nucleus displays dispersed chromatin and nucleoli can be distinguished easily. (×250)

neurons in each SON in the rat and that this number remains constant from newborn to adult animals (Crespo et al., 1990). The final number of cells in any region of the developing nervous system is determined by a combination of progressive and regressive processes (Cowan et al., 1984), such as cell proliferation and naturally occurring cell death. Cell death is a phenomenon that takes place in most regions of the CNS. In this regard, NSNs of the SON constitute a stable cell population during the postnatal period; in fact, the number of NSNs remains constant even in very old rats (McNeill, Clayton, and Sladek, 1980).

Immunocytochemical studies have shown that there are equal numbers of OT and VP neurons in the rat SON, but some topographic divisions can be made: VP neurons are predominantly situated ventrally (Figure 3.3) whereas OT neurons are located more dorsally in relation to other hypothalamic areas (Silverman and Zimmerman, 1983). Not all SON neurons are neurosecretory; 6 percent of the total neuronal population (Léránth et al., 1975) are small, nonse-

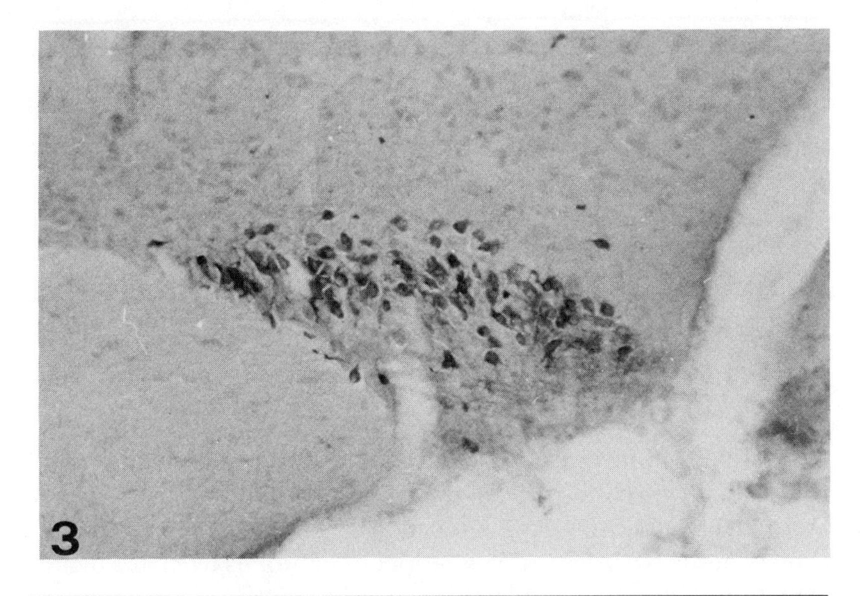

Figure 3.3 Light micrograph showing immunocytochemical staining with anti-oxytocin. Oxytocinergic neurons are predominantly located in a dorsal position in the SON. (×250)

cretory interneurons that play an important role in the local synaptic microcircuitry of this nucleus. A reduced number of dendritic branches emerge from the perikarya (1–3) and extend to the ventral glial lamina, where most of the synaptic input to these cells takes place. SON axons course medially to the neurohypophysis and are incorporated into the pathway coming from the PVN, forming the hypothalamo-neurohypophyseal tract. Autoradiographic studies (Alonso and Assenmacher, 1981) have demonstrated that both tracts keep quite an orderly arrangement while going down to the neurohypophysis through the pituitary stalk: the SON axons travel ventrally in the median eminence and the PVN axons are placed more caudally; these latter axons are restricted to the more peripheral layers of the neural lobe, while SON axons are scattered throughout the whole organ. A characteristic of these neurosecretory axons is the presence of swellings along the nerve fibers. These bead-like structures, termed "Herring bodies," contain aggregates of NSGs, and their number and size increase notably, with a concomitant reduction in the interswelling distance, during parturition and dehydration.

Hypophysis extirpation induces a drastic reduction (73 percent) in the number of neurons of the SON, mainly because of the elimination of those cells located in the most ventral region near the ventral glial lamina (Crespo et al., 1990). This reduction in the number of SON neurons after hypophysectomy is a consequence of retrograde degeneration; the maintenance of the remaining neuronal population (27 percent) is probably due to the fact that these neurons present axon collaterals before the median eminence or to the fact that they project directly to another extraneurohypophyseal system.

Nuclear Events

Several lines of investigation have demonstrated that the nucleolus is directly involved in ribosomal RNA (rRNA) synthesis and that its morphology and size vary as a function of the level of transcriptional activity of the rDNA genes. Transcription, in turn, is regulated by the cytoplasmic demand of rRNA for protein synthesis (Sommerville, 1986). Physiological conditions, such as the diurnal cycle (Armstrong and Hatton, 1978) and lactation (Russell, 1980), can modify the level of neurosecretory activity in SON neurons and, consequently, induce variations in the number, size, morphology, and location of nucleoli. In a pioneering work, Hatton, Johnson, and Malatesta (1972) observed that desert rodents, which live under short water supplies, excrete very concentrated urine and that their SON cells present higher proportions of multiple nucleoli than are found in laboratory rats (20–69 percent versus 0–20 percent). When desert animals are maintained under laboratory conditions with free access to water, the number of multinucleolated neurons decreases appreciably. Smear preparations of whole neurons impregnated with silver for preferential nucleolar staining have revealed that 68 percent of rat NSNs are mononucleolated (Crespo et al., 1988). During osmotic stress, induced by giving the animals a 2 percent NaCl solution in the drinking water for one week, most of these neurons (60 percent become multinucleolated; after one week of recovery the normal values are restored. Using quantitative *in situ* hybridization to investigate changes in levels of rRNA in neurons of control and dehydrated rats, Kawata, McCabe, and Harrington (1988) found that the amount of rRNA in these NSNs was significantly increased by osmotic stimulation. This transient increment both in nucleoli numbers and rRNA synthesis during dehydration represents an early cellular adaptive response to high hormone-synthesis demands. At the nucleolar level this represents an augmentation in rDNA gene transcription for the synthesis of rRNA precursors.

By the same token, nucleolar morphology reflects the functional activity of this structure. During dehydration, ribosomal cistrons of the nucleolar organizer regions (NORs) are very active (Goessens, 1984) producing an increment in the nucleolar granular and fibrillar components. There is an increment in the number of fibrillar centers and in their dense fibrillar component, where ribosomal DNA is located and transcribed (Wachtler et al., 1989). An increase is observed in the number of vacuoles or interstices in the nucleolar matrix, which have been related to the storage and transport of pre-rRNA precursors; consequently, the nucleoli display a reticulate appearance. These processes lead to an increment in nucleolar volume as a result of the activation of synthesis, processing, and transport of rRNA precursors. Concomitantly with these structural changes, the nucleolus moves from the center of the nucleus toward the periphery to establish contact with the nuclear membrane, thus occupying a privileged position for transferring rRNA precursors to the perikaryon (Figure 3.4).

The cellular content of mRNA, which encodes a protein, is an index of specific protein-synthesis activity. In this sense, several *in situ* hybridization studies (Majzoub et al., 1983; Sherman et al., 1986) have shown that osmotic stimulation increases both VP and OT mRNA content, and this increment is closely related with the duration of the dehydration period (for references, see Kawata, McCabe, and Harrington, 1988). As a direct consequence of the overactivation of neurohormone mRNA synthesis, the average size of the NSN nuclei also increases. However, our experiments (unpublished data) have revealed that, in contrast to the nucleoli, the nuclei do not return to their original volume in a recovery period of one week. This asynchrony in the recovery of the cell nucleus and the nucleolus might be attributable to a different RNA turnover rate, which in turn is due to the fact that newly synthesized rRNA is more stable and has a longer life than neurohormone mRNA.

Cytoplasmic Events

A great deal of work has been done on the cytoplasmic changes in NSNs under stimulation and on the concomitant hypertrophy and

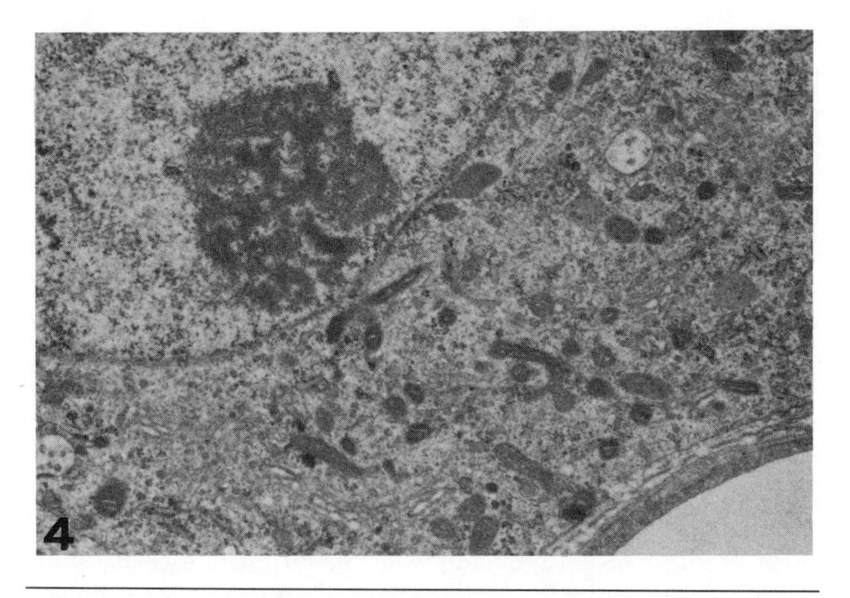

Figure 3.4 Immunoelectron micrograph of the SON of a dehydrated rat. The nucleolus displays a reticulate appearance and establishes contact with the nuclear membrane. (Immunocytochemical staining with neurophysins, ×30,000)

proliferation of perikaryal organelles. Extensive reviews of these events have been published (Defendini and Zimmerman, 1978; Castel, Gainer, and Dellman, 1984). For this reason we will limit our attention in this section to some of the most relevant aspects. Stimulated neuronal activity in the CNS frequently leads to cellular hypertrophy. When NSNs are osmotically stressed, they undergo a 1.5-fold increase in size, and this increment is related to volume increases in both nuclei and perikarya (Figure 3.5). This morphometric increment corresponds at the ultrastructural level with an augmentation in the cytoplasmic protein-synthesis machinery and the appearance of aggregates of secretory granules in the perikaryon to be transported through the axon. When rats that have been hyperosmotically stressed for one week are given normal drinking water for a period of seven days, NSNs recover their normal cytoplasmic volume. During this recovery period the lysosomal content of the cytoplasm, mainly secondary lysosomes, increases to ensure the recycling of cellular organelle components.

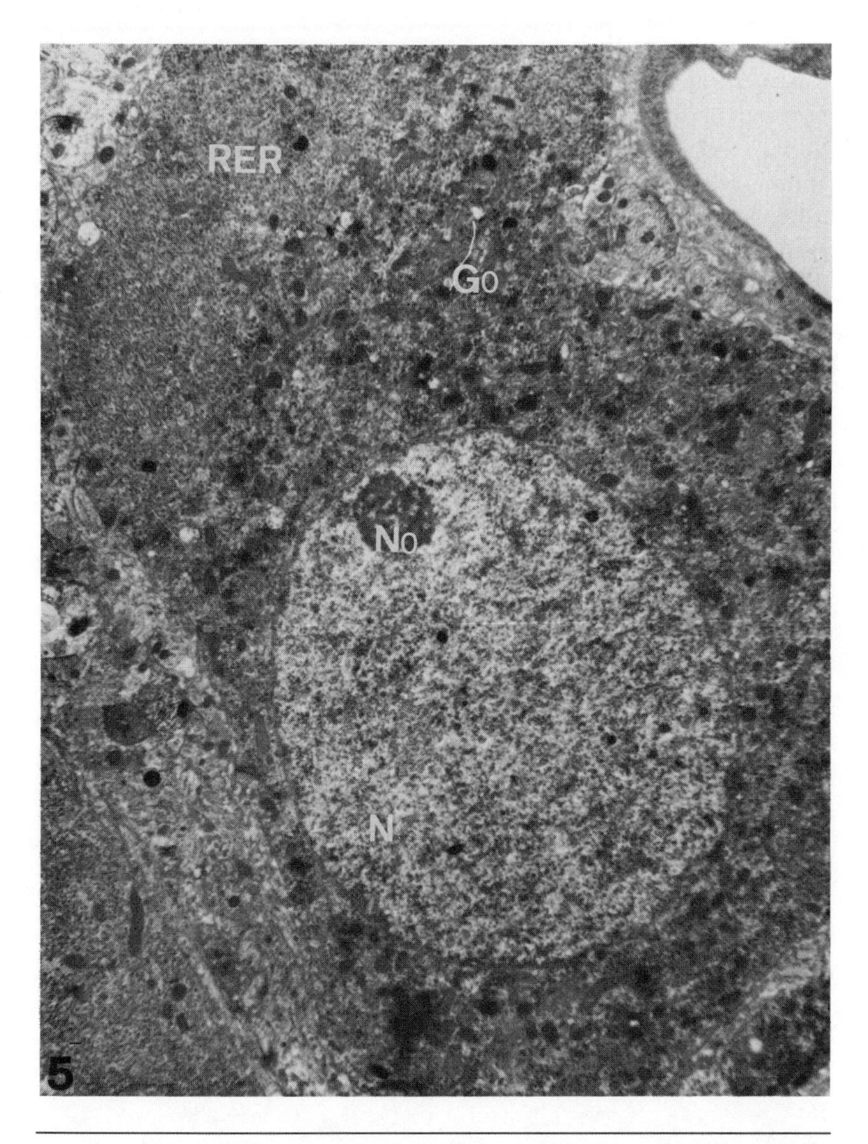

Figure 3.5 Low-power immunoelectron micrograph of the SON of a saline-drinking rat. *N*, nucleus; *No*, nucleolus; *Go*, Golgi complexes; *RER*, rough endoplasmic reticulum. (Immunocytochemical staining with neurophysins, ×15,000)

The Nissl substance in these NSNs, whose counterpart at the ultrastructural level is the RER, typically occupies a marginal position in the perikaryon, in close proximity to the cell membrane. Conventional electron microscopy reveals that the RER of control rats is composed of three to seven stacks of flattened laminar cisternae, but in water-deprived rats the RER is composed of a greater number of stacks (up to twenty), and the cisternae appear dilated and occupied by a moderately electron-dense content that represents the newly synthesized prohormone (Figure 3.6). In thick sections impregnated with heavy-metal salts and studied with the aid of high-voltage electron microscopy (Alonso and Assenmacher, 1979), most of the RER in osmotically stimulated NSNs displays a configuration of intricately anastomosed three-dimensional tubulae. Using pre-embedding immunocytochemistry with antibodies to neurophysin in mouse SON neurons, Broadwell and Brightman (1979) reported for the first time the presence of immunoreactive neurophysin in the RER lumens.

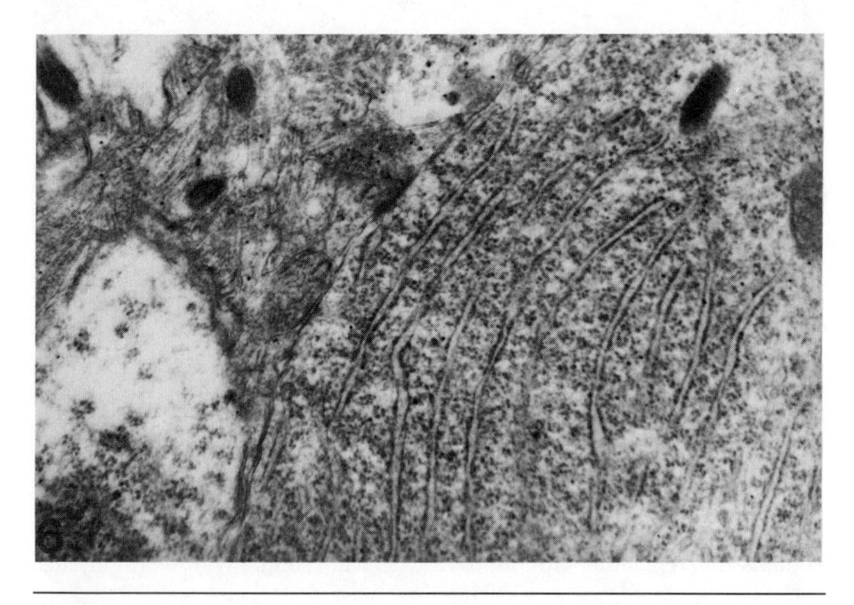

Figure 3.6 Electron micrograph of the SON of a dehydrated rat. The RER occupies a typical marginal position in close proximity to the cell membrane. The lumens of the cisternae appear to be occupied by a moderately electron-dense content. (×25,000)

The Golgi complexes have a perinuclear disposition and they receive the prohormones transferred from the RER (Figure 3.7). In their lumens the prohormones are condensed and sulfated; in addition the VP prohormone suffers terminal glycosylation (North, 1987). Several lines of investigation (Gainer, Sarne, and Brownstein, 1977; North et al., 1983) have proposed that in the Golgi the prohormones are associated with several enzymes, which at this level are inactive and packed into granules 100–200 nm in diameter, the neurosecretory granules (NSGs). In the course of axonal transport to the posterior lobe these NSGs suffer a change in internal pH. This change activates the proteases, leading to several post-translational prohormone cleavages and to the secretion of the final products by exocytosis (Brownstein, Russell, and Gainer, 1980).

Neuron-Neuron and Neuron-Glia Relationships

In normal conditions, adjacent NSNs are separated from each other by the interposition of fine astrocytic processes (Figure 3.8), which

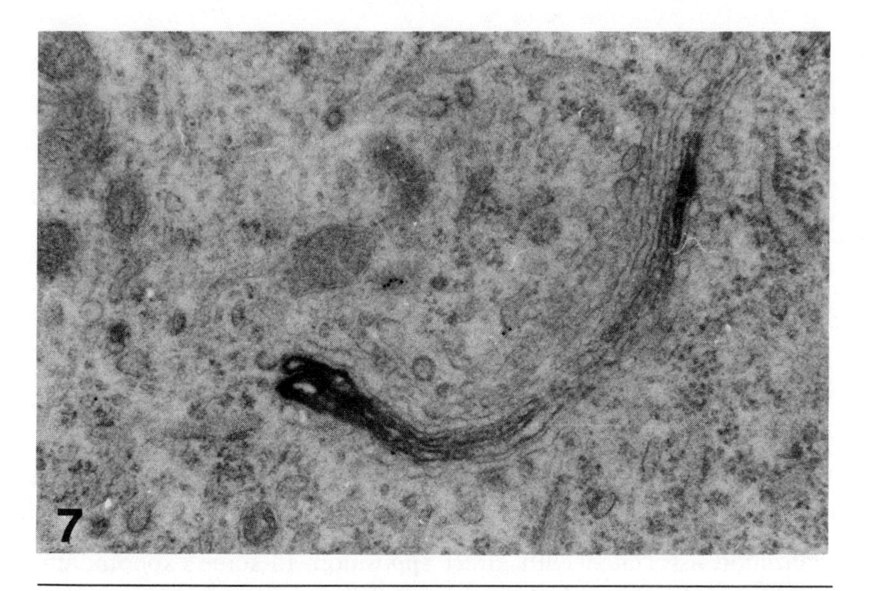

Figure 3.7 After double immunostaining (PAP for neurophysins and immunogold for OT), the Golgi complex shows PAP immunoreactivity and the neurosecretory granules (NSG) for OT. (×35,000)

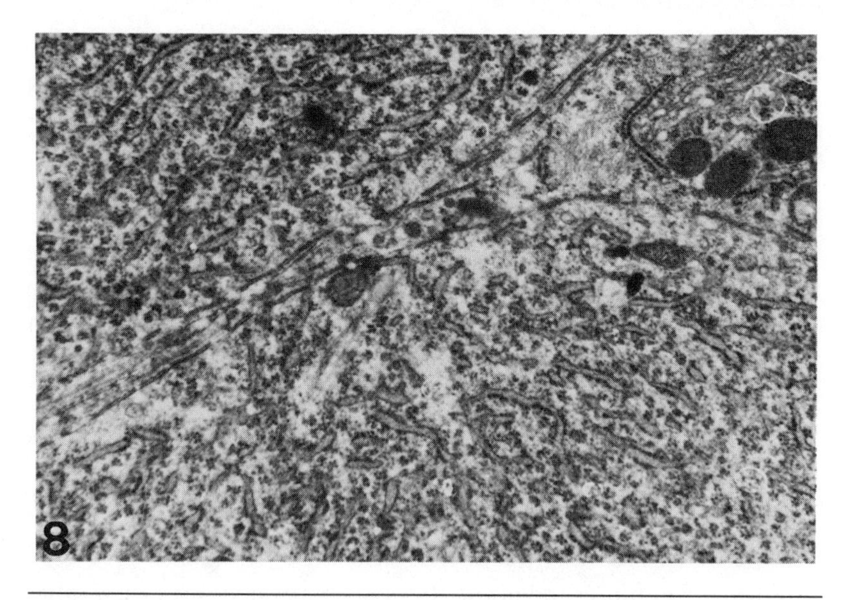

Figure 3.8 Two NSNs (from a control rat) separated from each other by the interposition of a fine astrocytic process. (Immunostaining for neurophysins, ×30,000)

cover 84 percent of the neuron surface, and direct neuronal juxtapositions are both rare and limited to small zones (Modney and Hatton, 1989). Several quantitative studies have established these appositions to represent 1–2 percent of the total cytoplasmic membrane (Theodosis et al., 1986; Chapman et al., 1986; Modney and Hatton, 1989). Significant differences have been found in the extent of direct juxtaposition among adjacent NSNs as a consequence of physiologic or induced conditions (such as lactation or dehydration, respectively). In dehydrated animals up to 8 percent of the neuronal surface area is in direct contact with neighboring neurons (Figure 3.9), a four- to eight-fold increment with respect to control values (Modney and Hatton, 1989).

From a morphological point of view there is no junctional specialization associated with direct apposition. In some experimental conditions, however, intermediate cell junctions may appear, and these are usually referred to as puncta adherentia. Their specific role has not yet been established (Mugnaini, 1982), although they are

Figure 3.9 Direct juxtaposition between two adjacent NSNs as a consequence of dehydration. (Immunostaining for neurophysins, ×30,000)

common in fetal SONs during the active synaptogenesis period (Sikora and Dellmann, 1980). A relationship has been established between these specializations and the amount of neuronal contact: their number increases in direct proportion to neuronal juxtaposition (Chapman et al., 1986). Andrew et al. (1981) used intracellular injections of the fluorescent dye Lucifer yellow into *in vitro* single NSNs to prove that there is dye transfer between these neurons. In another approach, using freeze-fracture replicas of the same material, these authors have occasionally observed the presence of gap junctions between NSNs. This type of junction is common between SON astrocytes. Although there were close neuronal body appositions, most of the cellular coupling was between different dendrites or between dendrites and somas.

When the stimulus that induced these modifications ceases, there is a gradual return to a normal state. In addition to recovery of normal neuronal size, direct membrane apposition diminishes and glial processes reappear between neurons. Several investigations have related the recovery time to the duration of the stress. Thus, after pregnancy and lactation recovery takes one month; after two consecutive pregnancies and lactations it takes two months (Theodosis and Poulain, 1984a). When lactation is suppressed after delivery, reorganization takes only two days (Montagnese et al., 1987). In the case of induced dehydration for ten days, normal values have been reported after five days of rehydration (Tweedle and Hatton, 1984).

Using anti-OT serum, Chapman et al. (1986) have observed at the ultrastructural level that these changes affect OT neurons selectively: they found no variations in the nonstained neurons, which were inferred to be VP neurons. At present, there is no explanation for this difference in response between VP and OT cells, and the mechanism(s) underlying these responses are unknown. A difference in growth rate between OT and VP neurons could provide a cellular mechanism. Thus, if OT neurons suffered greater volume increases than VP cells, some of these differences could be explained. However, quantitative studies have reported similar growth rates in both populations. Second, there may be a special relationship between astrocytes and OT neurons that leads to these differences. Astrocytes have an important function in regulating the ionic composition of the perineuronal space, mainly the extracellular content of Na^+ and K^+. As a result of synaptic activity, K^+ tends to increase in the extracellular space, so these astrocytic changes may represent a process of potassium buffering to reduce K^+ concentrations. At the same time, astrocytes are known to be involved in the metabolism of several neurotransmitters. One of them, γ-aminobutiric acid (GABA), has a very important inhibitory action and is found throughout the SON. Thus, specific glial changes could represent the morphological counterpart of an enhanced inhibitory action on OT neurons that regulates their firing responses. Changes similar to those found in cell somas have been reported in the dendrites of these NSNs. Dendro-dendritic membrane coupling occurs after both chronic and acute dehydration, but some differences have been observed in the recovery process. Longer dehydration requires longer recovery periods. This suggests that the capacity of recovery is directly dependent on the duration of dehydration. In any case, soma membranes recover earlier than their related dendritic membranes (Perlmutter, Tweedle, and Hatton, 1985). There are two possible explanations for this difference in recovery rate between soma and dendrites. In the first place, it could be related to a differential process starting at the soma and reaching the dendrites after a delay due to a slower turnover rate of the dendritic membrane. Second, it could be linked to the fact that most neural inputs to NSNs are through their dendrites, and synaptic reorganization at this level may provoke a delay in dendritic recovery.

Synaptic Organization and Plasticity

Electron microscopic studies (Léránth et al., 1975) revealed three types of synapses onto NSNs: axo-somatic, axo-dendritic (impinging onto a smooth dendritic membrane or a dendritic spine), and axo-axonic synapses. Quantitative analyses have established an average of 600 terminals per neuron (Léránth et al., 1975). Several different neurotransmitters have been immunocytochemically demonstrated in SON synapses (Swanson and Sawchenko, 1983). Among these, noradrenaline (NA) and adrenaline (A) terminals are well characterized, and the SON receives a dense catecholaminergic projection from several cell groups located in the medulla. The A1 group of the ventral medulla sends projections that make synaptic contacts primarily in the ventral region of the SON, where VP neurons are located. The A2 group from the medial part of the nucleus tractus solitari and the A6 group from the locus coeruleus send the same type of terminals to this nucleus. These catecholaminergic pathways are responsible for the transmission of visceroceptive information to the SON (Sawchenko and Swanson, 1981). Terminals containing serotonin have been reported in the dorsal region of the SON, where OT neurons are located (Loewy, Wallach, and McKellar, 1981).

The major inhibitory innervation of the SON is represented by GABA-ergic terminals. Quantitative studies by Theodosis, Paut, and Tappaz (1986b) estimated that half of the synapses onto NSNs are of this type. These GABA-ergic terminals are evenly distributed over both OT and VP neurons of the SON, and they terminate on cell bodies, axons, and dendrites (Buijs, Van Vulpen, and Geffard, 1987). No GABA-positive neurons have been reported in the SON, but GABA neurons have been observed in the dorsal vicinity of the SON (Theodosis et al., 1986) and in the diagonal band of Broca (Buijs et al., 1987). Although there is no conclusive evidence, these regions could represent the origin of these terminals in the SON.

Oxytocinergic terminals have been reported on OT neurons within the SON (Theodosis and Poulain, 1984), and Pow and Morris (1988) have presented evidence of exocytosis of OT NSGs in both soma and dendritic membranes without relation to any synaptic specialization. During milk ejection there is a significant increase in oxytocin release within the SON, but there is no change after osmotic stimulation. It has been suggested that this local increment in oxytocin release is a prerequisite for the onset and maintenance of the characteristic intermittent electrical activity of OT NSNs leading to milk ejection (Moos et al., 1989).

During suckling and parturition a pattern of periodic high-frequency bursts of action potentials is evoked (Summerlee, 1981), inducing a pulsating release of oxytocin into the bloodstream. In response to osmotic stimulation, a continuous tonic activation occurs in OT neurons, which produces a sustained release of OT in contrast to the phasic pattern of electrical activity of VP neurons (Brinble, Dyball, and Forsling, 1978). This local oxytocin release, both synaptic and paracrine, may have some physiological effect on the milk-ejection reflex, since it increases during the suckling but not during osmotic stimulation (Moos et al., 1989).

Synaptic terminals containing vasopressin have been described in the SON. It has been found that these inputs have two different origins. Occasionally, VP neurons that project to the neural lobe form recurrent axon collaterals to the SON, whereas VP neurons that present an intranuclear distribution of their axon and dendrites make local synaptic contacts (Ray and Choudhury, 1990). These terminals make contact with NSNs that project distally, and they could act as an intranuclear autoregulatory system. Further studies are necessary in order to elucidate their functional significance.

Colocalization of several neurotransmitters in the same terminal was first suggested after morphological studies revealed the appearance of synaptic vesicles of different size and content. More precise immunocytochemical techniques have demonstrated the colocalization of dopamine, galanin, and neuropeptide Y (NPY). Beroukas, Willoughby, and Blessing (1989) found NPY together with NA in some SON terminals in areas innervated by A1 projections, and these terminals formed synapses on the soma and proximal dendritic trunks of VP neurons. Functional studies (Willoughby and Blessing, 1987) have suggested that coliberation of NPY and NA may control

VP liberation, since NPY injections into the SON cause VP secretion in unanesthetized rats.

Tweedle and Hatton (1984) were the first authors to report the appearance of multiple synapses (one terminal contacting two or more postsynaptic elements) in the mature SON after chronic dehydration, but previously Sikora and Dellman (1980) had reported the presence of this type of synapse during the prenatal development of the SON (Figure 3.10). In quantitative ultrastructural studies of lactating rats, Theodosis and Poulain (1984) found that these synapses couple 20 percent of all OT neurons but only 3 percent of VP cells. In a recent study, Modney and Hatton (1989) have observed that dehydrated animals show a three-fold increase in the percentage of terminals forming multiple synapses in comparison with control animals, and a number of these terminals are "triple" connections, with a presynaptic bouton contacting two dendrites and a soma or two somas and a dendrite. Most of these multiple synapses are GABA-ergic (Theodosis, Paut, and Tappaz, 1986), but other neurotransmitters may be involved since not all these synapses present a

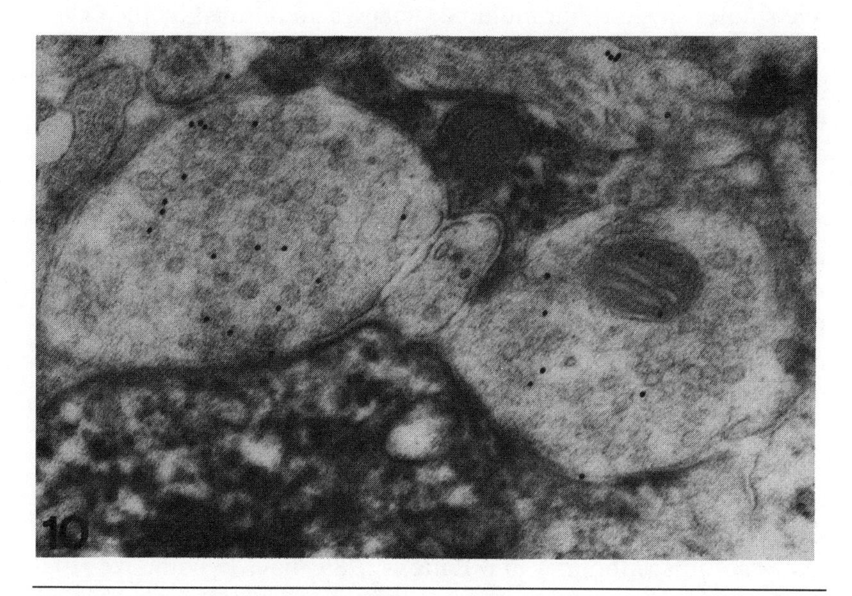

Figure 3.10 Electron micrograph of the SON of a dehydrated rat (with double immunostaining: PAP for neurophysins and immunogold for GABA) showing a double synapse. (×45,000)

positive reaction to GABA. In control animals, ICC electron microscopic studies have shown the presence of dopamine in some of these multiple synapses (Buijs et al., 1984). The fact that most of these multiple synapses are GABA-ergic and terminate onto OT cells makes this relation quite intriguing.

Neither the process of formation of multiple synapses nor their functional role is well understood yet. Synaptogenesis requires the participation of the pre- and postsynaptic elements for the formation of membrane specializations that are able to handle with neural transmission. At present there is a great deal of controversy about the cellular and molecular mechanisms that result in synaptogenesis—specifically, whether the presynaptic element induces the formation of its postsynaptic counterpart or vice versa. In the case of multiple synapses, the fact that different neurotransmitters have been found in the terminals indicates that multiple synapse formation is not closely related to neurotransmitter specificity.

The appearance during stress of neuron membranes that are normally covered by glial processes represents the transient exposure of new membrane areas that are potential postsynaptic loci. Quantitative studies suggest that multiple synapses are formed by the development of new presynaptic areas on previously existing single terminals, since the overall number of contacts does not increase (Modney and Hatton, 1989). Postsynaptic membrane specializations are very dynamic structures and can be formed very quickly. In this sense, it has been proposed that the appearance of membrane thickening during ontogenesis is the first morphological event of synapse formation (Larramendi, 1969), and thickening of the newly exposed membrane may influence the formation of a second presynaptic site. Desmosome-like structures similar to the puncta adherentia of the perikaryon appear on synapses during dehydration, but it remains uncertain whether or not these junctions are synapse precursors.

Electrophysiological studies of dehydrated rats (Poulain and Wakerley, 1982) have evidenced changes in the firing pattern of OT neurons, suggesting that this type of synapse may represent the morphological basis of firing synchronization during dehydration. Other authors (Mondney and Hatton, 1989) have hypothesized that these double synapses may act as a compensatory mechanism to maintain synaptic efficacy in response to the increase in neuronal membrane area after stress. Another explanation that seems quite plausible is that the preferential development of multiple synapses on oxytocinergic elements may be related to enhanced postsynaptic induction in these neurons.

Vascular Changes

An important characteristic of the magnocellular neurosecretory system is the richness of its microvasculature in comparison with other hypothalamic areas (Ambach and Palkovits, 1979). This rich capillary plexus is the major source of nonneural inputs to the SON. Sposito and Gross (1987) have reported that the rat SON has a capillary density 2–4 times greater than that of the gray matter and 8 times greater than that of the white matter. SON capillaries have a blood-brain barrier (BBB) structure (Figure 3.11): they are of the continuous type of capillaries and have endothelial cells joined by tight junctions, few plasmalemmic vesicles, and pericytes attached

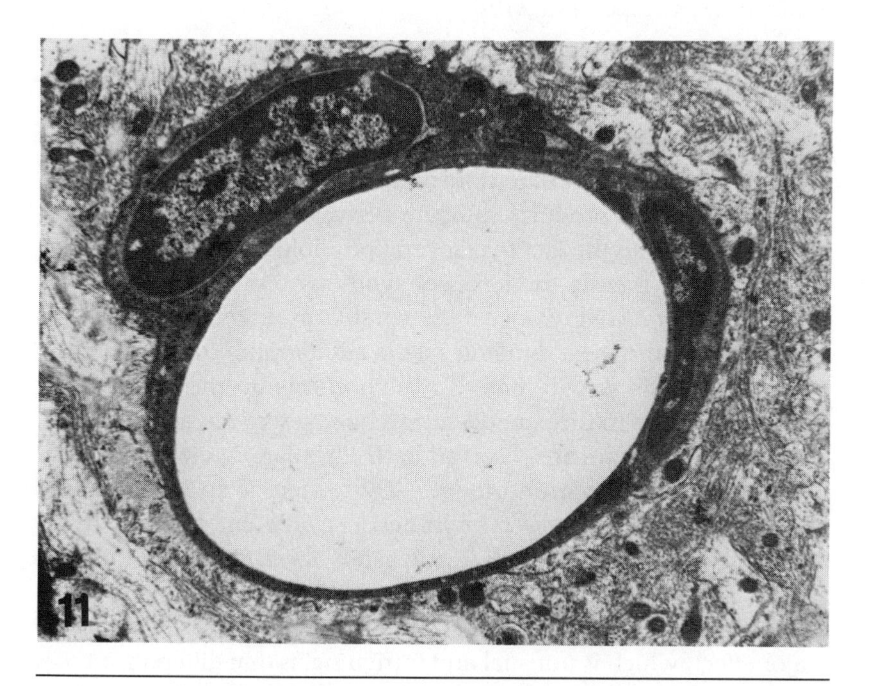

Figure 3.11 Electron micrograph of a blood vessel in the SON of a dehydrated rat. Capillaries of the SON have a blood-brain barrier structure. (×25,000)

to the capillary walls. Taking into account the structural organization of the capillary wall, the high degree of vascularization together with the large capillary surface area, and the low rate of transendothelial exchange, we can see that small changes in the plasma osmolality will produce drastic variations in osmotic plasma pressure on the capillary wall. This has led Gross et al. (1986) to postulate that the SON could act as an osmometer. Furthermore, SON neurons present solitary cilia that might function as receptors of the changes in pericellular osmolality (Lafarga et al., 1980).

It has been found that the SON of dehydrated rats contains more capillaries than are present in control tissue. Moreover, images suggesting vascular sprouting appear under osmotic stress. These vascular responses to dehydration may be a consequence of two factors: an increment in the plasma concentration of sodium chloride, which produces rapid changes in the osmolality of the SON perivascular space, and/or an increased demand in the local metabolism resulting from stress, as has been reported for other CNS regions (Siveraag et al., 1988; Crespo et al., 1989).

Conclusions

Peptidergic neurons represent a model for analyzing the cell biology of neurons that synthesize, transport, and release neurohormones. They release their products into the perivascular space of the neurohypophysis, and the fact that several physiologic and experimental conditions can modify the levels of synthesis activity, together with the property of recovery once these conditions disappear, makes this system one of the most dynamic of the adult brain. Enhancement of protein-synthesis activity has two main effects on these cells apart from the changes in the general cytoarchitecture of the nucleus. First, morphologic changes are observed in the cellular organelles involved in the synthesis of neurohormones. Thus, there is an increase both in the nuclear machinery for synthesis of different types of RNAs and in its counterpart at the cytoplasmic level for neurohormone synthesis. Although these changes are important, they are quite similar to those observed in other protein-secretory cells. The second major effect, which is unusual and intriguing, is that different stresses lead to complex rearrangements in the spatial relationships of these neurons; the increase in cell volume is accompanied by the formation of new neuronal membrane to allow for cell growth. This new membrane usually establishes a close relationship with a neighboring neuron, forming direct appositions between adjacent cells. At the same time, new neuronal connections onto these stimulated neurons are formed.

These dynamic events are adaptive changes that are reversible. This capacity of response and adaptation in the adult hypothalamus may be a model for analyzing the effect of environmental alterations on the maintenance of cell structure and function. In this context, there are several studies which indicate that sex hormones may play a role in this process. Sex hormones such as steroids modulate the electrical activity of magnocellular neurons (Akaishi and Sakuma, 1985), and increased plasma levels of these hormones occur during dehydration and lactation (Ezzarani et al., 1985). After ovariectomy, there are cellular changes that are similar to those observed after dehydration (Crespo et al., 1989). These findings suggest that the changes observed in these NSNs may be dependent, in part, on the concomitant influences of other factors, like estrogens, as has been described in other regions of the CNS (McEwen, 1981).

In summary, NSNs of the SON offer a useful model for studying functional neuronal adaptability in general and synaptogenesis in particular in the adult CNS. The formation of new afferent inputs in a preexisting framework of connectivity and their subsequent elimination suggest the possibility that functional activity and/or the presence of a putative synapse-inducing factor may lead to the preferential appearance of new synaptic contacts from preexisting ones in one type of SON neurons. A deeper knowledge of the mechanisms involved in this phenomenon will provide a better understanding of the rules that control this plastic restructuring.

Acknowledgments

This work was supported by Grant PM88-0168 and a Grant to Research Personnel for Overseas Study from DGICYT (Spain). The author thanks M. Castel and H. Gainer for the gift of antiserum and technical advice. S. Cohen, C. Viadero, and E. Hattab provided excellent technical assistance. Part of this work was carried out at the Institute of Life Sciences, Hebrew University of Jerusalem, under the sponsorship of M. Castel, to whom the author is deeply indebted.

4

Pathogenesis of Late-Acquired Leptomeningeal Heterotopias and Secondary Cortical Alterations: A Golgi Study

Miguel Marín-Padilla

The presence of glial cells and their processes, axonic fibers, and an occasional neuron within the leptomeningeal space overlying the cerebral cortex has been described in a variety of genetic and acquired encephalopathies, including dyslexia (Brum, 1965a,b; Friede and Mikolasek, 1978; Cavines et al., 1978; Clarren et al., 1978; Clarren and Smith, 1978; Galaburda and Kemper, 1979; Wisniewski et al., 1983; Williams and Cavines, 1984; Galaburda et al., 1985; Freeman, 1985; Larroche and Razavi-Encha, 1987; Choi and Matthias, 1987; Barth, 1987; Friede, 1989; Cavines et al., 1989; Evrard et al., 1989). Names used to describe these superficial lesions include: marginal cortical dysplasias, brain warts, status verrucosus, marginal glioneuronal heterotopias, glial nests, meningeal ectopias, and leptomeningeal heterotopias (LMHs). The last of these names will be used in this chapter.

Microcephaly, hydrocephaly, microgyria, megalencephaly, agyria-pachigyria, and hydranencephaly are among some of the developmental encephalopathies with LMHs. Acquired disorders associated with these lesions include multicystic encephalopathies, microgyria, porencephaly, perinatal infections, and fetal alcohol syndrome. The pathogenesis of these lesions remains poorly understood (see Chapter 5). Most investigators consider them abnormalities of neuronal migration (Barth, 1987; Sarnat, 1987; Larroche and Razavi-Encha,

1987; Choi and Matthias, 1987; Evrard et al., 1989; Friede, 1989). One group of LMHs, however, occurs late in gestation, when neuronal migration is completed, and these are not expected to interfere with that developmental process. Some of these late-acquired LMHs result from direct injury and subsequent rupture of the cerebral cortex's external glial limiting membrane (EGLM). This type of lesion is often found in the brain of prematurely born infants who have suffered from perinatal asphyxia, and it is usually associated with hypoxic-ischemic and/or hemorrhagic cortical damage (Marín-Padilla, personal observations). The histopathology and pathogenesis of these late-acquired LMHs will be analyzed and discussed in this chapter. The structural alterations of the cerebral cortex underneath these LMHs will also be analyzed and their clinical relevance discussed.

First, some of the major events that occur in the course of prenatal development of the human cerebral cortex will be outlined, and the structural composition and role of the CNS external glial limiting membrane will be briefly discussed. An understanding of these aspects of cortical development is crucial for a meaningful discussion concerning the nature, composition, pathogenesis, and clinical relevance of late-acquired cortical LMHs.

Prenatal Ontogenesis of the Human Cerebral Cortex

In the course of prenatal development, the human cerebral cortex undergoes a series of fundamental transformations. These could be grouped into three major events, namely: the establishment of a primordial cortical organization, the process of neuronal migration resulting in the inside-out formation of the cortical plate (gray matter), and the process of neuronal maturation resulting in the inside-out structural and functional organization of the neocortex. Each one of these events occurs during the course of a distinct period of development (Figure 4.1). Hence, three developmental periods, *the embryonic, the fetal, and the perinatal,* should be distinguished in the course of human prenatal development (Marín-Padilla, 1983, 1988a, 1990a).

The embryonic period covers the first eight weeks of gestation and includes the first 22 stages of human development (Figure 4.1A). The cerebral vesicles become recognizable in the telencephalon at

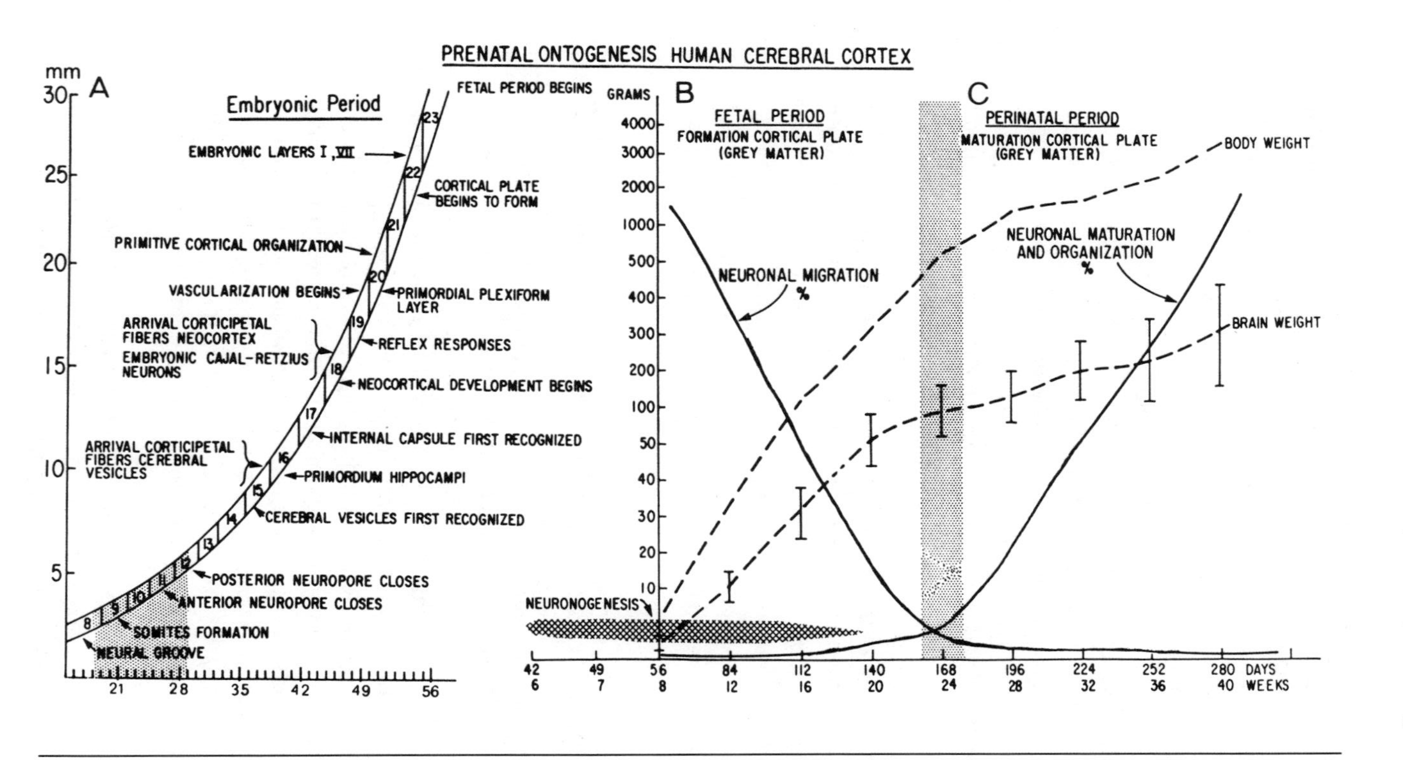

Figure 4.1 The major developmental events of the embryonic *(A)*, fetal *(B)*, and perinatal *(C)* periods of the prenatal ontogenesis of the human cerebral cortex. The embryonic period is characterized by the appearance of the cerebral vesicles, the arrival of the first corticipetal fibers, and the establishment of the primordial plexiform layer (PPL) prior to the formation of the cortical plate. The fetal period is characterized by the process of neuronal migration and the inside-out formation of the cortical plate. The perinatal period is characterized by the ascending neuronal differentiation and maturation of the cortical plate (gray matter). (Modified from Marín-Padilla, 1990a.)

stages 14–15, and neocortical development begins at stages 17–18 (Larroche, 1981; Marín-Padilla, 1983; O'Rahilly and Müller, 1987; Müller and O'Rahilly, 1988a,b). First, a superficial primordial plexiform layer (PPL) is established in the developing neocortex. The PPL is thought to represent a premammalian-like cortical organization that may be shared by amphibians, reptiles, and mammals (Marín-Padilla, 1971, 1972, 1978). It is superficial (external) plexiform lamina composed of primitive corticipetal fibers and neurons (associative and projective) scattered among the fibers. Its establishment is a prerequisite for the subsequent development and inside-out formation of the cortical plate (Marín-Padilla, 1978, 1983, 1988a 1990a,b; Luskin and Shatz, 1985; Bayer and Altman, 1990). The cortical plate, considered a more recent mammalian innovation, starts to form *within the PPL* at stage 22, marking the beginning of the fetal period. The PPL gives rise to layers I and VII (subplate zone) and the cortical plate to layers VI, V, IV, III, and II of the adult cerebral cortex.

The fetal period starts with the appearance of the cortical plate, about the eighth week of gestation ("w-g" hereinafter), and continues throughout its progressive inside-out formation (ca. 24th w-g) (Figure 4.1B). This period is characterized and dominated by the process of neuronal migration. Human cortical neuronogenesis starts, throughout the ependymal (ventricular) surface, immediately after the formation of the cerebral vesicles (stages 14–15). Neuronal migration starts following the arrival of primitive corticipetal fibers and continues through the 16th–20th w-g. Although cortical neuronogenesis may be completed by mid-gestation (ca. 18 w-g), the actual migration of neurons continues for a longer period. The distance between the neurons' ependymal origin and their final cortical destination varies from region to region and, consequently, the time spent in their migration varies. In general, neuronal migration is completed earlier throughout the anterior regions of the developing cortex than throughout its posterior or occipital regions. The greater percentage of cortical neurons have probably completed their migration by the 24th w-g.

The perinatal period (Figure 4.1C) is a distinct stage in the prenatal development of the human cerebral cortex (Marín-Padilla, 1990a). The 24th w-g has been arbitrarily chosen as the demarcation between the fetal and perinatal periods. This age generally represents the lower limit for the possible survival of prematurely born infants, and neuronal migration throughout the cortex should already be completed. The perinatal period is essentially characterized and dominated by the ascending differentiation and maturation of cortical neurons in response to the progressive inside-out penetration of corticipetal fibers. Consequently, this period is also characterized by the ascending structural and functional organization of the cortex's gray matter. During the perinatal period, neuronal plasticity is at its maximum. Cortical neurons respond to incoming fibers by an exuberant formation of complex but distinctive dendritic arborizations and by the establishment of an uncountable number of synapses, which eventually diminish as cortical development progresses (Easter et al., 1985; Rakic et al., 1986; Huttenlocker, 1982; Cowan et al., 1984; William and Herrup, 1988; Zecevic et al., 1989; Rakic, 1989).

During the perinatal period, the actively growing cerebral cortex is particularly vulnerable to perinatal asphyxia (Volpe, 1987). Infants born during this period often become patients in our hospital, and many of them require respiratory assistance. Neonatal intensive care units (NICU) have been established throughout the world for the care of these infants. Hypoxic, ischemic, and/or hemorrhagic brain damage often occurs in premature infants after periods of respiratory difficulty. A better understanding of the development and maturation of the human brain during this crucial perinatal period will undoubtedly improve the management and, we hope, help prevent perinatal brain damage.

The External Glial Limiting Membrane

The CNS, as an ectodermally derived tissue, is separated from the surrounding tissues by a distinct basal lamina (BL) that covers its entire surface (Figure 4.2). The surface of the developing and adult CNS and of layer I in the cerebral cortex is composed of the endfeet processes, in close and tight apposition, of glial cells. Together, these endfeet constitute an external glial limiting membrane (EGLM) that produces and maintains the CNS superficial BL (Andres, 1967; Krisch et al., 1982, 1983; O'Rahilly and Müller, 1986; Marín-Padilla, 1985, 1987, 1988b). The EGLM, as the anatomical demarcation or barrier of the CNS, plays an important role in maintaining both the structural and functional integrity of the CNS.

In the course of development, the EGLM is perforated only at

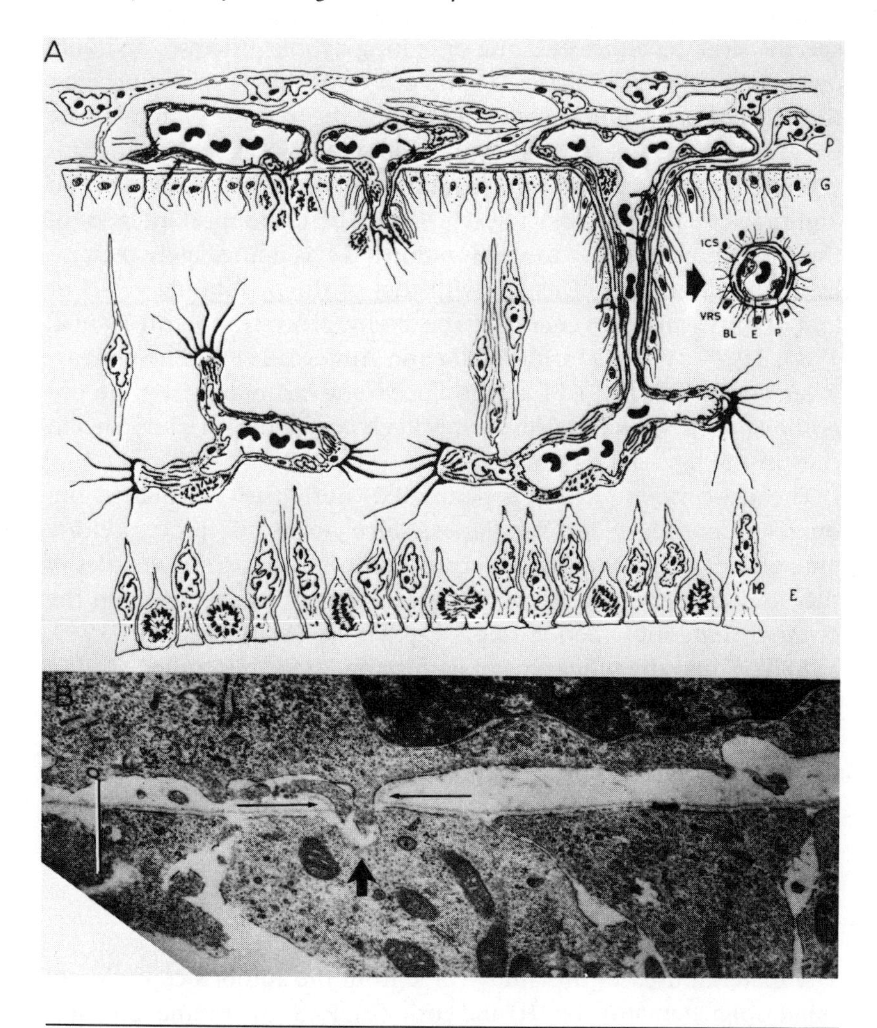

Figure 4.2 Ink drawing of the embryonic mammalian cerebral cortex *(A)* showing its external glial limiting membrane *(G)* and basal lamina, which is being perforated by leptomeningeal capillaries. Also illustrated is its ependymal (ventricular) surface *(E)* with the mitotic cells of early neuronogenesis. (Modified from Marín-Padilla, 1987.) Visible in the electron micrograph *(B)* is the close apposition of the glial endfeet processes forming the external glial limiting membrane covered by the CNS external basal lamina. Also illustrated are the perforating filopodia *(thick arrow)* of a capillary endothelial cell and the fusion *(thin arrows)* of the CNS basal lamina with that of the perforating capillary. Scale = 1 μ.

specific sites, by either entering or exiting axonic processes and only by entering blood capillaries (Figure 4.2A). At these perforation sites, the CNS basal lamina is perforated by the release of proteolytic enzymes by the axonic and/or the endothelial cells' advancing tips (Folkman, 1982; Marín-Padilla and Amieva, 1989; Farbman and Squinto, 1985; Dodd and Jessel, 1988). The anatomical integrity of the CNS, interrupted at the perforation sites, is immediately reestablished by the fusion of its BL with that of the perforating vessel or that of the sheath cells encircling the axonic processes (Marín-Padilla, 1985, 1987, 1988b; Marín-Padilla and Amieva, 1989). The perivascular and perineural BLs are produced and maintained by the endothelial and Schwann cells, respectively, or by the olfactory sheath cells in the first cranial nerve.

The neocortex's EGLM is perforated, during its vascularization, only by entering blood capillaries, since no axonic process either enters or exits from it. The entering arterioles and exiting venules of the adult cerebral cortex evolve by circulatory mechanics from the original entering (perforating) capillaries (Marín-Padilla, 1985, 1988b). Thus, any other type of perforation of the neocortex's EGLM is always a pathological condition. Such a perforation interrupts the cortex's anatomical integrity, permits the escape of nearby glial and/or neuronal elements and approaching fibers into the leptomeningeal space, and, consequently, allows a focal leptomeningeal heterotopia to become established at the perforation site.

Material and Method

The material used in this study came from the author's collection of rapid Golgi, hematoxylin (H) and eosin (E), PAS, and Holmes preparations of the cerebral cortex of premature infants (ranging in age from 20 to 36 weeks of gestation), newborns, and young children who came to autopsy for a variety of reasons. Immaturity, respiratory distress syndrome, bronchopulmonary dysplasia, intracerebral (matrix) hemorrhages, necrotizing enterocolitis, perinatal infections, aspiration of meconium, and congenital (cardiac, renal, CNS) malformations were the most common causes of death (Marín-Padilla, 1970a,b, 1988a, 1990a,b; Marín-Padilla and Marín-Padilla, 1982).

Seven cases have been selected to illustrate the histopathology of acute, subacute, and chronic forms of late-acquired LMHs. Acute forms are illustrated by three cases. Cases #8869 (Figure 4.3B) and

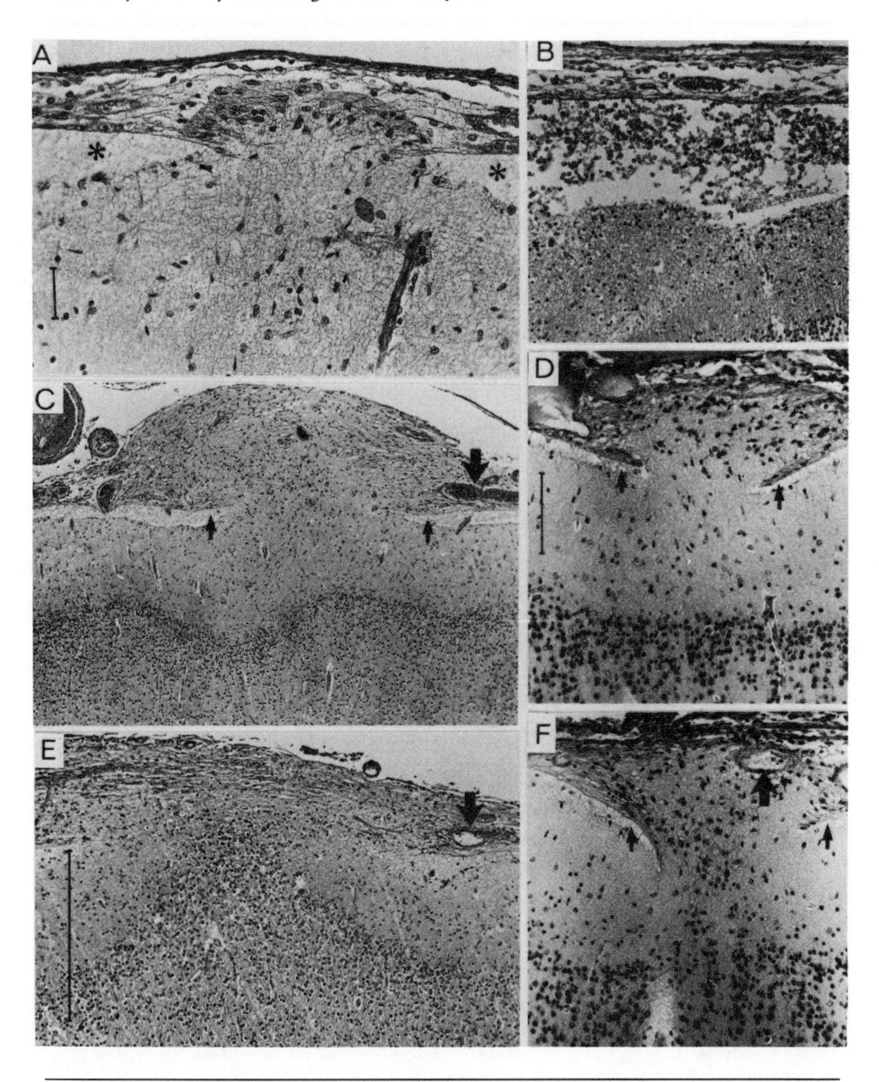

Figure 4.3 The neuropathologic features of acute *(A, B, D)* and subacute *(C, E, F)* late-acquired leptomeningeal heterotopias. Shown here are the severity *(asterisks)* and the persistence *(thin arrows)* of the subpial edema, the presence of acute subpial hemorrhages *(B)*, the separation of the nervous tissue from the pial surface *(A, B, C)*, and the presence of post-inflammatory vessels *(thick arrows)* that characterize these heterotopias. Scales = 100 μ.

#804493 (Figure 4.3A), both born at 29 w-g, suffered from severe perinatal asphyxia due to lung immaturity and respiratory distress syndrome, lived a few hours, and had severe acute brain damage (matrix and intraventricular hemorrhages). Case #9054 (Figure 4.3D), born at 38 w-g, had severe respiratory difficulty due to prenatal aspiration of meconium (premature rupture of membranes). At 5 days of age, a computerized tomography (CT) scan showed right frontal and left parietal hypoxic-ischemic damage of the white matter, and another CT scan 5 days later showed early cavitation at those regions. The infant lived 12 days and had extensive multicystic encephalopathy and extensive white-matter infarcts with gliosis, liquefaction, and early cavitation.

Subacute forms are illustrated by three cases. Case #9054 (Figure 4.3F) has already been described above. Case #722143 (Figure 4.3C) was born at 35 w-g with congenital jejunal atresia requiring corrective surgery. At 5 days of age, severe respiratory failure occurred due to infection and persistent intestinal obstruction. Respiratory difficulties persisted, leading to progressive bronchopulmonary dysplasia. This infant lived 26 days and had diffused gray-matter neuronal necrosis and white-matter infarcts. Case #907 (Figure 4.3E), born at 28 w-g, had respiratory distress syndrome leading to progressive bronchopulmonary dysplasia, lived 3 months, and had extensive brain damage, white-matter infarcts, periventricular leukomalacia, and focal occipital and parietal microgyria.

The chronic forms are illustrated by two cases of severe prenatal brain damage. Case #88175 (Figures 4.4 and 4.5) was an eight-year-old child, mentally retarded, partially blind, and epileptic, born with extensive damage to the left occipito-parietal region and a single large porencephalic cyst confirmed by postmortem magnetic resonance imaging (Figures 4.4A,B). This child died of accidental drowning and had old (chronic) LMHs in the attenuated but surviving cortex over the large porencephalic cyst. Case #794190 (Figures 4.6–4.8), born at 34 w-g with extensive brain damage, multicystic encephalopathy, focal microgyria, and microcephaly, lived 18 days and had chronic LMHs in the cortex over the damaged regions.

Results

The *differentiating and maturing* cerebral cortex of the prematurely born infant is especially vulnerable to asphyxia. It is not uncommon

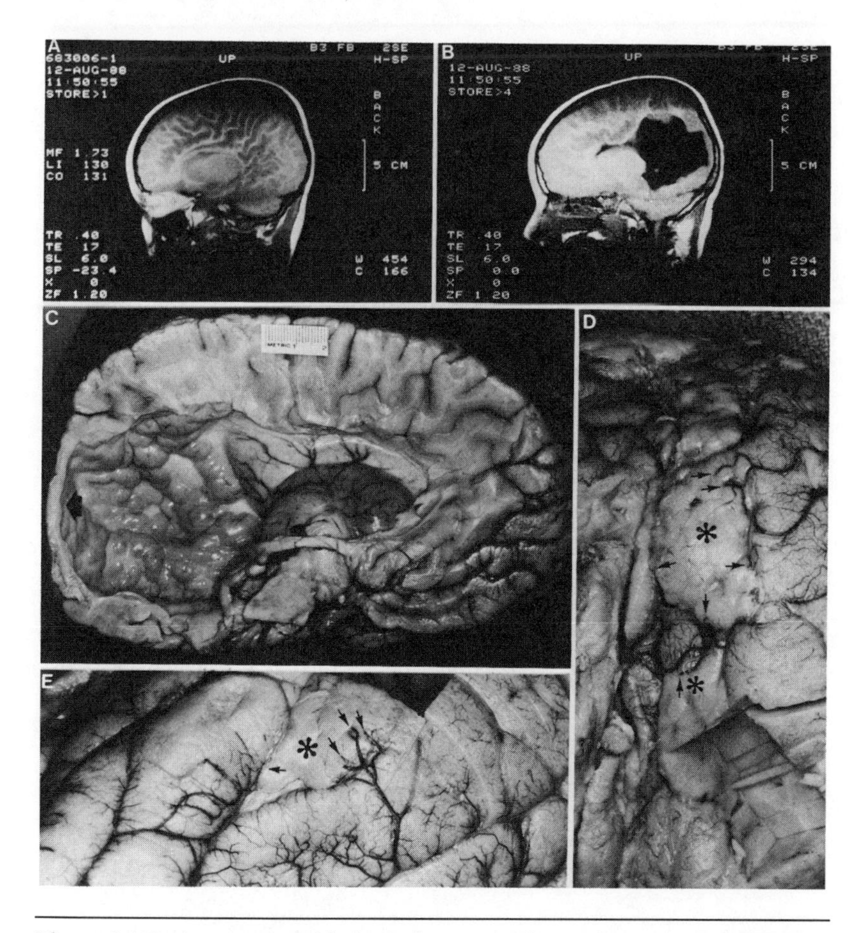

Figure 4.4 Various neuropathological aspects of case #88175, including the postmortem magnetic resonance images (MRI) of the normal right cerebral hemisphere *(A)*, the single porencephalic cyst of the left hemisphere *(B)*, the autopsy specimen of the left hemisphere showing the extent and size of the porencephalic cyst *(C)*, and the still-surviving thin cortex above it *(thick arrow)*. The cerebral cortex away from the damaged area is unaffected and has developed normally. Two anatomical views *(D, E)* of the cortical surface above the cystic damaged area show large chronic leptomeningeal heterotopias *(asterisks)* and the characteristic disappearance of pial capillaries under the ectopic tissue *(thin arrows)*.

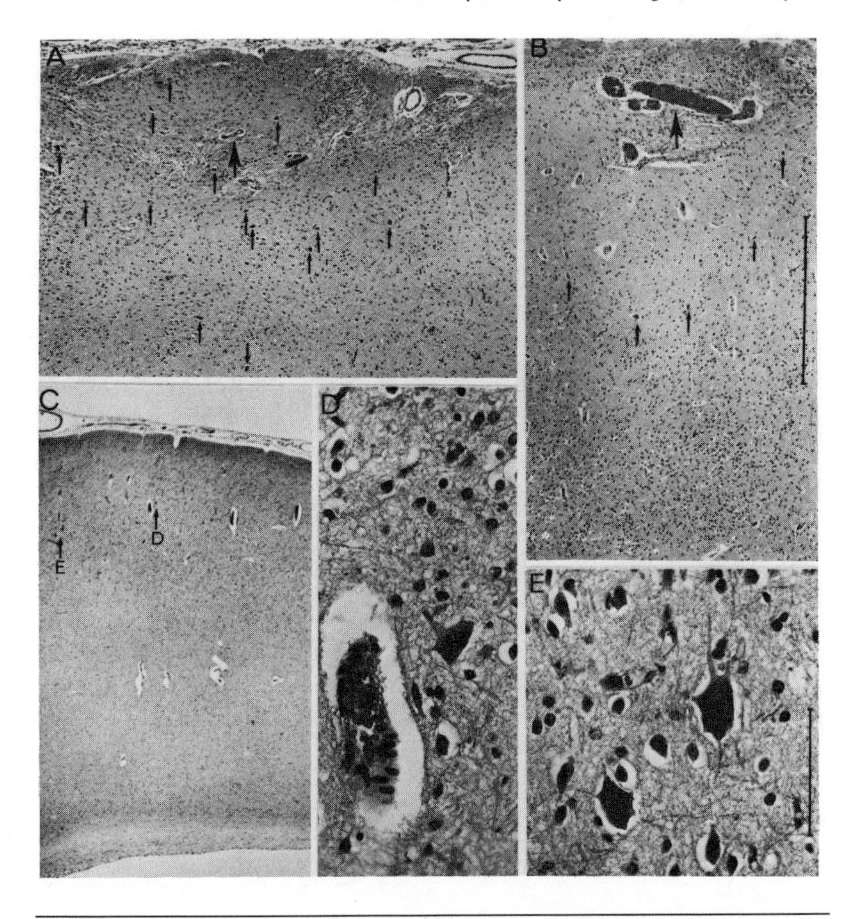

Figure 4.5 The histopathologic features of chronic leptomeningeal heterotopias and of the underlying altered cortex of case #88175. The horizontal post-inflammatory vessels within the heterotopia *(thick arrows)*, the disorganization and absence of laminations of the underlying cortex, and the presence of a few meganeurons *(thin arrows)* are illustrated in *A* and *B*. The thin surviving and altered cortex (gray matter) above the damaged area, its reduced amount of white matter, the paraventricular gliosis, and the location of three meganeurons *(arrows)* are illustrated in *C*. These three stellate meganeurons are illustrated at a higher magnification in *D* and *E*. Scales in *A*, *B*, *C* = 100 μ; scales in *D*, *E* = 50 μ.

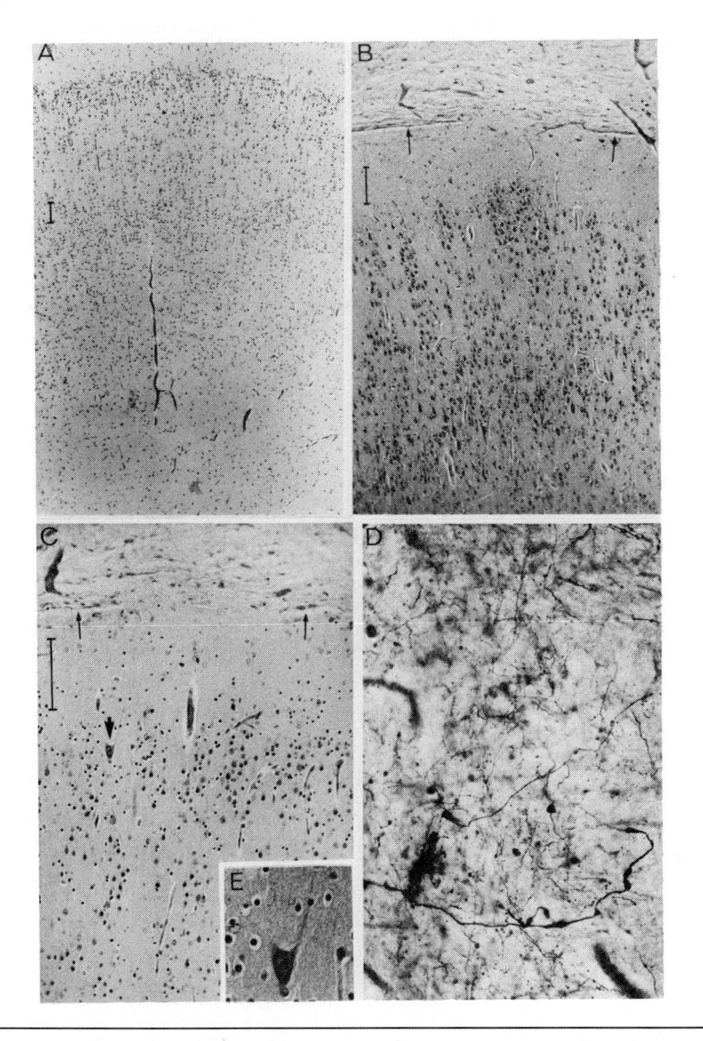

Figure 4.6 Preparations from case #794190, illustrating the normally developed cerebral cortex away from damaged areas *(A)*, the altered cortex *(B)* beneath the chronic leptomeningeal heterotopia, and its characteristic cellular aggregates separated by cell-free areas. The ruptures of the cortex's external glial limiting membrane (EGLM) are clearly illustrated between the thin arrows in *B* and *C*. These arrows also illustrate the persistence of the subpial glial endfeet edema. A closer view of a heterotopia with a rupture in the cortex's EGLM and a layer II stellate meganeuron are illustrated in *C*. This stellate meganeuron is reproduced at a higher magnification in the inset. A Golgi preparation *(D)* reveals the numerous thin and thick fibers trapped within the heterotopia. Afferent fibers that enter into the heterotopia through the EGLM gaps obviously miss their neuronal targets. Scale = 100 μ.

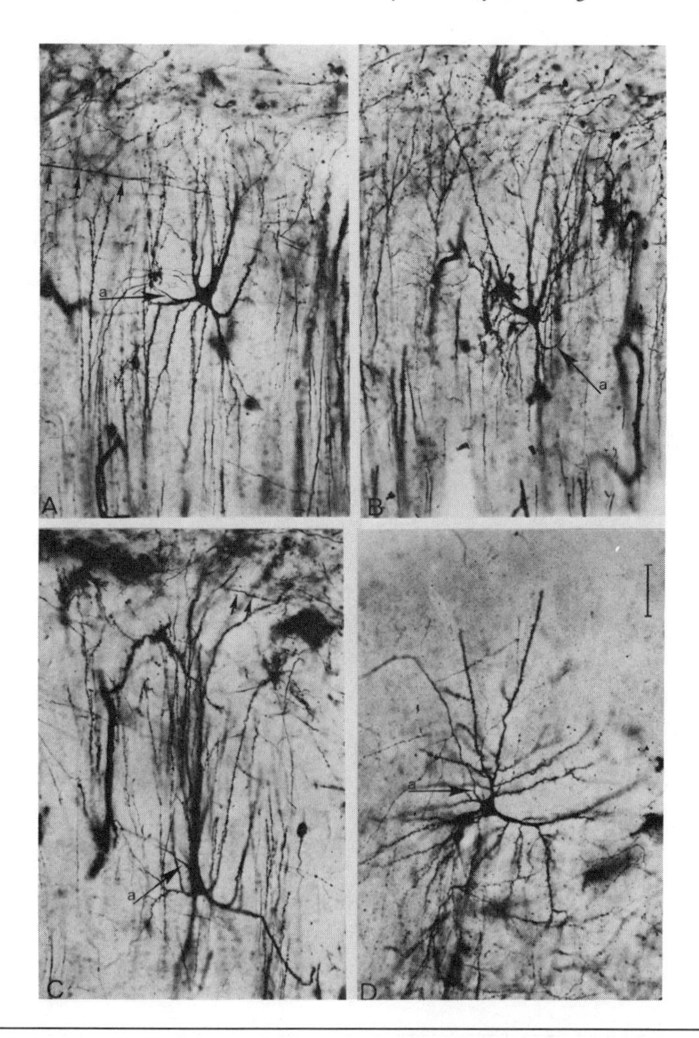

Figure 4.7 Rapid Golgi preparations of case #794190, illustrating the stellate dendritic morphology of four layer II meganeurons under a leptomeningeal heterotopia. The descending *(A)*, ascending *(B, D)*, or transverse *(C)* axons *(long arrows)* of these meganeurons branch and terminate locally. The ascending dendrites of some of them *(B, C)* penetrate the heterotopia. These meganeurons are considered to represent hypertrophic and restructured local-circuit interneurons that have overresponded to the abnormal circuitry created by the heterotopia. Many different fibers are seen entering the heterotopia from below. The presence of Cajal-Retzius horizontal axons within the obliterated layer I are marked with small arrows. Scale = 100 μ.

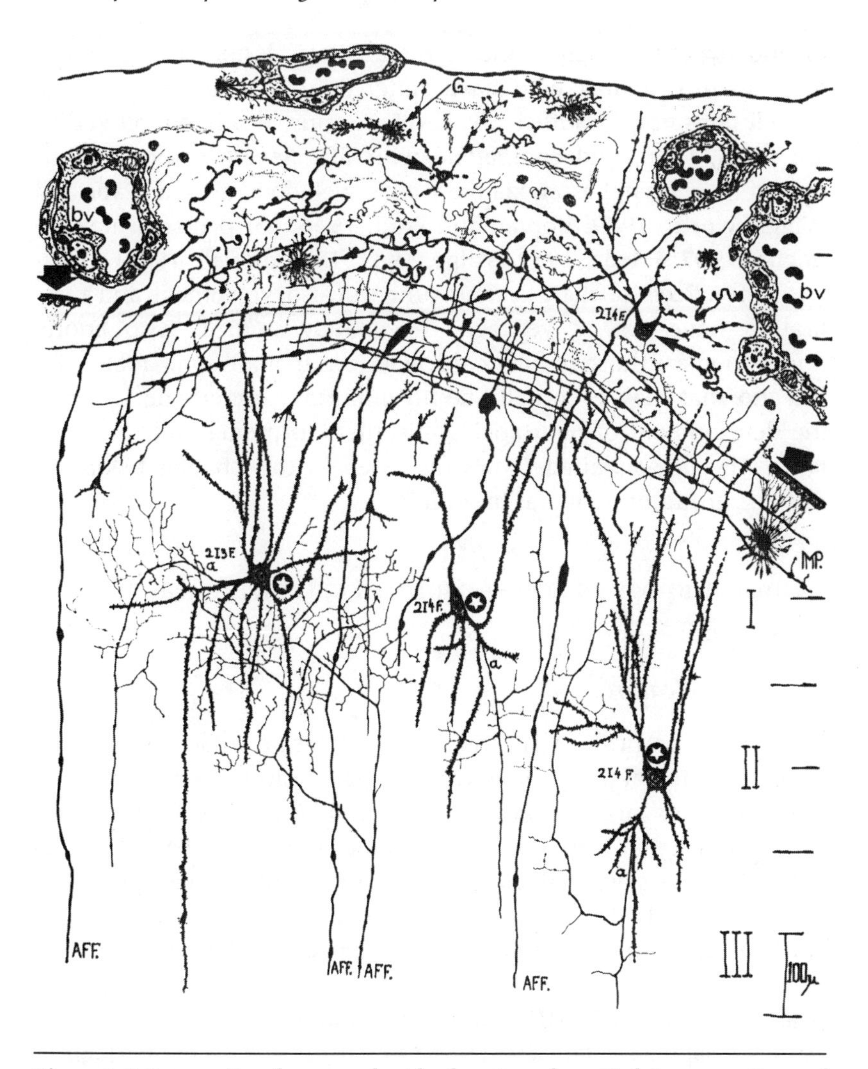

Figure 4.8 Composite of camera lucida drawings from Golgi preparations of case #704190, illustrating the overall structural organization of a chronic leptomeningeal heterotopia, the altered structure of the underlying cortex, and the stellate morphology of three layer II meganeurons *(asterisks)*. The ascending dendrites of some of them penetrate the heterotopia. The fibrillar composition, the presence of glial *(G)* and neuronal *(small arrows)* elements, the large EGLM's gap or rupture *(thick arrows)*, and the post-inflammatory vessels *(bv)* that characterize acquired leptomeningeal heterotopias are illustrated. Some afferent *(AFF)* fibers with anomalous dilatations penetrate the heterotopia while others trapped within the altered cortex branch profusely throughout its upper region. Scale = 100 μ.

to find hypoxic-ischemic and/or hemorrhagic lesions in the brains of premature infants who have suffered from perinatal asphyxia. Such lesions may be sufficiently severe to cause death, but often they are small, focal, and compatible with life. Severe cortical lesions caused by perinatal asphyxia have been well documented (Pape and Wigglesworth, 1979; Kaplan and Ford, 1966; Larroche and Razavi-Encha, 1987; Friede, 1989; Volpe, 1987; Evrard et al., 1989). However, the effects of these lesions on cortical development and maturation have not as yet been adequately studied. Also poorly understood are those smaller cortical lesions that are compatible with life and that may eventually be an important underlying factor in the development of neurological disorders in prematurely born infants. The late-acquired LMHs discussed in this chapter belong in this last group of perinatal brain injuries.

Neuropathology of Late-Acquired Leptomeningeal Heterotopias

Anatomical features

The size of cortical LMHs found in premature infants who have suffered perinatal asphyxia is variable. They can be very small, medium-sized, or large. Small ones are usually invisible to the naked eye, requiring microscopic confirmation (Figures 4.3A,D,F). Medium-sized ones are visible as discrete verruca-like surface protuberances apparently deprived of visible superficial blood vessels (Figures 4.3C,E). Large ones are easily recognized as surface lesions of variable sizes that could cover large portions of a gyrus (Figures 4.4D,E). The pial blood vessels of LMHs seem to terminate abruptly at their edges as they disappear from view, buried under the ectopic tissue (Figures 4.4D, E, arrows). This vascular peculiarity, undoubtedly the most distinctive feature of medium-sized and large LMHs, facilitates their gross identification. Large LMHs are most frequently found in the attenuated but surviving cortex overlying areas of hypoxic-ischemic white-matter damage (Figures 4.4–4.8).

Histological features

The microscopic study of late-acquired LMHs has disclosed that at least three types—acute, subacute, and chronic lesions—can be recognized. Acute LMHs are lesions that have occurred recently, perhaps a few hours prior to the infant's demise. Subacute lesions may have occurred days or weeks before, whereas chronic ones may have occurred months before death. Subacute and especially chronic LMHs undergo secondary changes compatible with a continued reorganization, including gliosis and fibrosis, and subsequently affect the structural organization of the underlying and still-developing cortex (Figures 4.3C,E,F, 4.5, and 4.6).

Acute LMHs are characterized by rupture of the EGLM, severe glial edema, lack of layer I obliteration, and absence of recognizable alterations in the underlying cortex (Figures 4.3A,D). The superficial glial edema may be quite severe and could cause an actual separation of the nervous tissue from the overlying leptomeninges (Figure 4.3A, asterisk). Furthermore, it is not uncommon in acute LMHs to find, in addition to the superficial glial edema, recent subpial hemorrhages (Figure 4.3B). The edematous separation of the nervous tissue from the overlying pial surface undoubtedly damages both the EGLM and the pial-perforating vessels. Damage to perforating pial capillaries explains the subpial hemorrhages, and the disruption of the edematous glial endfeet certainly ruptures the cortex's EGLM (Figure 4.3A). Through these superficial gaps, nearby glial and neuronal elements could easily escape into the leptomeningeal space and form LMHs (Figure 4.3A). It should be emphasized that some degree of edema is invariably found in glial endfeet around disruptions of the cortex's EGLM in acute, as well as in subacute, LMHs (Figures 4.3C,D,F, thin arrows).

Subacute LMHs are characterized by persistent superficial glial edema around the EGLM's rupture, obliteration of layer I, and alterations of the structural organization of the underlying developing cortex (Figures 4.3C,E,F). Some of the surface blood vessels are already buried within the heterotopic tissue, rendering them invisible if viewed from the surface (Figures 4.3C,E,F thick arrows). These vessels, rather than coursing perpendicular to the pial surface as they normally do, are horizontal to it. They are, therefore, components of the heterotopia, not typical CNS vessels. They are post-inflammatory vessels that have evolved in response to the original injury.

The extraneural heterotopic tissue is composed of glial cells, numerous fibers and glial filaments, and an occasional small neuron. In older heterotopias, there are in addition to these neural-derived elements mesodermal-derived structures, including fibroblasts, collagen, and macrophages (Figure 4.3E). The amount of gliosis and/or fibrosis varies, according to the age of the lesion and degree of secondary reorganization.

In subacute LMHs, layer I may be partially (Figure 4.3F) or completely obliterated (Figure 4.3E), and the structural organization of the underlying cortex may be mildly (Figure 4.3F) or severely altered (Figure 4.3E). The most common cortical alteration observed is the formation of columnar cellular aggregates of various sizes surrounded by cell-free areas (Figures 4.3C,F). These columnar aggregates are most prominent in the upper cortical layers, are often displaced toward the heterotopia, and could even enter into it. The upward displacement of these aggregates obliterates layer I and causes disorganization of the underlying cortex. Although these alterations resemble abnormalities of neuronal migration, they are in fact focal lesions that have occurred late in gestation, when migration is either complete or nearly so. They do not represent anomalies of neuronal migration in the strict sense, because the radial glia is invariably damaged at the EGLM's rupture. Furthermore, while there are no indications elsewhere in the cortex of anomalies of neuronal migration, there are evidences of perinatal brain damage in these infants.

Chronic LMHs are characterized by prominent gliosis and fibrosis, evidences of progressive reorganization, and significant alterations of the structural organization of the underlying cortex (Figures 4.5A–C, 4.6B–C). These old lesions are invariably associated with extensive encephaloclastic brain damage involving primarily the white matter (Figures 4.4A–C). They are found at the surface of the attenuated but still-surviving cortex (gray matter) overlying the damaged area or residual (porencephalic) cyst (Figures 4.4D–E). The damaged area is characterized by a thick layer of paraventricular gliosis (leukomalacia), a narrow and significantly reduced white matter, and a disorganized gray matter without appreciable or clear laminations (Figures 4.5A–C). The fact that the cerebral cortex away from the damaged area appears unaffected suggests that the original injury was a locally destructive process, more likely a vascular accident (Figures 4.4C, 4.5A). The presence of viable but disorganized cortex over the damaged areas suggests that developing neurons of the gray matter were able to survive the original injury and have been secondarily reorganized (restructured) in response to the local abnormality.

The structural alterations found in the cortex beneath chronic LMHs vary from mild (Figure 4.6B) to severe (Figures 4.5A–B). In chronic LMHs the degree of gliosis is pronounced, and often only small aggregates of cortical neurons are recognizable within the gliotic scar (Figures 4.5A–B). The surviving cortical neurons within these areas have probably undergone progressive alterations in their structure and connectivity. In this respect, it should be pointed out that it is not uncommon to find large anomalous neurons (meganeurons) in the altered cortex beneath chronic LMHs (Figures 4.5D–E, 4.6C insert). These meganeurons are easily recognizable because of their large size. They are mostly multipolar cells with several thick dendrites arriving from their body (Figures 4.5A–E, arrows; 4.6C). Although they are found at any cortical level, those found at the upper layers are particularly prominent because at this level they are surrounded by smaller neurons. The nuclear size of layer II meganeurons has been quantitatively analyzed and compared with that of pyramidal cells of layers II and V of the same location (Figure 4.9). Their mean nuclear perimeter is roughly twice that of layer II and similar to that of layer V pyramids. The nuclear surface area of layer II meganeurons is comparable to that of the large layer V pyramids. These meganeurons are considered to represent restructured and hypertrophic neurons that have responded and readapted to the local abnormal circuitry created by the lesion. To elucidate the structural organization of subacute and chronic LMHs and the morphology and connectivity of the altered underlying cortex and meganeurons, a special procedure—namely, the Golgi method—is necessary.

Rapid Golgi Studies

Rapid Golgi studies of chronic LMHs have confirmed the disruption of the cortex's EGLM (Figure 4.8, thick arrows) and have demonstrated their structural organization and composition. They have also demonstrated the type of structural alterations that characterize the underlying altered cortex and have established the morphology of its surviving neurons, including that of meganeurons (Figures 4.7 and 4.8). In addition, they have given support to the idea that an injury to the cortex's EGLM may be followed by the formation of an LMH, which subsequently could alter the structural organization of the underlying and still-developing cortex.

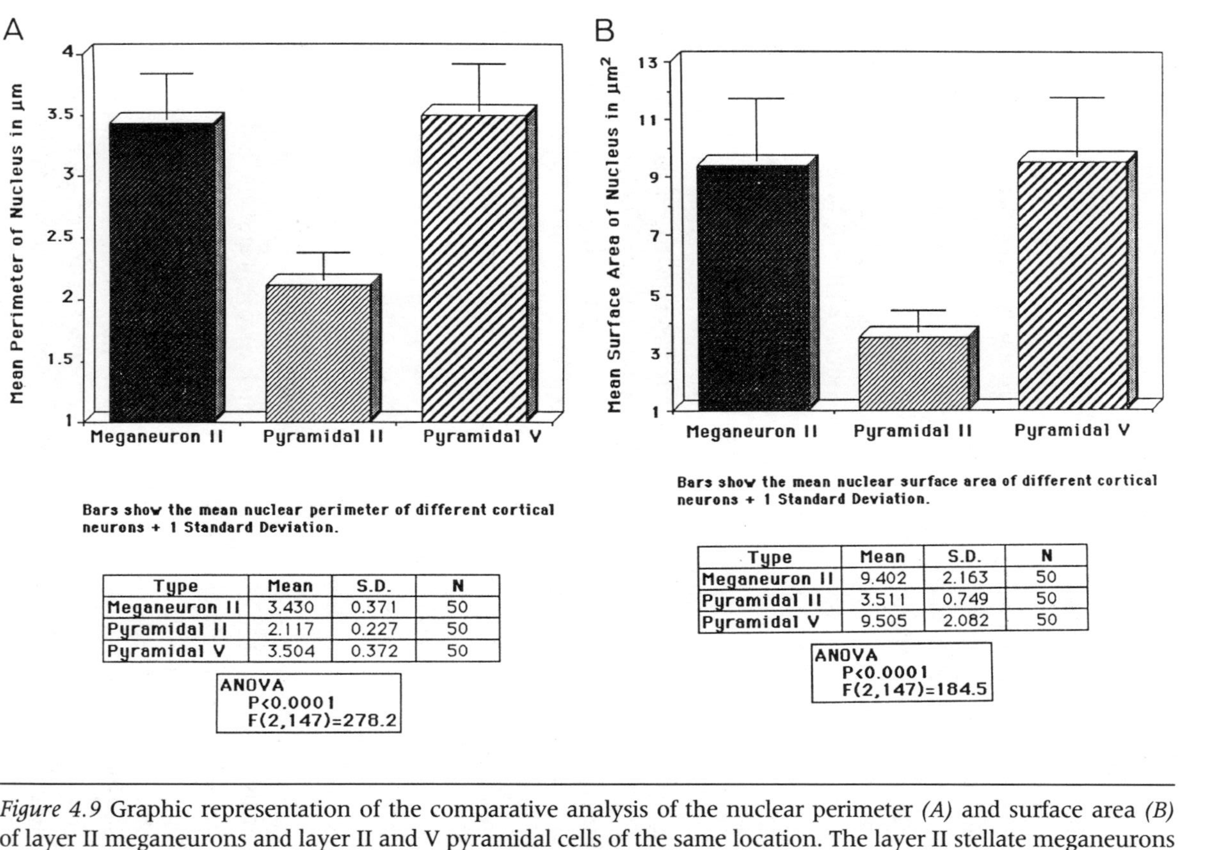

Type	Mean	S.D.	N
Meganeuron II	3.430	0.371	50
Pyramidal II	2.117	0.227	50
Pyramidal V	3.504	0.372	50

Type	Mean	S.D.	N
Meganeuron II	9.402	2.163	50
Pyramidal II	3.511	0.749	50
Pyramidal V	9.505	2.082	50

Figure 4.9 Graphic representation of the comparative analysis of the nuclear perimeter *(A)* and surface area *(B)* of layer II meganeurons and layer II and V pyramidal cells of the same location. The layer II stellate meganeurons are comparable in size to the large pyramidal cells of layer V.

A chronic LMH is a glial and fibrovascular scar composed of both neural- and mesodermal-derived elements (Figure 4.6D). Neural-derived elements include glial cells, a few poorly developed neurons, and many fibers of different caliber (Figure 4.8). The fibers' caliber ranges from fine and beaded (cathecolaminergic?) to thick afferent terminals, with many intermediate types. Afferent fibers enter into the heterotopia from below, lack any recognizable target, and terminate by freely giving off tortuous collateral fibrils (Figures 4.6D and 4.8). Afferent fibers that have entered and terminated in the heterotopia have obviously missed their destination. Consequently, there must be differentiating cortical neurons under the LMH that have been deprived of their normal inputs. Such deprived developing neurons could become underdeveloped and, hence, atrophic. On the other hand, they could instead respond by overgrowing to other local fibers trapped under the LMH. The meganeurons observed in the altered cortex beneath the LMH could represent such overdeveloped or hypertrophic neurons.

Some of the afferent fibers within and below the heterotopia show signs of degeneration, including unusual dilatations, the formation of large beads, and incomplete staining (Figure 4.8). There are also many fiber terminals scattered throughout the heterotopia whose origin remains undetermined. Fibrillary and protoplasmic astrocytes and many of their filaments are recognizable within the heterotopia, as is the occasional small and poorly developed neuron (Figure 4.8). Mesodermal-derived elements include post-inflammatory, rather than CNS, blood vessels, fibroblasts, collagen, and macrophages (Figures 4.6D and 4.8).

In Golgi preparations, the Cajal-Retzius cells' horizontal axons (Retzius' tangential fibers) are still recognizable within the obliterated layer I. These thick horizontal fibers are slightly bent toward the heterotopia but do not penetrate it (Figures 4.7A and 4.8). Their upward deviation suggests that the lesion occurred late in gestation and caused a secondary bending of the fibers as they passed under the heterotopia. Remnants of the EGLM are still recognizable in some areas (Figure 4.8, thick arrows). Their presence determines the size or length of each surface disruption through which the heterotopias are formed.

The cytoarchitecture of the cerebral cortex beneath the heterotopia is abnormal; essentially, it is disorganized (Figures 4.5, 4.6B,C, and

4.8). It should be emphasized that the cortex around but away from the heterotopic area has developed normally and has typical neuronal assemblages. On the other hand, the cortical neurons below an LMH do not respond to the normal developmental constraint. They seem to respond to the local constraints caused by the heterotopia, and they therefore develop abnormally. This altered cortex lacks recognizable laminations, it has many anomalous neurons, and the neuronal assemblages are abnormal. Although the morphology of some neurons beneath the LMH is still recognizable (for example, the pyramidal cells), that of others is not. There are poorly developed or atrophic neurons with short dendrites with a few spines, and fine axons with a few collaterals. Contrarily, there are also a few hypertrophic meganeurons with overdeveloped dendritic and axonic arbors (Figures 4.7 and 4.8). Many other neurons are unclassifiable. A peculiar finding is the disorientation of dendrites among some neurons. These neurons have dendrites that are oriented predominantly in a particular direction, as if they were responding only to incoming axons arriving from that direction. Other neurons have very long and tortuous dendrites. Neuronal interrelationships are abnormal. Bundles of glial filaments often separate and isolate some of the neuronal aggegates. Glial scars, probably replacing areas of cellular death, are common. Nests of poorly developed and small neurons are found within gliotic scars (Figures 4.5D,E, 4.6C, 4.7, 4.8).

The presence of large hypertrophic meganeurons in the altered cortex beneath the heterotopia is undoubtedly the most interesting finding (Figures 4.7 and 4.8). Meganeurons are found at various depths throughout the cortex. Their size is enormous and comparable to that of the large pyramids of layer V (see Figure 4.9). They are invariably stellate cells with several thick dendrites that radiate from the soma in all directions. Their ascending and descending dendrites are quite long, reaching lengths of more than 500 μ. The dendritic arborization of a single meganeuron could cover ca. 1 mm^2 surface area. Ascending dendrites of meganeurons in layers II and III could reach and even penetrate into the heterotopia (Figures 4.7 and 4.8). Both dendrites and soma are covered by numerous typical spines, suggesting a considerable degree of activity. The overall number of spines per neuron is extraordinary. Their axons frequently originate from one of the main dendrites away from the soma. The dendrite that carries the axon could be ascending (Figure 4.7D), transverse (Figure 4.7A), or descending (Figure 4.7B). The axon may rise from an anomalous site of the soma (Figure 4.7C). The axon may ascend or descend, and it gives off many collaterals that branch and terminate locally. Although the nature of these meganeurons remains unexplained, it is proposed that they represent hypertrophic, readapted, and restructured local-circuit interneurons. Obviously, further studies will be necessary to elucidate the nature, development, and structure of these meganeurons.

Discussion

The histopathology and pathogenesis of the late-acquired LMHs found in the developing cortex of premature infants who have suffered from perinatal asphyxia have been analyzed. These surface lesions are considered to be caused by hypoxic-ischemic and/or hemorrhagic injury and subsequent rupture of the cortex's external glial limiting membrane (EGLM). Glial and neuronal elements close to the ruptured site escape from the nervous tissue into the overlying leptomeninges and form the lesions. Incoming fibers approaching these surface gaps could also enter into the heterotopia, becoming trapped in it and, hence, failing to reach their neuronal targets. Eventually, the LMHs affect the normal development of the underlying and still-developing cortex and, subsequently, could be the cause of cortical dysfunction. A better understanding of these acquired lesions and of their effects on cortical development would undoubtedly improve the management and, possibly, the prevention of perinatal brain damage. It should also lead to a better understanding of their often devastating neurological effects.

Morphologically, LMHs resemble anomalies of neuronal migration. They are not abnormalities of neuronal migration in the strict sense, however, because the radial glia is invariably damaged at the site of the lesion. LMHs are acquired neuropathological lesions that occur late in gestation after neuronal migration is completed (perinatal period) and that are associated with evidence of injury to the developing brain. An injury to the endfeet processes of the EGLM, followed by its disruption, is considered an important cause of late-acquired LMHs. Glial and neuronal elements appear to continue their migration into the leptomeningeal space, thus resembling abnormalities of migration. However, these elements are not able to utilize the radial glia to continue their extraneural migration because the radial processes are invariably injured at the site of the EGLM's rupture.

An injury to the EGLM after neuronal migration has been completed (perinatal period) is not expected to interfere with that developmental process, but it could certainly interfere locally with the structural organization of the underlying and still-developing cortex. Essentially all the structural alterations observed in the developing cortex below a late-acquired LMH could be explained as secondary or adaptative changes in response to the local disturbances created by the lesion. It should be emphasized that the altered cortex below an LMH is usually surrounded by unaffected cortex that has developed normally. Neuronal modifications observed with the rapid Golgi method in the altered cortex beneath a LMH include: atrophic or underdeveloped neurons, hypertrophic or overdeveloped meganeurons, monodirectional orientation of dendrites in some neurons, and recognizable as well as unclassifiable neuronal types. Anomalous neuronal interrelationships, gliosis, and focal glial scars are also observed in this altered cortex.

Of all the cortical alterations observed under late-occurring LMHs, the presence of hypertrophic meganeurons is perhaps the most interesting. In addition to having been detected in tuberous sclerosis (Huttenlocher and Wollmann, 1980) and storage diseases, large anomalous neurons have been described in a variety of disorders associated with epilepsy (Bignami et al., 1968; Taylor et al., 1971; Townsend et al., 1975; Marchal et al., 1989; Vigevano et al., 1989). The presence of these meganeurons has been attributed to neuronal dysplasia caused by genetic or developmental disturbances. Meganeurons have also been described in acquired disorders caused by perinatal brain damage, also associated with epilepsy (Marín-Padilla, 1979, 1988c; Mervis and Yates, 1980).

The meganeurons found in the altered cortex below late-acquired LMHs are believed to represent modified and restructured local-circuit interneurons. Their location, stellate morphology, ascending, descending, or transverse axon, and the intracortical distribution of the axon are characteristic features of these interneurons. Local-circuit neurons would have a better chance to survive a cortical injury than would projective ones (Cajal, 1968). Surviving local-circuit neurons may be receptive to incoming axons trapped by the heterotopia, becoming overtargeted by them and responding with overgrowth and hence with hypertrophy. Cajal (1968) was probably one of the first to describe structural modifications of cortical neurons following experimental surgery (undercutting of) of the white matter. Some pyramidal neurons whose axon has been severed at the white-matter level survive, and their proximal axonic collaterals persist and overgrow, establishing new connections with neighboring neurons. As a result, inputs reaching them are deviated toward a more local circuitry. Consequently, these restructured neurons cease to function as projective pyramids as they become transformed into local-circuit neurons and, hence, assume a different functional role. A similar situation probably occurs in the surviving cortex above hypoxic-ischemic white-matter infarcts. The structural organization of the attenuated but surviving cortex above large white-matter infarcts has not been adequately studied.

The idea, although not a new one, that a focal injury to the developing cortex could lead to structural and, consequently, to functional modifications is corroborated and supported by the findings presented here. During the perinatal period (24 w-g to the time of birth), the human cortex is particularly vulnerable because its neurons are undergoing significant structural and functional transformations in response to incoming fibers. To achieve its complex structural and functional organization, the cerebral cortex must be subject to significant developmental constraints. An injury to the EGLM, disrupting the cortex's anatomical integrity, could disturb the normal developmental constraints of the underlying and still-growing cortex. Thus, the neuronal abnormalities found below LMHs are considered to represent secondary modifications or re-adaptations to the locally disturbed cortical development. Needless to say, these acquired neuronal alterations could explain the neurological disorders observed in some premature infants who survive periods of perinatal asphyxia. The need to study, to evaluate, and to follow-up these children clinically is of primary importance, and all the necessary human and material resources should be made available for these studies.

Acknowledgments

This work has been supported by the National Institute of Neurological and Communicative Disorders and Stroke, Grant #NS-22897, NIH.

5

Dyslexia and Brain Pathology: Experimental Animal Models

Glenn D. Rosen
Gordon F. Sherman
Albert M. Galaburda

Developmental dyslexia has commonly been thought to be a primary disturbance in the acquisition and achievement of reading skills. This diagnosis requires not only that there be evidence of a reading disability but also that there be no overt cognitive, emotional, neurological, or socio-economic explanations for this deficit. Although reading ability is the hallmark of the disorder, dyslexic children and adults often exhibit other cognitive problems, such as difficulties with oral comprehension of syntactically complex sentences, awareness of the phonological structure of words, and working memory for linguistic objects (Kean, 1984). It is therefore apparent that developmental dyslexia can be considered more as a generalized disturbance of language function.

Dyslexia affects approximately 5–10 percent of the school-age population, is diagnosed more commonly in boys (Finucci et al., 1983), and is sometimes seen more often among left-handed and ambidextrous individuals (Geschwind and Behan, 1982). It is often seen in members of the same family (DeFries, Julker, and LaBuda, 1987; Urion, 1988) and may be a genetically linked trait (Smith et al., 1983; see also Chapter 10). Other disorders (including stuttering and attention deficit disorders) and ailments that implicate dysfunction of the immune system (such as allergies and autoimmunity) are more common in dyslexics and their families (Geschwind and Behan, 1982).

In the following section, we will discuss our findings based on examinations of brains obtained at autopsy of dyslexic individuals. These findings—namely, symmetry of language-related regions of the brain and various forms of developmental neuropathology—have been successfully modeled in rodents. We will discuss these animal models and how further understanding of the mechanisms associated with these anomalies may lead to better understanding of the deficits associated with dyslexia.

Neuroanatomical Findings in Dyslexia

Ever since dyslexia was first identified it has been hypothesized that structural abnormalities in the brain may underlie the behavioral disorder. It was not until 1968, however, that the brain of a dyslexic was microscopically examined. William Drake (1968) reported the results of the examination of the brain of a fourteen-year-old severely dyslexic boy who had died of a hemorrhage in the cerebellum. There was a family history of left-handedness, learning problems, migraine headaches, and vascular malformations.

Gross examination revealed that the brain contained a large number of small gyri and a thin corpus callosum. The cortex of the brain was large, and microscopic examination revealed ectopic neurons (neurons that are out of place) in the subcortical white matter.

Since this time we have examined the brains of five male and three female dyslexics (Galaburda and Kemper, 1979; Galaburda et al., 1985; Humphreys, Kaufmann, and Galaburda, 1990). The five males ranged in age from 20 to 56 years, and the three females were 20, 36, and 88 years of age. Two consistent findings were seen: developmental neuropathology and symmetry of language-related regions of the brain.

Neuropathology

The brains of all five male dyslexics and two of the three females (Galaburda and Kemper, 1979; Galaburda et al., 1985; Humphreys, Kaufmann, and Galaburda, 1990) showed a high rate of focal microdysgenesis of the cerebral cortex. In the males, the abnormalities consisted of nests of neurons in layer I (ectopias), subjacent or adjacent disorganization of cortical lamination (dysplasias), and abnormal vasculature. Ectopic neurons, which often distorted the upper surface of the cortex, were seen in large numbers (ranging from 30 to 100 per brain), mostly in perisylvian and anterior vascular border-zone territories and in the left more than in the right hemisphere. In addition, abnormal, poorly laminated small gyri and sulci (microgyria) were present in two of the cases.

A recent study of ten male control brains (age matched), which were prepared and examined exactly as the dyslexic brains were, revealed only three brains that showed abnormalities similar in type to those of the dyslexic brains. But the abnormalities were found in far smaller numbers (one or two per brain) and were located in different areas of the cortex (Kaufmann and Galaburda, 1989). These findings contrast sharply with the large number of ectopias seen in the male dyslexics.

Two female dyslexics showed multiple instances of focal myelinated cortical infarction, with neuronal loss, gliosis, and myelination of the scars affecting perisylvian and cerebral arterial border-zone territories (Humphreys, Kaufmann, and Galaburda, 1990). The presence of myelin in the scars suggested that the injury preceded the second or third year of postnatal life (Malamud, 1950; Borit and Herndon, 1970; Norman, 1981; Larroche, 1986). One of the males showed, in addition to microdysgenesis, a small number of these myelinated glial scars. The brain of the twenty-year-old female had no scars in the cortex but did have a small number of ectopias distributed equally between the hemispheres.

Other abnormalities included a bilobar hippocampal oligodendroglioma and a frontal arteriovenous anomaly in one female case; one male and another female case also showed arteriovenous anomalies; and a male showed architectonic abnormalities in the lateralis posterior and medial geniculate nuclei of the thalamus. Instances of subtle hippocampal disorganization also were seen, but sporadically.

Symmetry

The five male and three female dyslexic brains showed symmetry of the planum temporale, a language-related cortical area located on the superior surface of the temporal lobe. The plana in all cases not only were equal in size, but were also large (equivalent to the normal left side) on both sides. In control brains (presumably normal readers), 85 percent show asymmetry, the larger side usually being the left. Only 15 percent have plana of equal size on the two sides

(Galaburda et al., 1987). Recent imaging studies on the brains of living dyslexics also have shown similar differences in the asymmetrical organization of the brain (Larsen et al., 1990; for a review, see Hynd and Semrud-Clikeman, 1989.

Research issues

Given that these two categories of anatomical abnormalities are seen in dyslexic brains, the overwhelming question remains how might each of these anomalies, either alone or in concert, cause the cognitive deficits. Are both the developmental neuropathological changes and symmetry necessary for dyslexia to be manifested, or is one of the two more directly associated with the cognitive problems? Are these two types of anomalies related directly to one another? For example, does the brain develop symmetry of the language areas and ectopias and glial scars as the result of the same etiologic process? Or are the two the results of independent events? Why do the types of developmental neuropathological changes seen in the male brains differ from those of the females? Are there more widespread neuroanatomical changes associated with the developmental neuroanatomical anomalies that cannot be viewed with current techniques?

Although there are no definitive answers to these questions as yet, we are beginning to gain valuable insight from work with various animal models. In the following section, we detail our findings on animal modeling of developmental pathology and brain symmetry. For the former, we use strains of mice that spontaneously develop autoimmune disease as well as ectopic collections of neurons in the cerebral cortex that are remarkably similar to those seen in dyslexics. In addition, we have induced focal microdysgenesis in newborn rats, which allows closer examination of the etiology of microgyria. For the study of symmetry and asymmetry in the brain, we are able to use various experimental model systems, including both the mouse and the rat.

Spontaneous Neuropathology in Animal Models

An animal model for the study of spontaneous microdysgenesis in the neocortex grew out of the results of studies on the relation between the immune and nervous systems. Geschwind and Behan (1982) found an increase in both learning disabilities and autoimmune disorders in strongly left-handed individuals and their families. In an initial study, 500 strongly left-handed and 900 strongly right-handed individuals were compared on the incidence of autoimmune diseases (rheumatoid arthritis, celiac disease, thyroiditis, ileitis, colitis, and uveitis), migraine headaches, and developmental learning disorders (dyslexia and stuttering). The left-handers had more than twice as high a rate of immune disorders and ten times the rate of learning disabilities found in right-handers. Migraine headaches were also more commonly seen in the left-handers. This same relationship was present for first- and second-degree relatives. In a follow-up study, left-handers were found to have more allergies and skeletal malformations, thyroid disorders, migraine headaches, and learning disorders. Approaching the same question from a different angle, the researchers also found that individuals with immune problems had higher rates of left-handedness (Geschwind and Behan, 1984).

Other studies have supported Geschwind and Behan's findings (Pennington et al., 1987; Schachter, Ransil, and Geschwind, 1987; Lahita, 1988; Searleman and Fugagli, 1988; Urion, 1988). It has also been reported that learning disorders are more often associated with right-sided congenital limb malformations (Dlugosz et al., 1988). Others have reported no connection among handedness, immune disorders, and dyslexia. For example, Satz and Soper (1986) suggested that because most dyslexics are right-handed there was no connection between handedness and dyslexia. It is clear, however, that even with a shift toward left-handedness and ambidexterity in dyslexics, the majority of dyslexics would still be right-handed (Galaburda, 1989). Salcedo et al. (1985) and Schur (1986) did not find a correlation between the autoimmune disorder systemic lupus erythematosus and left-handedness. However, both failed to study a large enough group of subjects to show a relationship if one existed.

Altogether these studies point to a link among left-handedness, autoimmune disease, and learning disabilities, and their findings stimulated us to examine the brains of certain mouse strains that spontaneously develop autoimmune disease.

Autoimmune mice

The New Zealand Black (NZB) and BXSB mouse strains spontaneously develop severe autoimmune disease that curtails life expectancy. In NZB mice, the autoimmune disease is characterized by the presence of autoantibodies and abnormalities of stem cells, macrophages, and T and B lymphocytes. The NZB develops hemolytic anemia and dies prematurely at about 16–17 months. The development of autoimmune disease in the NZB is under the control of multiple genes (Steinberg et al., 1981). The BXSB strain, which was developed from a C57BL/6J female and an SB/Le male, develops immune disease characterized by the development of autoantibodies, lymphoproliferation of B cells, and immune complex glomerulonephritis. BXSB males are more severely affected than females and die much younger. These sex differences are not under the control of sex hormones but instead appear to be determined by an accelerator gene located on the Y chromosome (Theofilopoulos and Dixon, 1981).

About 20–50 percent of these mice have ectopias in layer I of the neocortex and dysplasia of the underlying layers. Typically, only one ectopic nest is seen in each affected brain, although 25 percent of the affected brains of the BXSB strain have multiple ectopias (Sherman et al., 1987). In a small number of cases there are neuron-free patches extending through the entire depth of the cortex. The anomalies are usually present in only one hemisphere, with no preference for the right or left sides overall. More mice (40–50 percent) from the BXSB strain have brain pathology than from the NZB (20–30 percent), and more multiple anomalies are present in the BXSB. The location of the anomalies also differs between strains: most anomalies in the NZB mice are present in the sensorimotor cortices, whereas most of those in the BXSB are in the frontal-motor cortices (Sherman et al., 1987). It is important to note that each litter of the NZB and BXSB strains contain at least one affected mouse.

To further examine the incidence of brain abnormality associated with a defective immune system, we have surveyed other strains that develop various abnormalities of the immune system (Sherman et al., 1990a). These were the MRL/1, MRL +/+, NZB/W, Snell dwarf, SJL, and two mouse strains that contain the nude (*nu*) gene (BALB/cByJ-*nu*/*nu*, C57BL/6J-*nu*/*nu*), as well as seven control stains without immune disorders. The brains of mice from these strains were examined in Nissl-stained serial sections under a light microscope for the presence of abnormalities; specific attention was given to examining the samples for ectopic collections of neurons in layer

I of the neocortex, as reported in the autoimmune NZB and BXSB strains.

The highest number of brain abnormalities (20–40 percent) were seen in the C57BL/6J-*nu/nu* and Snell dwarf strains. The anomalies in the C57BL/6J-*nu/nu* mice consisted of ectopic neurons in layer I of the neocortex, whereas the Snell dwarf mice had either neuron-free areas in the cortex or rippling of cortical layers II–IV, and one case had agenesis of the corpus callosum. Four to eight percent of the mice from the SJL, MRL/1, and MRL +/+ strains had either neuron-free areas in the cortex or ectopic neurons in layer I. The BALB/cByJ-*nu/nu* and control strains did not have any cortical abnormalities. Studies are under way to determine whether immune-based alterations to the developing brain, as suggested by the correlation between brain abnormalities and immune disorders, are responsible for the brain anomalies present in immune-disordered strains.

Anatomical characterization of the ectopias

We have begun a number of anatomic studies of these ectopias. In order to assess the neurochemical composition of the neurons contained within the ectopias, we stained for VIP (vasoactive intestinal polypeptide), NP-Y (neuropeptide Y), GABA (γ-aminobutyric acid) and somatostatin, all of which robustly stain in the cerebral cortex. VIP, NP-Y, GABA and somatostatin-like staining was seen in neurons within the ectopias in the NZB mouse. VIP and somatostatin stained the largest number of neurons in the ectopias.

Furthermore, by counting the number of VIP neurons in the affected and unaffected hemisphere of the brains of NZB mice containing focal cerebrocortical microdysgenesis, we addressed the question of whether these abnormalities represent abnormal placement of neurons, abnormal numbers, or both, and thus we now better understand the mechanism underlying this form of cerebral malformation. We found an increase, in comparison to the opposite hemisphere, in the total number of VIP neurons in the cortical region containing an ectopia as well as in the area medial to the ectopia. No cortex neighboring the area of ectopia actually had fewer VIP-stained neurons than the corresponding areas in the opposite hemisphere. This would suggest that there is an absolute increase in VIP-stained neurons in the affected hemisphere, rather than a relative increase in the area of ectopia caused by neurons that have migrated laterally from adjacent cortex to populate the ectopic region. Thus, we concluded that in this form of cortical malformation VIP-stained neurons are more numerous than in normal cortex, as well as abnormally positioned in layer I (Sherman et al., 1990c).

Neurofilament staining

The relatively restricted and subtle nature of the brain anomaly as demonstrated on routine cell stains is not totally persuasive that the focal dysgenetic changes seen in dyslexics and in autoimmune mice could account for learning failure. We have proposed, therefore, that there exists a more widespread disruption of the cortical architecture that involves the neural networks participating in the behaviors that fail. In a further attempt to determine the magnitude of cellular and fiber disruption associated with the ectopias, we stained cortical sections containing ectopias in frontal-motor and somatosensory cortex from NZB and BXSB mice with an antibody directed against the 68 kDa subunit of neurofilament protein, which is contained within neuronal cell bodies, dendrites, and axons.

There was a dramatic change in neurofilament staining in the cortical layers underlying the ectopic neurons, as a result of dense, radially oriented fiber bundles spanning the thickness of the cortex underlying the ectopias. In all instances, these bundles distorted the highly stratified neurofilament lamination seen in normal mouse cortex and present in adjacent and homologous contralateral cortex of the autoimmune mouse. The frontal ectopias were especially interesting because it was possible to see the multifascicled fiber bands underlying the ectopias join the lateral aspect of the corpus callosum. It is also important to note that even in a small ectopia, where standard stains show no associated cortical dysplasia, dense neurofilament staining was present in layers II and III (Sherman et al., 1990b).

The results of neurofilament staining support the hypothesis that the apparently delimited cytoarchitectonic disturbance seen in Nissl stains of the neocortex of autoimmune mice, illustrated by isolated foci of ectopic neurons in layer I, is accompanied by more extensive cortical disorganization. Thus, nests of misplaced neurons in the molecular layer may be associated with abnormal connectivity, as exemplified by the abnormal patterns of neurofilamentous fibers and dendrites in subjacent cortex.

Radial glial fibers

In order to investigate the contribution of the glial-fiber network in guiding neurons to ectopic positions in the cortex, we stained radial glia in newborn autoimmune mice for evidence of a disturbance in the organization of these fibers. Ectopias were identified in thionin-stained sections and the adjacent section was immunohistochemically stained for radial glial fibers with Rat-401 (provided by S. Hockfield, Yale University).

The pial-glial membrane, which was enhanced by antibody staining, was interrupted in the area of the ectopias, and there was aberrant organization of radial glial fibers in association with this breach. The cortical column underlying the ectopias contained an increased density of glial fibers that were radially oriented toward the ectopias. At the breach in the pial-glial membrane, however, most fibers curved outward toward the edges of the ectopias and seemed to be anchored here to the intact pia. In addition to the fibers attached at the edges, a matrix of disorganized glial fibers, oriented both radially and horizontally, was present within the ectopias.

A likely scenario for the formation of the ectopias is that an insult to the developing brain produced a breach in the pial-glial membrane and neurons migrated along radial glia into the space created by this interruption. The neurons are contained within that space by the overlying pia and the intact edges of the membrane. As the breach is formed the anchoring points of the glial cells are pulled aside; this explains the presence of the curved glial fibers at the edges of the ectopias. Most neurons probably migrated into the ectopias along these glial fibers, but the disorganized fibers seen within the ectopias also may play a role in this regard.

Behavioral effects of ectopias

Behavioral studies have shown that the NZB mouse is deficient on certain learning tasks. NZB mice learn poorly a conditioned avoidance task where a noxious stimulus must be associated with a tone (Nandy et al., 1983). Control mice learn this task easily. In addition, NZB mice have difficulty learning a passive avoidance response (Spencer and Lal, 1983; Spencer et al., 1986) and a water escape task (Denenberg et al., 1988). The deficit in the NZB is not due to a general sensorimotor impairment but to a problem in formation or retention of stimulus-response associations (Spencer and Lal, 1983).

In studies of the effects of ectopias on behavioral performance and learning, NZB and BXSB mice have been tested on a variety of tasks designed to show not only differences between strains but also differences between mice with and without brain anomalies (Denenberg et al., 1991). The findings support the earlier studies showing that autoimmune mice are slower in performing certain learning tasks, and they also show that NZB and BXSB mice are more asymmetrical in paw use than are DBA mice. On a water escape task it was seen that right-pawed male and female NZB and male BXSB mice with ectopias were slower than left-pawed mice with ectopias. But no paw effect was seen for the mice without ectopias. Thus, ectopias interact with paw preference to alter learning ability.

Induced Neuropathology in Animal Models

As mentioned previously, instances of micropolygyria were found in the brains of two male dyslexics, and this type of developmental neuropathology has been produced in the rat brain by applying a freezing probe to the skulls of newborn rat pups (Dvorák and Feit, 1977; Dvorák, Feit, and Juránková, 1978). We have replicated and extended this work, using more specialized stains, to determine the true extent of damage incurred during the early stages of neocortical development and to reveal the mechanism by which this abnormality occurs.

On the day of birth (P0) or the day after (P1), rat pups were anesthetized with hypothermia and a freezing probe (−70°C) of 2 mm diameter was placed directly on the skulls overlying the left hemisphere for five seconds. After various lapses of time, the animals were sacrificed, their brains removed and sectioned coronally, and series of sections were stained for Nissl substance with Thionin, for myelin using the Loyez method, and for glial fibers using the phosphotungstic acid hematoxylin (PTAH) method. In addition, a series of sections were immunohistochemically stained for neurofilament (NF), VIP, glial fibrillary acidic protein (GFAP), glutamatergic neuronal processes (Glut), and radial glial cells (Rat-401) for the purpose of examining specific disturbances of immunoarchitecture of the neocortex (Humphreys et al., 1991).

The results of our studies have extensively confirmed the findings of Dvorák and colleagues (1977; 1978) in that we can reliably produce a focus of four-layered cortex that substantially resembles the polymicrogyric cortex (a pathological event of neuronal migration) seen in humans with a variety of neurologic dysfunctions (Richman, Stewart, and Caviness, 1974; McBride and Kemper, 1982). Severe freezing lesions typically showed evidence of myelin staining in the zone of selective neuronal loss. Adjacent sections stained for glial fibers by the PTAH method showed small foci of enhanced glial staining within the zone of neuronal loss. In addition, sections stained for GFAP showed clusters of GFAP-positive fibers within the zones of neuronal loss. These findings, taken together, indicate myelinated scarring in the area of the freezing lesion, a characteristic of lesions acquired early, before intracortical myelogenesis is complete (Myers, 1969; Borit and Herndon, 1970). In addition, our immunohistochemical studies suggest that regional changes in neuronal organization may be a far more important consequence of neocortical lesions induced in day-old rats than any minor alteration in neuronal migration. Stains for neurofilament clearly indicated that neurofilament-containing structures were markedly reduced in the external neuronal layer of the microsulcus and on the periphery of the freezing lesions.

Neurofilament-containing structures were also seen in elaborate sprouts similar to those seen in the neocortical neuronal ectopias of the immune-disordered mouse (Sherman et al., 1990b). Stains of glutamatergic processes also revealed widespread organizational abnormalities—processes were markedly reduced in zones of neuronal loss and adjacent neuronal areas. In contrast, VIP cortical neuronal bodies in the area of the freezing lesion appear to have been spared, although occasional disturbances of their processes were noted (Humphreys et al., 1991).

Thus, early induced injury to the cerebral cortex results in neuronal loss, laminar reorganization, and focal disturbances in neuronal migration. Immunohistochemical staining reveals that the alteration affects anoxia- and ischemia-sensitive elements and variably extends beyond the borders of the injury as demonstrated by routine cell stains.

Connections

One question posited above had to do with the relationship between developmental cortical anomalies and behavior. That such focal abnormalities as ectopias, glial scarring, and microgyria should be related to widespread functional difficulties seems surprising. However, the research reported in the previous sections suggests that what appears to be focal damage may well have more widespread effects. But in order to fully consider this question, one must explore the changes in the inherent connectivity of the damaged areas.

We had previously reported a case of spontaneous microgyria occurring in a rat who had subsequently received a corpus callosum section (Rosen, Galaburda, and Sherman, 1989). After examining the brain for callosal connectivity, we found exuberant, termination-dense areas in the region of the microgyria. In order to replicate this study, we severed the corpus callosum of adult rats who as newborns had received a freezing lesion to the skull. A week following surgery the animals were sacrificed, their brains removed and cut coronally, and one series of sections was stained for Nissl substance with Thionin and the adjacent series with the Fink-Heimer method (1967) for axonal termination. It was found that there were profound differences in interhemispheric connectivity in those animals with freezing lesions. These differences could be categorized in three ways: (1) exuberant patches of callosal termination in and around the area of induced microdysgenesis, (2) a decrease in callosal termination in homologous regions of the opposite hemisphere, and (3) a decrease in callosal termination in heterologous regions of the opposite hemisphere.

These findings are consistent with the observation that damage to the brain during development has widespread effects on connectivity and behavior. In hamsters, for example, unilateral neonatal lesions of the superior colliculus result in a profound restructuring of afferent connections and subsequent disturbances in orienting responses (Finlay, Wilson, and Schneider, 1979; Schneider, 1979; Schneider, 1981). In monkeys, the effects on behavior and connectivity of prenatal lesions of the frontal cortex are different from the effects of similar lesions in adulthood (Goldman and Galkin, 1978; Goldman-Rakic and Rakic, 1984).

Not all pathological events occurring during brain development affect the eventual patterns of connectivity, however. In the reeler mouse, whose cerebral cortex is characterized by the inversion of the normally inside-out placement of the cortical layers because of a disturbance of neuronal migration, the efferent connections arise from appropriate neuronal types irrespective of their laminar location (Caviness, 1976; Caviness and Yorke, 1976; Caviness, 1982; Caviness and Frost, 1983).

Yet it is not necessary to suggest that the mechanisms for the cortical anomalies in the reeler mouse and for induced microgyria are equivalent and must therefore have the same effects on cortical connectivity. One major distinction between the two phenomena might be their differential ability to cause the release of trophic factors capable of altering local circuitries. Nerve growth factor and other trophic factors can, in fact, alter the survival of neurons and axons during ontogenetic cell and axonal pruning (Hamburger and Yip, 1984) and have been reported to be released after injury to nervous tissue (Nieto-Sampedro et al., 1982; Needels, Nieto-Sampedro, and Cotman, 1986; Kromer, 1987). Thus, since the production of microgyria in this case is clearly the result of injury, unlike in the case of the reeler mouse, it is possible that trophic molecules are released in cases of induced microgyria but not in the reeler mouse. These trophic factors could diminish the normally extensive ontogenetic axonal pruning (O'Leary, Stanfield, and Cowan, 1981), could interfere with the normal ontogenetic correction of targeting errors (Cowan et al., 1984), and/or could stimulate sprouting of novel terminations not normally found.

In support of these findings of connectional reorganization following developmental injury are the results reported elsewhere in this volume (Chapters 1 and 22). Using unilateral carotid ligation in the neonatal cat, Finlay found that the resultant hypoxic visual cortex had significantly greater numbers of efferent connections from the homologous, nonhypoxic cortex. Afferent connections from the hypoxic area were, on the other hand, decreased. Similarly, Innocenti induced infragranular laminar necrosis in the developing cat visual cortex through the injection of ibotenic acid and found increased efferent connections from intrahemispheric laminar nuclei (Innocenti and Berbel, 1991b; Innocenti and Berbel, 1991a). These results, taken together with our own, suggest an increased survival of efferent connections to an area that is damaged during development.

Development

We have examined the brains of rats soon after they had received freezing lesions and been stained for GFAP-positive and radial glial cells. During the first few postlesion days, the area of induced damage shows an intact layer I subjacent to which there is a wide area containing pyknotic cells and tissue necrosis. In contrast, the neocortical ectopia in the autoimmune mouse is characterized by a more obvious break in the external glial limiting membrane. By P5 the typical infolding of the cortical surface is visible, and by P10 the appearance of adult-like microdysgenesis is present.

Many but not all radial glial cells appear damaged during the first two days after the freezing lesions. Bulbous varicosities at their foreshortened terminations are seen lying within the area of necrosis. By P5 some radial glial-like fibers traverse the damaged area and reach through overlying cell-dense areas up to the external glial limiting membrane. From P5 to P10, there is increased density of radial fibers within the area of damage, which may represent a collapse of surrounding radial fibers toward the center of the lesion. Between P1 and P5, GFAP-like immunoreactive cells are present mostly as reactive astrocytes in the area of damage. On P7 and P10 some radial glial fibers express GFAP, but mostly within the area of neocortical damage. By P15, there is evidence of some radial-glial fiber immunoreactivity in the area of the damage and nowhere else.

These results are similar in some way to those seen with the autoimmune mice. Radial glial cells play an important role in the eventual disposition of neurons in the region of damage. Unlike the case with the autoimmune mouse, however, after an induced injury layer I and the external glial limiting membrane remain intact despite the severity of the insult. This points to the probability that there are distinctly different mechanisms involved in the formation of these regions of cerebrocortical abnormality.

Symmetry

As mentioned above, one distinctive feature of the dyslexic brain is the symmetry of the planum temporale, a region that is ordinarily asymmetrical in favor of the left hemisphere (Galaburda and Kemper, 1979; Galaburda et al., 1985). Understanding the mechanisms involved in the development of symmetrical and asymmetrical brains would lead to a greater understanding of the ways in which the development of the brain of a dyslexic differs from that of normal readers. Toward that end, we have attempted to delineate some of the underlying biological mechanisms that distinguish symmetrical from asymmetrical brains in adulthood—specifically the gross and cellular morphometric, developmental, and connectional characteristics of symmetrical and asymmetrical brains. Before we discuss these issues, however, we will briefly review the evidence for anatomical asymmetry and functional lateralization in humans and animals.

Human and nonhuman anatomic asymmetries

Lateralization of cerebral cortical functions is well established in adult humans (Sperry, 1974; Springer and Deutsch, 1981) and is present from birth (Molfese, Freeman, and Palermo, 1975; Entus, 1977; Michel, 1981). The first modern attempt at deriving the anatomical basis of functional asymmetries was by Geschwind and Levitsky (1968), who examined the outside border of the planum temporale (an area on the posterior superior temporal plane bordered anteriorly by Heschl's gyrus and laterally by the sylvian fissure and purported to be involved in language processing). They found that 65 of the 100 brains they examined had a larger left planum whereas only 11 had the reverse asymmetry and 24 had no bias. They also found that the planum was, on the average, 33 percent longer on the left than on the right. These results have since been replicated by a number of investigators using direct and indirect techniques in adults (LeMay and Culebras, 1972; Teszner et al., 1972; Witelson and Pallie, 1973; Wada, Clarke, and Hamm, 1975; Rubens, Mahowald, and Hutton, 1976; Chi, Dooling, and Gilles, 1977; Galaburda et al., 1978; Pieniadz and Naeser, 1984; Galaburda et al., 1987) and fetuses and children (Witelson and Pallie, 1973; Wada, Clarke, and Hamm, 1975). The left sylvian fissure also is longer and the right is more posteriorly angled (Gundara and Zivanovic, 1968; LeMay and Geschwind, 1975; Rubens, Mahowald, and Hutton, 1976), and cytoarchitectonic area Tpt tends to be larger on the left than the right (Galaburda, Sanides, and Geschwind, 1978). Asymmetries have also been reported in the frontal opercula (Galaburda, 1980), the parietal lobe (Eidelberg and Galaburda, 1984), and in white-matter bundles like the pyramidal decussations (Yakovlev and Rakic, 1966; Kertesz and Geschwind, 1971).

Nonhuman species are generally less biased to the right or left as a population, yet evidence indicates that individual animals are lateralized from birth to adulthood (Nottebohm and Nottebohm, 1976; Nottebohm, 1977; Denenberg et al., 1978; Petersen et al., 1978; Denenberg, 1981; Ross, Glick, and Meibach, 1981; Ross, Glick, and Meibach, 1982; Rosen et al., 1983, 1984; Sherman and Galaburda, 1984; Hamilton and Vermeire, 1988). Anatomically, sylvian fissure lengths are longer on the left in chimpanzees (Yeni-Komshian and Benson, 1976), rhesus macaques (Falk et al., 1986), and some great apes (LeMay and Geschwind, 1975). In the rat, the right posterior neocortex is thicker in males (Diamond, Johnson, and Ingham, 1975) whereas females have a slightly (though not significantly) thicker left neocortex (Diamond et al., 1981). We have shown that the volume of the right neocortex was larger than that of the left in male rats but not in females (Sherman and Galaburda, 1984), and others have found the right hemisphere of the rat to be heavier, longer, wider, and taller in both adults and 15-day-olds (Kolb et al., 1982). During development, one subarea of the prefrontal cortex of the rat has been shown to be asymmetric as early as postnatal day 10 (van Eden, Uylings, and van Pelt, 1984).

Substrate size and asymmetry

Symmetrical brain substrates can result from (1) an increase in size of the normally smaller side, (2) a decrease in the usually larger side, or (3) a combination of a decrease in the larger side and an increase in the smaller side. In the first case, the total substrate area (right plus left) of symmetrical brain regions would be larger than their asymmetrical counterparts, whereas in the second case the opposite would be true. Total area size would be similar if the third scenario were true.

To test these possibilities, we examined the same 100 brains used by Geschwind and Levitsky (1968) and measured instead the total planum area in the left and the right side (Galaburda et al., 1987). We found a leftward-biased asymmetry in 63 percent of the plana, while 21 percent were rightward biased and 16 percent had no bias. When we plotted the total planum area (right plus left) against a measure of directionless asymmetry, a significant negative correlation was found, indicating that as asymmetry increased, the total planum area decreased. Moreover, the area of the smaller of the two plana significantly predicted asymmetry, whereas there was no correlation between the degree of asymmetry and the larger planum area. These results supported the hypothesis that symmetry was due to the development of two large planum areas in each hemisphere—as asymmetry increased, the total planum area decreased because of the decrease in size of the smaller of the two plana.

Histological basis of substrate asymmetry

The previous study demonstrated that a symmetrical brain region is made up of two large areas (as compared with the asymmetrical brain of a large area and a small area). When considering the morphometric histological basis of this difference, we must examine two parameters of cellular arrangement that determine architectonic volume: cell-packing density and cell numbers. Given this fact, we are left with three hypothetical possibilities to explain the difference between asymmetrical and symmetrical brains: smaller volumes may be the result of (1) a decrease in cell numbers without any change in cell-packing density, (2) changes in cell-packing density with the numbers of cells constant between the larger and the smaller area, or (3) a combination of an increase in cell-packing density and a decrease in the number of cells.

For methodological reasons, it was decided that the rat would be the proper animal model to study this problem. Before we could use the rat, however, we had to determine whether characteristics of asymmetry and symmetry generated from the previous experiment—that there was an inverse relationship between total area volume and degree of asymmetry—held true for this species. We therefore measured the volume of cortical area 17 (primary visual cortex), an area previously shown to be anatomically asymmetric. We found, analogous to the results obtained from the human planum temporale, that there was a negative correlation between the total area 17 volume and the asymmetry coefficient. In addition, the smaller of the two sides was also inversely related to the degree of asymmetry, whereas there was no relationship between the larger side and total volume asymmetry. Thus, rats with asymmetrical brain regions have, like the human, brain regions smaller than those that are symmetrical.

We next counted the cells within area 17 of these animals and found that there was no relationship between cell-packing density and volumetric asymmetry. Because any changes in cortical volume can be due only to differences in cell-packing density or cell numbers, it was clear that any change in asymmetry must be due to changes in total cell numbers (Galaburda et al., 1986). That changes in cell numbers and not cell-packing density are related to asymmetry is not surprising. For example, changes in cell-packing density large enough to account for the volumetric asymmetries would clearly disturb the cytoarchitectonic appearance of the areas and thus make recognition of these areas impossible. Others have shown (Rakic and Williams, 1986; Rakic, 1988) that changes in cytoarchitectonic intrahemispheric borders are subject to the same mechanism—specifically, changes in cell numbers, not cell-packing density.

Development of symmetric and asymmetric substrates

Symmetric regions are on the whole larger than asymmetric regions, which have fewer neurons on one side. In order to consider how an asymmetric region might develop, it would be important to consider the ways in which neuronal numbers in the cerebral cortex are regulated during development. According to the radial unit hypothesis of Rakic (1988), neurons that eventually make up the cerebral cortex originate as proliferative units of progenitor cells (neuroblasts) in the ependymal layer of the cerebral ventricles. Guided by radial glial fibers, migrating neurons from these zones are arranged in a columnar manner in the cortex with a one-to-one correspondence between proliferative unit and ontogenetic column. The numbers of neurons within a cortical area are regulated by: (1) the number of early progenitor cell divisions (which affects only the number of radial units), (2) the number of later divisions (which will determine the number of neurons within a radial unit), (3) the contacts of migrating neurons with thalamocortical and corticocortical afferents (which may have effects on architectonic boundary placement), and (4) ontogenetic neuronal death.

Using this hypothesis as a guide, we concluded that asymmetry in cytoarchitectonic volume (and hence neuronal number) could be the result of side differences in: (1) the number of proliferative units (which are the result of early progenitor-cell divisions), (2) the rates or numbers of later neuronal division irrespective of the initial state of the germinal zones, (3) the loss of neurons to adjacent architectonic areas, or (4) ontogenetic neuronal death. These possibilities are not mutually exclusive and all could play some role in the ontogenesis of neocortical asymmetry.

The investigation of these issues is not straightforward, however. Even at the earliest stage of development, when neocortical areas can be accurately parceled (postnatal day 10, or P10, in the rat), most if not all of the potential contributory events have already occurred and cannot be directly measured. We therefore devised an indirect method to investigate the relative effects of early division within the proliferative units versus the effects of later divisions. Specifically, we injected pregnant rats with [3H]-thymidine at various gestational ages in order to label in their pups the neurons undergoing their last mitosis at the time of injection (Rosen, Sherman, and Galaburda, 1991). The pups were then sacrificed at P10, P30, and P60, and the number of labeled and unlabeled neurons counted in area 17 and area 18a (a visual association area). Because the proportion of labeled to unlabeled neurons within a brain region is sensitive only to neuroblast division occurring during the later stages of brain development, this allowed an indirect measure of the importance of neuroblast division in determining side differences in architectonic volume. We found that while there was no difference in this ratio between the large and small side in asymmetric brains, there was a significant difference between architectonic regions: the ratio of labeled to unlabeled neurons was higher in area 17 than in area 18a.

These results suggest that late neuroblast division, while important for cytoarchitectonic differentiation, plays little or no role in determining side differences in the numbers of neurons. What of the possibilities of neuronal disappearance either through ontogenetic neuronal death or through changes in architectonic boundary placement? Both of these options seem unlikely. The only type of ontogenetic neuronal death that could explain the previous data would involve loss of entire columns of cells—something that has never been seen. Although thalamic deafferentation affects architectonic border placement differences (Rakic, 1988), we consider that an unlikely mechanism for the production of asymmetry—if the differences we report are related to asymmetric neuronal assignment to bordering architectonic areas, then one would justifiably predict that the shrinking of area 17 would be accompanied by the enlargement of one or more of its neighbors and that, for instance, a leftward asymmetry of area 17 would be accompanied by a rightward asymmetry of areas 18a, 18, and/or 7. This has been shown not to be the case. Additionally, if a shifting of borders could account for the emergence of asymmetry, then some areas would be larger when asymmetric while their asymmetric neighbors would be smaller. Yet in all of our previous research we have shown that asymmetric areas are consistently smaller than their symmetric counterparts (Galaburda et al., 1986; Galaburda et al., 1987; Rosen, Sherman, and Galaburda, 1989; Rosen et al., 1989).

We are therefore left with the possibility that asymmetry in neuronal number is the result of events occurring before the birth of the first neuron in the progenitor zones. It could therefore be the case that there are inherent asymmetries in the number of proliferative units on the two sides. Alternatively, there could be an initial symmetry of proliferative units but eventual side differences in proliferative unit production and/or proliferative unit death. What is becoming clear, however, is that asymmetries in substrate volume may result from asymmetries occurring very early in brain development, in the case of the human before the end of the second month of gestation (Rosen, Sherman, and Galaburda, 1991).

Connectivity in symmetric and asymmetric brain regions

It has long been speculated that the mechanism for cerebral dominance might lie in the callosal connectivity—that the dominant hemisphere exerts control over its homologue through the corpus callosum (Weiskrantz, 1977). By this reasoning, one might expect that the larger, dominant brain region, which has more cells than its homologue, would send more projections across the corpus callosum. Alternatively, it could be that symmetric and asymmetric brains have *patterns* of connections that are both qualitatively and quantitatively different.

To investigate this issue, we severed the corpus callosum of rats and examined the pattern of axonal terminal degeneration using silver degeneration techniques (Fink and Heimer, 1967). The somatosensory-somatomotor areas (SM-I; areas 3, 1, and 2) were parceled on the Nissl-stained sections and these sections were then overlaid with the adjacent silver-stained sections and the borders traced onto the latter. The percent degeneration (a measure which corrects for background staining and artifact) of SM-I, the number of patches of termination, and a directionless asymmetry coefficient were computed.

There was a negative correlation between the asymmetry coefficient and total (right plus left) volume of SM-I, indicating that as the degree of asymmetry of the architectonic region increased, the total volume of the region decreased—a finding previously reported for other brain regions, as noted above. In addition, there was a significant inverse relationship between the asymmetry coefficient and percent of callosal terminal degeneration. This indicated that in SM-I, symmetric regions had a greater percentage of callosal terminations than did more asymmetric regions. Finally, we found a negative relationship between the number of patches of callosal termination and asymmetry coefficient, indicating that there are more patches of callosal termination in the symmetrical, as opposed to asymmetrical, brain regions (Rosen, Sherman, and Galaburda, 1989).

Thus, symmetric brain regions have relatively greater numbers of callosal fibers and they contain more patches of termination. These findings introduce an additional dimension to the comparison of symmetric and asymmetric regions. If the detailed architecture of connections, as well as their number, affects functional capacity, symmetric and asymmetric brains may therefore differ in their preferred cognitive strategies as well as in their extent of hemispheric lateralization.

Variability in the pattern and number of callosal terminations may also be considered with regard to their ontogeny. Developmentally, callosal cells of origin are diffusely represented throughout the cerebral cortex (Wise and Jones, 1976; Ivy, Akers, and Killackey, 1979; Ivy and Killackey, 1981; Innocenti and Clarke, 1983). As the brain matures, callosal cells of origin are seen only in discrete laminar and columnar locations (Jacobson and Trojanowski, 1974; Zaborszky and Wolff, 1982). Likewise, the axonal terminations of these cells are distributed diffusely beneath the cortical plate until they penetrate it in discrete bundles and terminate on their appropriate targets (Zaborszky and Wolff, 1982; Olavarria and van Sluyters, 1985). The progressive restriction of callosal cells of origin and terminations is thought to result from axonal pruning rather than from the death of neurons (Ivy, Akers, and Killackey, 1979; Ivy and Killackey, 1981; O'Leary, Stanfield, and Cowan, 1981). The present finding of decreased callosal connections in asymmetric SM-I areas is consistent with an increased pruning of callosal axons and maintenance of ipsilateral cortical projections during development of this region, as has been previously demonstrated (Ivy and Killackey, 1982). Thus, the relative deficit of callosal connections with increasing hemispheric asymmetry may reflect the possibility that some neurons withdraw their callosal axons during development while those within the ipsilateral hemisphere are maintained. Alternatively, there may be a disproportionate loss of callosally related cells, as compared to noncallosal cells, in the asymmetric case. We have no evidence for this and indeed, as mentioned above, neuronal death is not a major factor in the development of callosal projections in SM-I (Ivy and Killackey, 1982; O'Leary et al., 1981).

Conclusions

We have reported two different types of neuroanatomical anomalies in the brains of dyslexics: neuronal ectopias and symmetry of the planum temporale. From our animal models, we have gained a better understanding of the ways in which these differences may be related to the functional deficits of dyslexia. For example, we have seen that apparently focal areas of microdysgenesis probably have more widespread effects—both in terms of the extent of the damage locally and in terms of connectivity. In addition, we have gained a greater understanding of the development and callosal connectivity of symmetrical brain regions through the study of animal models. Whether these two anomalies separately and distinctly contribute to the behavioral manifestations of dyslexia, or whether they act in concert, is not known. Complicating the matter further, it may well be that dyslexics have anomalies in other regions, not related to language, of the central nervous system (see Chapter 12). In spite of the questions that remain unanswered, however, it is clear that a uniquely human disorder such as dyslexia can be successfully studied in an animal model.

6

Anatomical and Functional Aspects of an Experimental Visual Microcortex That Resembles Human Microgyria

Giorgio M. Innocenti
Pere Berbel
Frederic Assal

In development, the fate of juvenile axons is probably controlled by interaction with their targets. Whether an axon will be maintained or eliminated may be decided by its intrinsic chemical features, the kind of activity code it relays, or its "vigor" at competing for some kind of trophic support with other axons afferent to the same target or with other axonal branches of the same neuron. The existence of transient axonal projections (for a review, see Innocenti, 1991) might indicate transient dependence of neurons on targets other than their final one.

In order to explore these issues at the cortical level, we performed early lesions of a brain site, the visual areas 17 and 18 in the cat. These areas receive three prominent transient projections, one from the contralateral area 17 and the others, respectively, from the ipsi- and the contralateral auditory areas. In an effort to avoid as much as possible the potentially confounding effects of axotomy, an "axon-sparing" excitotoxin, ibotenic acid ("ibo"; Eugster, 1968), was injected into areas 17 and 18 of newborn kittens.

The "microcortex" unexpectedly created by the excitotoxin resembles microgyria, a congenital dysplasia of human cerebral cortex, and provides a new way to study structural-functional relations and cell-cell interactions in a cortical structure. This article summarizes recent investigations on the ibo-induced microcortex (Assal et al., 1989; Hornung et al., 1989; Innocenti and Berbel, 1991a,b).

Cytoarchitectonics and Cell Composition of Areas 17 and 18 after Neonatal Injection of Ibotenic Acid

One month or longer after a neonatal injection of a sufficient dose (30 μg or more) of ibo on P2 or P2.5 (P = postnatal day), the lateral and postlateral gyri are narrower (Figure 6.1A). Although the normal sulci are present, in many of the experiments shallow and short new sulci have appeared near the injection site.

Nissl-counterstained coronal sections show that the narrowing of the lateral and postlateral gyri is due to a thinning of the gray and white matter. The cytoarchitecture is very characteristic (Figure 6.1C). Near the injection center the cortex is 500–700 μm thick. Layer I and two other cellular layers can be recognized. The outer cellular layer consists of small, round cell bodies (grains); the inner one, of small or medium-size pyramids. These two layers are continuous with layers II and III, respectively, of the surrounding intact cortex. Away from the injection center intact layers reappear in a characteristic outside-in sequence. Layer V and part of layer IV, however, are destroyed to a greater distance from the center of the injection than is layer VI, and this creates a neuron-free space between supra- and infragranular layers (Figure 6.1D). Still further away from the injection center, only the few neurons normally found in the adult white matter have disappeared. Occasionally, near the center of an injection, the cortex acquires uneven thickness (Figure 6.1E).

Impregnation with the Golgi method revealed that the outer cell layer of the ibo-injured cortex consists mainly of small pyramids with short, apical dendrites. The inner layer contains medium-size pyramids. In addition, there are throughout the cortex small stellate cells with smooth dendrites and locally arborizing axons, generally considered to be inhibitory interneurons (Figure 6.2).

Preliminary results indicate that neurons containing neuropeptide Y (NPY) and substance P (SP), which are a normal constituent of the deep layers in area 17, have been eliminated, although a rich

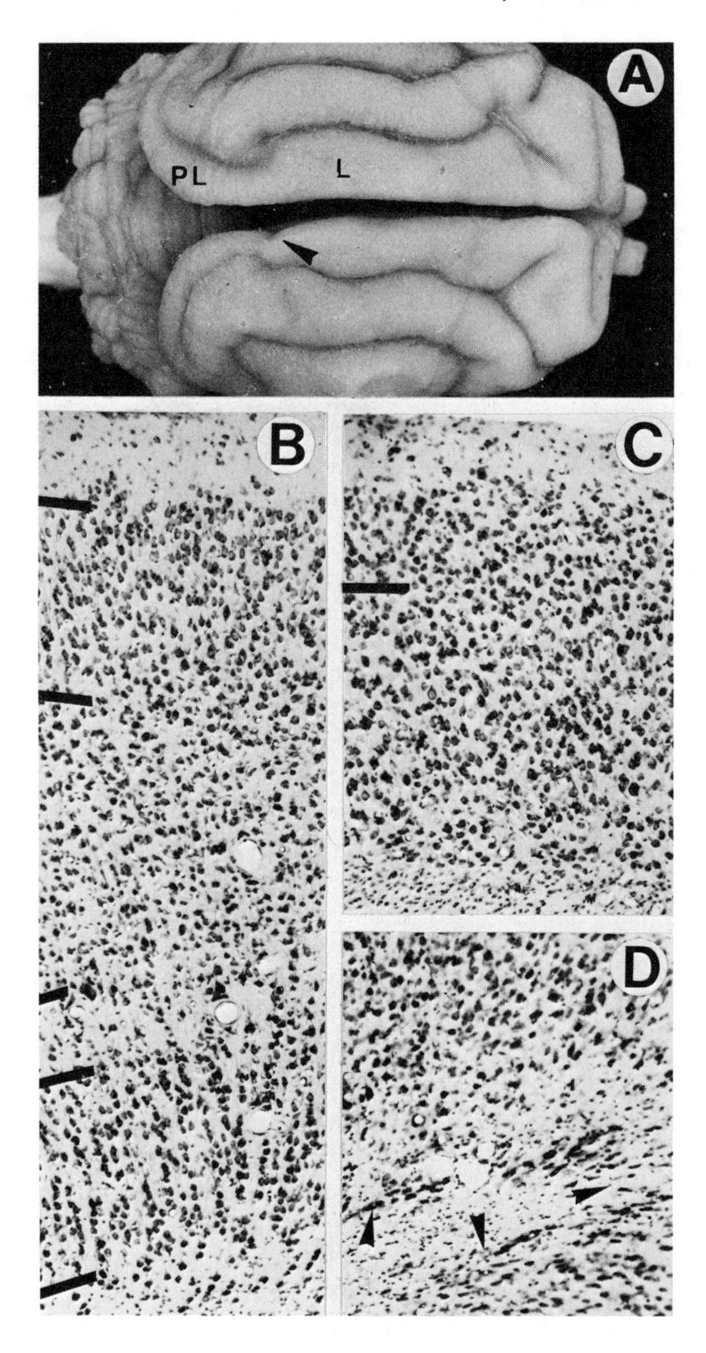
A
PL
L
B
C
D

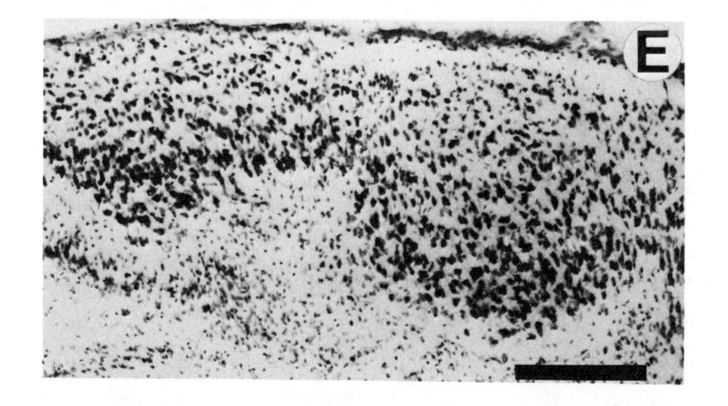

Figure 6.1 A: Dorsal view of a bran injected with ibotenic acid at P1. The narrower lateral *(L)* and postlateral *(PL)* gyri and abnormal sulcus *(arrow)* on the right are due to the ibo injection. *B–E* Photomicrographs of Nissl-stained coronal sections (40 μm thick) of area 17. *B:* Normal hemisphere, bottoms of layers I, III, IV, V, and VI are marked. *C:* Severe lesion; the border between outer (granular) and inner (pyramidal) cell layers is marked. *D:* Bottom of the cortex at the convexity of the postlateral gyrus at some distance from the ibo injection; arrowheads point to the bottom of the gray matter; layer VI is thinned, layer V and part of layer IV are destroyed. *E:* Very severe lesion; notice the undulating bottom of gray matter. Thick bar *(B–E)* = 200 μm.

network of NPY axons is found in the microcortex, presumably originating from neurons located in the normal cortex (Hornung et al., 1989).

In its cytoarchitectonics, this kind of ibo-injured cortex resembles microgyria, a congenital dysplasia of the human cortex (for references, see Evrard et al., 1989). Typically, human microgyria is localized, often to a vascular territory. It consists of four layers. The most superficial one is cell poor and continuous with layer I of the normal cortex. The layer underneath can be subdivided into two sublayers continuous with layers II and III of the normal cortex. Still deeper there are, in this order, a cell-poor layer continuous with the layers IV and V of the normal cortex and a cellular layer continuous with layer VI. This architecture resembles that found at the periphery of the ibo lesion. On the other hand, in some microcortical cases the two deeper layers can be missing and the cytoarchitecture then closely resembles that found near the center of an ibo lesion (for a case of this kind, see Innocenti and Berbel, 1991a). Unlike ibo-

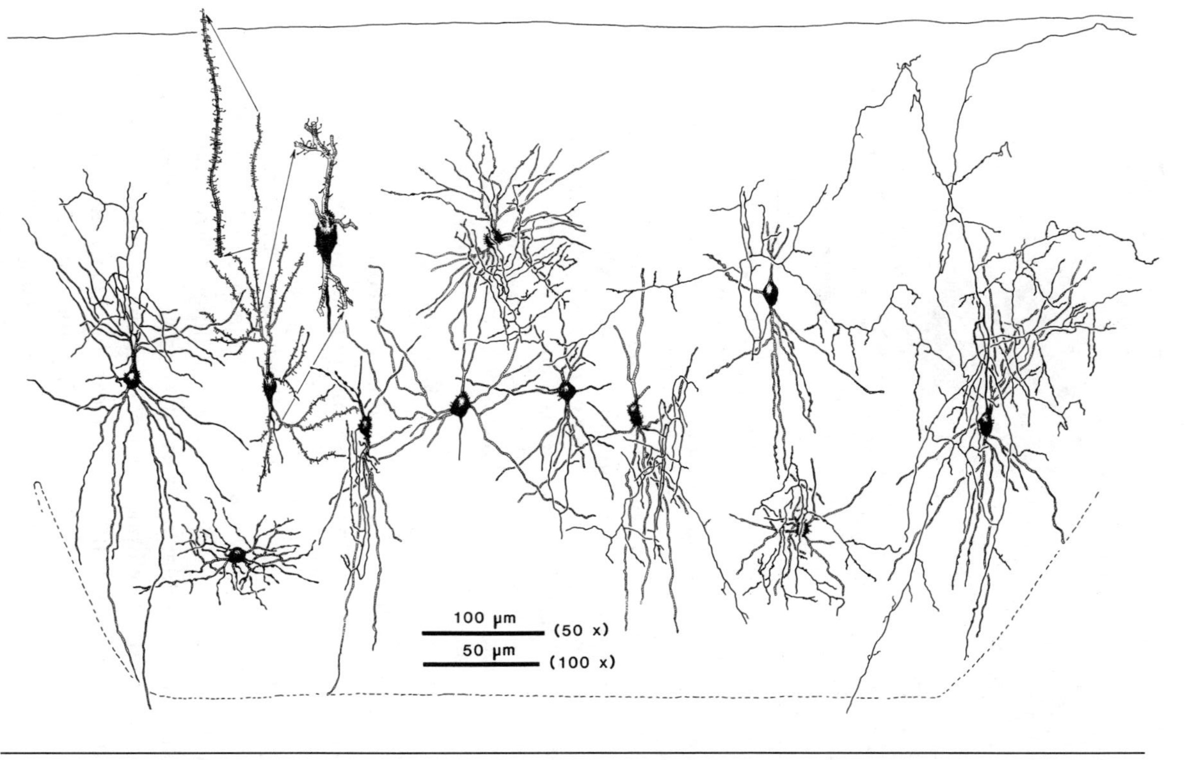

Figure 6.2 Composite drawing of Golgi-impregnated Golgi type II neurons, including both spiny and smooth stellate cells, from ibo-injected areas 17 and 18. Relative radial position of neurons is preserved. An interrupted border marks the bottom of the gray matter. Location and shape of spines are accurately drawn on enlargements of parts of some neurons, as indicated by arrows. Calibration bars refer to magnification of the main drawings and of the enlargements. Power of the objectives used for the drawings is also indicated. (From Innocenti and Berbel, 1991a.)

injured cortex, human microgyric cortex forms numerous shallow gyri and sulci. Because of its similarity with microgyria, we refer to the ibo-injured cortex as microcortex.

Ibo-induced microcortex and microgyria may share a common pathogenetic mechanism. Although the origin of microgyria is debated, one possible cause is ischemic events around the end of the period of neuronal migration. Excitotoxic components, which can be prevented by blockade of the NMDA receptor, appear to contribute to the neuronal damage caused by cerebral ischemia in the adult (Simon et al., 1984).

In order to check whether in the kitten the ibo excitotoxicity operates through NMDA-receptor activation, the following experiment (inspired by Schwarcz et al., 1979) was performed. In two kittens, ibo was injected in the visual areas in one hemisphere at the usual dose and concentration; in the other hemisphere, it was added with equimolar amounts of APV, in order to block the NMDA receptor. The presence of APV prevented the appearance of microcortex, suggesting that, as is true in the adult, the activation of the NMDA receptor may be crucial for ibo neurotoxicity.

The cytoarchitectonic features and cell composition suggest that microcortex may consist exclusively of supragranular layers. This is confirmed by several lines of evidence.

Cytochrome-oxidase staining shows that the darker band corresponding to layer IV becomes progressively deeper and finally remains as a small, inconstant rim at the gray matter–white matter border.

Anterograde axonal tracers injected in microcortical areas 17 and 18 fail to label projections to the superior colliculus, which normally receives afferents from layer V. Projections to the lateral geniculate nucleus (LGN), which normally arise from layer VI, are also absent. The lack of cortical projections to the LGN was confirmed by injections of tracers in this nucleus, which failed to label neurons in the microcortex (see Figure 6.4, below).

Development of Microcortex after Neonatal Injections of Ibotenic Acid

In order to understand what causes the differences in survival of infra- and supragranular layers in the microcortex, we studied the effects of ibo injections 24–48 hours or 6–8 days after the injection.

Twenty-four to forty-eight hours after an injection of ibotenic acid at P2 or P2.5, regions of the lateral and postlateral gyri are in the process of degenerating. The severity of degeneration varies characteristically with the distance from the injection center. Near the center, the cortex consists almost exclusively of layer I and of a few-cell-thick layer of apparently intact neurons. This layer is continuous with the cortical plate of the surrounding cortex, which at this age probably corresponds to part of the prospective layers II–III. Progressively deeper cortical layers reappear in an inside-out sequence with increasing distance from the injection center. In addition, at the periphery of the lesion a layer devoid of neurons is interposed between a thinned layer VI and the upper cortical layers and is continuous with the surrounding intact layers IV and V.

In electron micrographs, the injured cortex and its underlying white matter show wide empty spaces (Figure 6.3A) and scattered degenerating cells (Figure 6.3D), most of them probably neurons, with pyknotic nuclei and severely vacuoled or disintegrating cytoplasm. Many phagocytes with the morphology of "gitter cells" are interspersed among the degenerating elements. Astrocytes can be recognized neither by light microscope nor by ultrastructural criteria; nor are they immunohistochemically detectable from their positivity to antibodies against the glial fibrillary acidic protein (GFAP). Since astrocytes are present in the normal cortex surrounding the ibo-injected site or in the contralateral hemisphere, they may have degenerated in the ibo-injected cortex. This is somewhat surprising, even though it is known that astrocytes have functional receptors to excitatory amino acids (Pearce et al., 1986).

Even more surprising is the presence of intact radial glial cells, which can be recognized by their positivity to a vimentin antibody or electron microscopically by their typical processes rich in intermediate filaments and terminating with typical endfeet in layer I (Figure 6.3). The density of radial glia, however, may be lower than that in the surrounding normal cortex. Intact, elongated cells with the cytological features of neurons can be seen along the radial glia and are probably migrating neurons (Figure 6.3D).

Although many axons are vacuolated or swollen, some appear intact and occasional synapses can be seen.

Six to eight days after ibo injections on P2 or P2.5, a large fraction of the degeneration debris has been eliminated and the injected part

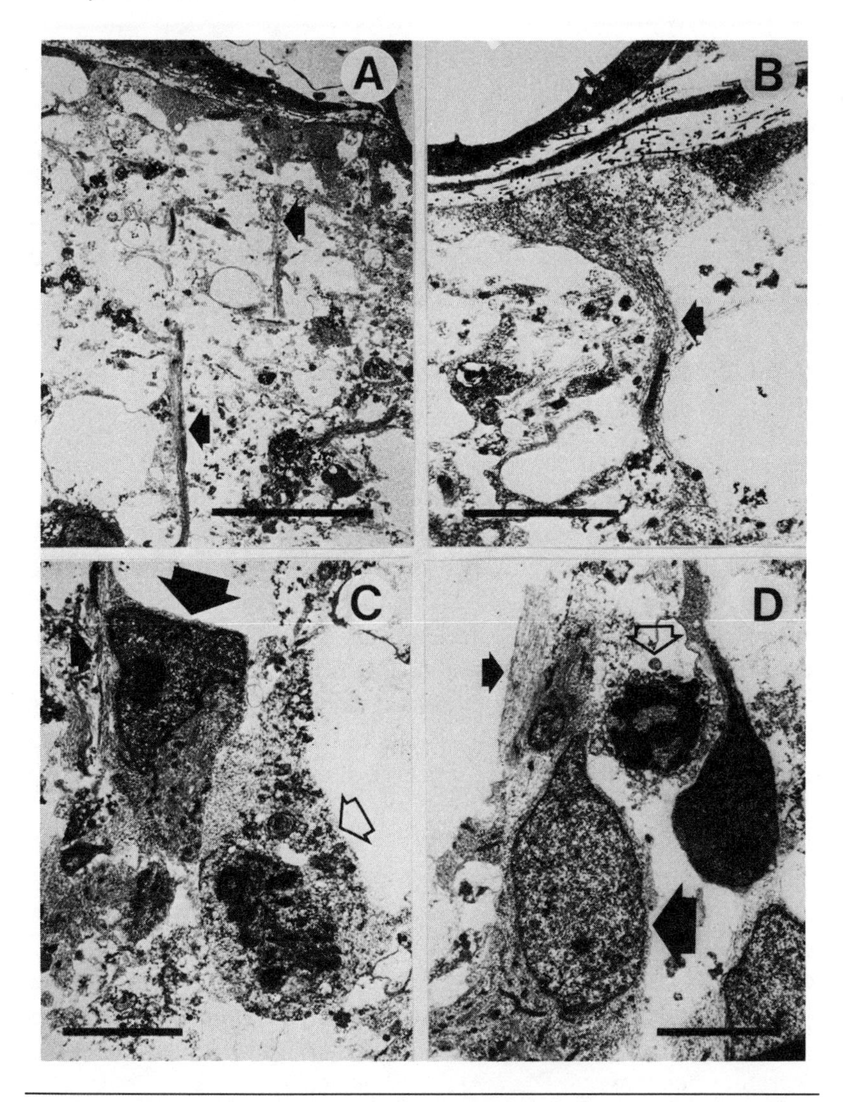

Figure 6.3 Electron photomicrographs through an injected portion of area 17, 24 hours after ibo injection. *A–B:* Layer I; processes of radial glia are marked by filled arrows; notice also the broad extracellular spaces and the degeneration debris. *C–D:* Degenerating cell layer, probably layer III; small filled arrows point to processes of radial glia, open arrows to degenerating cells, probably neurons; migrating neurons are denoted by large filled arrows. Calibrations are 10 μm in *A,* 5 μm elsewhere. (From Innocenti and Berbel, 1991a.)

of the lateral and postlateral gyri have been reorganized. The center of the lesion, presumably corresponding to the point of the injection, consists of a cell layer that is a few hundred μm thick and continuous with the supragranular layers of the surrounding cortex. Underneath, the tissue contains sparse, degenerating cells, "gitter cells," and a few intact neurons. It also contains accumulations of GFAP-positive astrocytes, more numerous than in the intact cortex and concentrated near the interface between the white and the gray matter.

Possible Mechanisms of Microcortex Formation and Effects of Age at Injection

The results described above indicate that at the doses, rates of delivery, and concentrations used here, ibo seems to affect indiscriminately most cortical neurons. At least some early postmigratory and/or migrating neurons, however, may be insensitive to ibo. Some neurons appear to terminate their migration "normally" and undergo "normal" development despite exposure to ibo, and these may be responsible for the recovery of cortical structure in the days following the injection. In addition, differential laminar sensitivity to ibo or differential spread of the toxin in different layers may contribute to the characteristic outside-in reappearance of the cortical layers at the periphery of the lesion.

This hypothesis is compatible with the finding (Shatz and Luskin, 1986) that, in kittens, neurons destined to supragranular layers of area 17 have not completed migration at birth. One would predict that if the migrating and early-postmigratory neurons are responsible for the recovery of cortical structure, the damage would become progressively more severe the older the age at injection (because with increasing age the cortex becomes progressively depleted of migrating neurons).

Consistent with the hypothesis, injections of ibo on P6 or P7 reduces the cortex to a cell-poor superficial layer, wider than the normal layer I, and a deeper layer that is only one or few cells thick. The superficial layer is continuous with layers I and II of the normal cortex while the deep layer is continuous with layer III. Unlike in the normal cortex, this layer III is abnormally undulating. This type of cortex resembles ulegyria, another cortical dysplasia, more than microgyria. Human microgyria and ulegyria may therefore have a common pathogenesis but may arise at different stages of cortical maturation (see Humphreys et al., 1991).

Injections of ibo on P14 or P20 result in even more severe lesions. The cortex appears completely destroyed and is replaced by a layer containing glia and apparently no neurons. The transition from the intact to the injured cortex tends to become sharper the older the animal at injection. Nevertheless, the destruction of the deep layers still extends further away from the injection center than that of the superficial layers.

Connectivity of Ibo-Induced Microcortex

The cytoarchitecture of the microcortex raises specific questions related to the problem of the origin of cortical connectional specificity.

One question is whether the microcortex still receives thalamic input in the absence of layer IV, the main target of projections from LGN.

Injections of retrograde tracers restricted to the microcortex invariably label neurons in the LGN and in the lateralis posterior complex. In the LGN they extend over all laminae delineating a column whose size and location are compatible with the maintenance of normal retinotopic connections between LGN and cortex. Large (20–30 μm in diameter) and medium-size (10–20 μm in diameter) neurons are retrogradely labeled in the LGN. Their size and morphology are compatible with the possibility that both Guillery's class 1 and class 2 neurons still project to the visual areas. Since the number of labeled neurons was not counted, and the topography not quantitatively assessed, these findings do not exclude the possibility of changes in the thalamocortical projection.

Injections of horseradish peroxidase (HRP) in electrophysiologically identified portions of the LGN (Figure 6.4) show, in the normal area 17, retrogradely labeled neurons in layer VI and a combination of anterogradely filled geniculocortical axons and initial axon collaterals of layer VI neurons in layers I, III–IV, and VI. Moving from the normal to microcortical area 17, the band of retrogradely labeled neurons becomes thinner and disappears. The band of anterograde transport in the intermediate cortical layers acquires progressively deeper positions and then disappears almost completely. Near the center of the microcortex only a light termination, resembling that

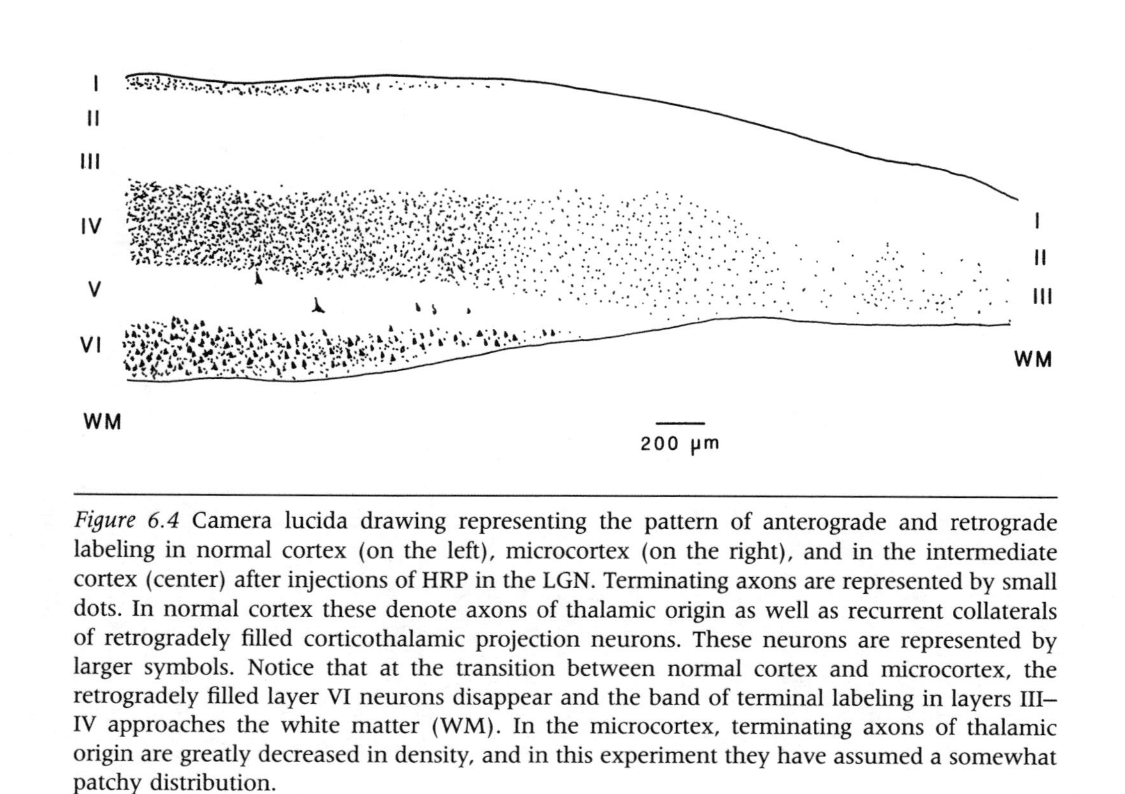

Figure 6.4 Camera lucida drawing representing the pattern of anterograde and retrograde labeling in normal cortex (on the left), microcortex (on the right), and in the intermediate cortex (center) after injections of HRP in the LGN. Terminating axons are represented by small dots. In normal cortex these denote axons of thalamic origin as well as recurrent collaterals of retrogradely filled corticothalamic projection neurons. These neurons are represented by larger symbols. Notice that at the transition between normal cortex and microcortex, the retrogradely filled layer VI neurons disappear and the band of terminal labeling in layers III–IV approaches the white matter (WM). In the microcortex, terminating axons of thalamic origin are greatly decreased in density, and in this experiment they have assumed a somewhat patchy distribution.

extending in normal cortex in layer III, remains. Occasionally this termination appears patchy. From the experiments performed thus far it cannot be excluded that some portions of the microcortex may be completely devoid of LGN afferent.

These results stress the robustness against epigenetic perturbations of two aspects of cortical connectivity, both related to the radial organization of the cortex. The supragranular layers, of which the microcortex consists, do not establish projections to the structures that normally receive afferents from the missing layers. Neither the change in their relative radial position in the cortex nor the absence of competitors from the deep layers seems to be sufficient to induce neurons in layers II and III to establish projections to subcortical areas. Elsewhere, one of us has argued that where a cortical axon grows may be initially determined by specific affinities between the axon and the substrate along which it grows (Innocenti, 1991). The thalamic axons, on the other hand, in the absence of their normal target in layer IV do not invade the supragranular layers but seem to remain confined to the layers they would normally grow to. Thus, the thalamic axons may recognize specific cellular targets in the cortex; this conclusion is compatible with the described "waiting" of thalamic axons in the subplate, which appears to be terminated by some unknown aspect of layer IV maturation (Shatz and Luskin, 1986).

Other aspects of the connectivity of the microcortical areas 17 and 18 are at least qualitatively normal. The microcortex receives projections from all the visual areas in the ipsi- and the contralateral hemisphere, which would normally project to areas 17 and 18. Particularly interesting, in the light of what will be discussed below, is the fact that although callosal projections to areas 17 and 18 at the time of ibo injection originate from the whole of area 17, the transient exuberant projection from medial area 17 to the microcortex is not stabilized; the projection therefore becomes restricted, in the normal hemisphere, to the region near the 17–18 border. Preliminary results stress the notion that even in the microcortex, afferent callosal axons behave differently at the border between areas 17 and 18, which they enter, and in medial area 17, which they seem to avoid. This behaviors suggests that callosal axons recognize different parts of areas 17 and 18 (discussed in Innocenti, 1991).

Still other aspects of the connections of the microcortex emphasize the plasticity of cortical connectivity and the role of transient or exuberant projections in this plasticity.

Afferents to the microcortex are found to originate outside the visual areas, in a portion of the ipsi- and contralateral ectosylvian gyrus and sulcus corresponding to AI and possibly AII (Figure 6.5). The origin of the projections occupies a similar extent and location in the three animals in which it was studied. It is larger and consists of more neurons in the ipsilateral than in the contralateral hemisphere. The neurons at the origin of the projection are usually small and occasionally medium-size pyramids in layers II and III. This projection is absent in normal adult cats, in which only a few neurons, most of them in layer VI, are found in auditory cortex following injections in areas 17 and 18. The projection is present in kittens until P30 (Innocenti et al., 1988) and its elimination involves axonal loss (Clarke and Innocenti, 1990). It is therefore possible that the auditory projection to microcortical areas 17 and 18 represents an abnormally stabilized exuberant projection.

In an effort to study the pattern of termination of the auditory projection in the microcortical visual areas, and possibly to learn whether it represents a stabilized juvenile projection, AI and AII were injected with WGA-HRP in 12 kittens. In all these animals the precision in the location of the injection was critical, since the region where the auditory projection originates is only about 4 mm2 and relatively close to the deep layers of the lateral suprasylvian cortex that project to areas 17 and 18 in normal cats (Symonds and Rosenquist, 1984). Criteria suggesting that these neurons were not involved in the injection site included the pattern of thalamic labeling and the absence of labeling in the contralateral lateral suprasylvian areas. In animals that fulfilled these criteria (the 12 mentioned above), labeled axons can be found that reach areas 17 and 18 through a characteristic trajectory along the roof of the lateral ven-

Figure 6.5 Top left: Sketch from a photograph of a neonatally ibo-injected brain; site of injection was on the right lateral gyrus (notice its narrowing and the appearance of an abnormal sulcus); levels of section shown below are denoted by horizontal bars. *Top right:* Coronal section through the site of a WGH-HRP injection; density of hatching roughly corresponds to density of precipitate. The injection is restricted to the microcortex. *Middle:* Distribution of retrogradely labeled neurons in the left and right (as shown) areas

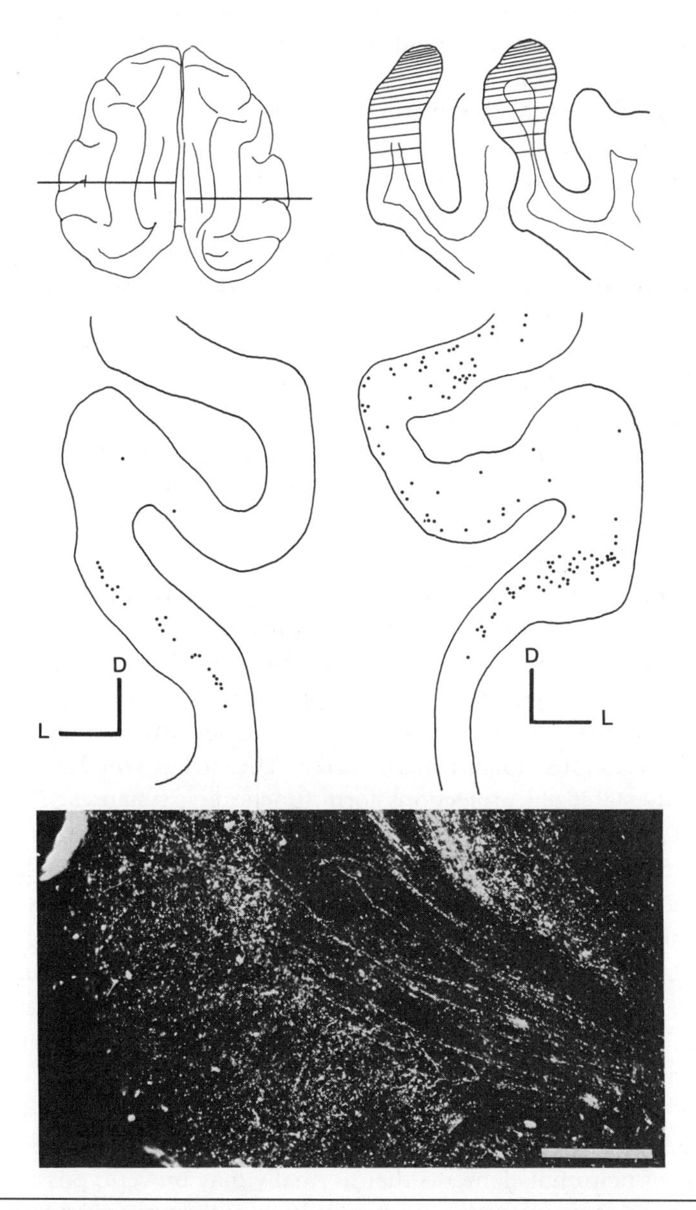

AI and AII in the same experiment. Bars = 1 mm; D = dorsal, L = lateral. *Bottom:* Polarized light photomicrograph of axons anterogradely labeled by a WGA-HRP injection in areas AI and AII and entering microcortical area 17 (on the left) of the same hemisphere, at the base of the postlateral gyrus. Top is up; bar = 200 μm.

tricle. This trajectory is similar to that of the juvenile transient projection mentioned above. The axons terminate particularly densely in the deep layers of the microcortex (Figure 6.4).

The development of the auditory projection to the microcortical areas 17 and 18 was studied in 5 of the kittens mentioned above at different times after the ibo injection (on P2–3). The results were compared with those obtained in age-matched controls from a parallel study (Innocenti et al., 1988). In this series of experiments, the earliest consequences of ibo injection were found in a P14 kitten. In this animal, axons labeled from the auditory cortex were restricted to the microcortical parts of areas 17 and 18 and were more numerous than anywhere in areas 17 and 18 of normal animals of comparable age. In older, normal animals (killed on P27 or P41), the projection had practically been eliminated, although it persisted in the microcortex, where axons seemed actually to have elaborated their terminal arbors.

In conclusion, layer of origin, trajectory, and termination of the auditory cortex projections to microcortical areas 17 and 18 are similar to those of the transient exuberant projections that existed at the time of ibo injection; the developmental analysis also favors the hypothesis that these projections are at least partially maintained. Several questions remain unanswered. The first is whether the abnormally stabilized projections form functioning synapses. The second is about the mechanism of stabilization. In the present study axons from auditory cortex entered exclusively the microcortical part of area 17 and 18. The stabilization, therefore, probably has to do with altered local conditions at the target of the projection. More indirect causes, such as changes in the connectivity of AI and AII due to the lesion of areas 17 and 18 (for a somewhat similar condition, see Frost, 1986), are unlikely and might be ruled out by the analysis of the connections of auditory areas.

The mechanism of stabilization may be related either to the changes in the astrocytes at the site of ibo injection or to the elimination of neuronal elements that normally may prevent, possibly via a competitive mechanism, the stabilization of the projection from the auditory cortex. Since some axons from auditory cortex are seen to enter microcortical regions where only layer VI was destroyed partially, neurons in this layer or in the subplate may have this function in normal development. In the present experiment the neurons of origin of the projections were restricted to a small region at nearly identical location in the three cases studied. Why precisely this part of the projection should be maintained is unclear.

The projections from auditory to visual areas may not be the only transient projections that are stabilized in the microcortex. Microcortical area 17 seems to possess long intracortical projections of a sort not found in normal adults but described in young kittens (Luhmann et al., 1990a).

Functional Properties of Microcortex

The origin of the characteristic functional properties of visual cortical neurons—namely, the mechanism producing orientation and directional specificity—remains mysterious. Concerning orientation specificity, the discussion is focused on the role of the spatial organization of the retinal receptive fields that feed onto a cortical neuron (Hubel and Wiesel, 1962), on local inhibition in the cortex (Benevento et al., 1972), and on the orientation bias present at the level of the geniculate body (Shou and Leventhal, 1989). On the other hand, orientation- and direction-specific receptive fields were recently described in the somatosensory area when retinal projections were rerouted to the ventrobasal thalamic nuclei (Métin and Frost, 1989).

It may be speculated that since orientation specificity is a property of visual cortical columns in area 17, its generation may involve interaction of neurons within one column and/or of neurons of neighboring columns. Anatomical and electrophysiological evidence exists for the latter hypothesis (Gilbert and Wiesel, 1979, 1989; Matsubara et al., 1985; Bolz and Gilbert, 1989; Luhmann et al., 1990a,b), though there is evidence that orientation-specific responses may be independently generated at different radial levels in a cortical column (Malpeli, 1983). In development, interaction between neurons in one column and in neighboring columns may be needed for the acquisition of orientation specificity.

The microcortex allows a new approach to the question of the genesis of response properties in visual cortex. For example, we performed a set of electrophysiological investigations in microcortical areas 17 and 18 using conventional recording and stimulation techniques. The study so far is based on the analysis of 261 single units and 127 multiunits recorded in 37 penetrations through the convexity or the medial bank of the lateral and postlateral gyri, and more rarely from its lateral bank.

Ad hoc anatomical definitions were adopted to classify the sites of recording. Because the microcortex has lost layer IV and the large layer III pyramids, the border between areas 17 and 18 cannot be defined on cytoarchitectonic criteria. We considered recording sites in the medial bank of the caudal lateral and postlateral gyri as having been in area 17. Sites near the convexity of the lateral and postlateral gyri were considered to be near the 17–18 border. More lateral sites were considered to be in area 18. This subdivision is in agreement with the electrophysiological maps of the visual areas in normal cats (Tusa et al., 1978) as well as with differences in the response properties of neurons in areas 17 and 18 and with the progression of receptive-field locations in the two areas. Recording sites were also grouped according to the severity of the cytoarchitectonic changes induced by ibo. Microcortex was three-layered (layer I, II–III, see above). A special type of microcortex, called supermicrocortex, was found in two experiments; it consisted of a thin, abnormally undulating pyramidal layer, probably a remnant of layer III and of a broader cell-poor layer above it. The transitional cortex was characterized by a thinned or nearly completely absent layer VI, nearly invariably accompanied by a clearing of neuronal cell bodies in layer V. Layer IV can also be partially destroyed.

Most of the penetrations (29) turned out to be in area 17; 52 single units and 35 multiunits were recorded from microcortex or supermicrocortex; 35 single units and 14 multiunits were recorded from transitional cortex. For comparison, 92 units and 30 multiunits were recorded from normal cortex adjacent to the microcortex or from intact animals.

Visual responses could be readily evoked from most of the recording sites and, as described in detail below, they were qualitatively indistinguishable from those found in normal cortex. Nevertheless, in several experiments responses either could not be evoked or were sluggish and inconstant along certain penetrations or parts of penetrations through microcortex. The possibility that the thalamocortical input may be greatly reduced in these regions is being investigated.

Figure 6.6 illustrates an interesting penetration through the microcortex. Responses were usually unequivocal. The contours of receptive fields could be identified by monocular stimulation. Most

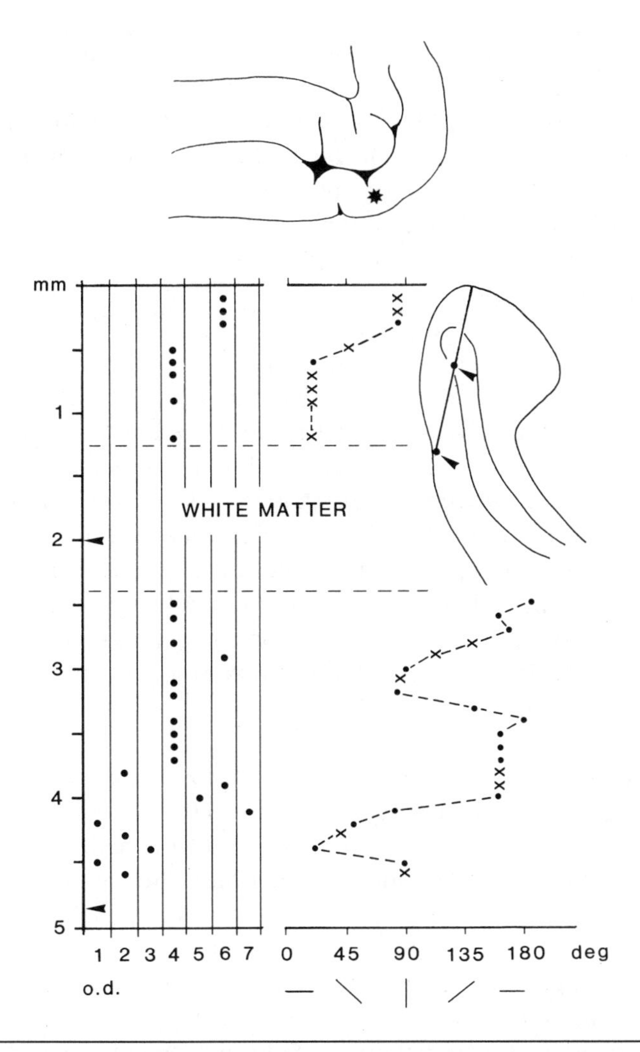

Figure 6.6 Reconstruction of a microelectrode track through the microcortical area 17 (asterisk on surface view, above graph). Ocular dominance and orientation selectivity of single and/or multiunit activity were recorded along the penetration shown in the inset to the right of the graph. *Left:* Ocular dominance (o.d.) classes identify neurons exclusively or preferentially responding to contralateral (1, 2, 3), ipsilateral (7, 5, 6), or either (4) eye. *Right:* Orientation preference is in degrees from the horizontal. Loci responding selectively to oriented stimuli are represented by dots. Crosses are loci unselective for orientation; their position in the diagram is arbitrary but in the tangential part of the penetration corresponds approximately to the orientations that would have been found in an abnormal animal. Arrowheads denote electrolytic lesions.

responses were specific for the orientation and direction of stimulus motion. Nevertheless, high proportions of nonoriented receptive fields were found in this and a few other penetrations through microcortex. The relevance of this finding is currently being investigated. Single units are distributed more or less normally within classes of ocular dominance (Hubel and Wiesel, 1962).

Quantitative comparison of the response properties in the different types of cortex is limited by the scatter of the sample. In general, the spatial distribution of response properties resembled that found in the normal cortex. In particular, along tangential penetrations traversing area 17 from lateral to medial, the receptive field of neurons moved laterally and ventrally in the contralateral visual field, indicating that the microcortex retains a retinotopic order similar to that of normal cortex. However, some change in the magnification factors of the retinal representation cannot be excluded on the basis of the available data. Along tangential trajectories through area 17, the regular and expected gradual changes in orientation specificity were usually found. Again, the possibility must remain open that these changes differ in some quantitative way from the normal. Shifts across ocular dominance columns were also observed in some penetrations.

Additional response properties were analyzed in a limited number of cases. Whenever it was tried, the usual classification of receptive fields into simple and complex types could be achieved by using stationary stimuli (Hubel and Wiesel, 1962). Frequently, units that required long stimuli, indicative of a need for spatial summation, were found. In a few cases units selective for high stimulus velocity were met, suggesting the need for temporal summation at the geniculocortical input.

A more global functional analysis of the microcortex involved the use of C14-deoxyglucose and monocular stimulation with gratings of all orientations in one case; binocular stimulation with gratings of vertical orientation was performed in another case. Figure 6.7 illustrates the results obtained in the monocularly stimulated animal. In the intact hemisphere a band of high deoxyglucose activity was found in layers III–IV, and it showed a moderate degree of waxing and waning, possibly related to the ocular dominance. Consistent with the absence of the normal thalamic input to layer IV, this band was absent in the microcortex. But there was still a

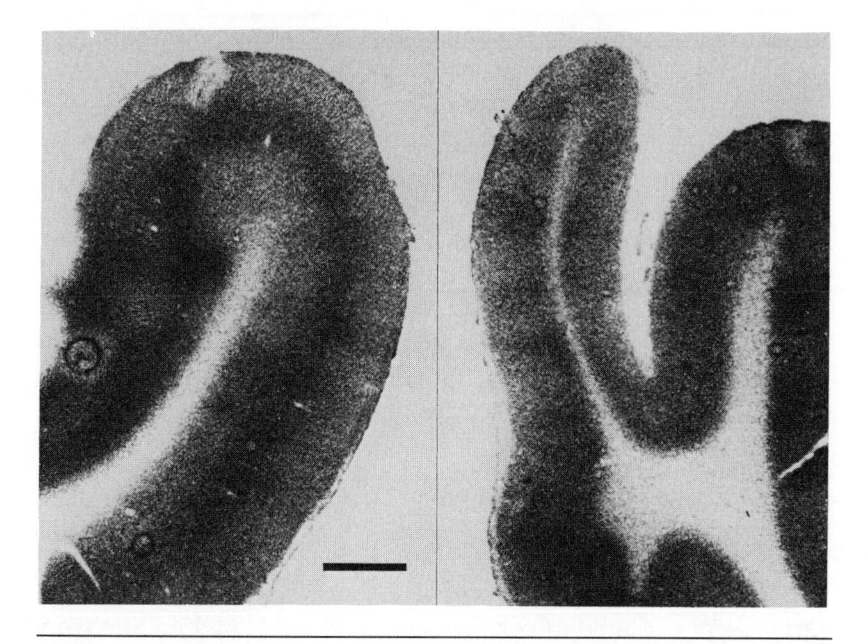

Figure 6.7 Photomicrographs of autoradiograms of coronal sections through the normal cortex *(left)* and the ibo-induced microcortex *(right)*. This kitten was injected intravenously with 16.5 μCi/100 g b.w. 2-deoxyglucose; its left eye (contralateral to the lesion) was stimulated with hand-moved gratings of different spatial frequencies and orientations, projected on a tangent screen 50 cm in front of the animal. Following deep Nembutal anesthesia and perfusion with 3.3% formalin in 0.1 M Sörensen buffer (pH 7.4), the brain was extracted, frozen, and processed for autoradiography. Notice the striking difference in size between the hemispheres; furthermore, the microcortex lacks the continuous band of activity corresponding to layer IV. Deoxyglucose uptake is distributed in columns of similar width (530 μm) and peak-to-peak distance (650 μm) in the two hemispheres. Bar = 1 μm.

hint of a columnar organization, possibly representing ocular dominance domains.

Conclusions

In research efforts devoted to human developmental abnormalities, animal experiments are useful as explanatory or predictive devices. The work we discussed may be relevant as a model for specific human pathologies, such as microgyria and ulegyria. More fundamentally, there is hope that it may provide insights into the basic rules governing the development of cortical connectivity. The evidence accumulated thus far stresses the crucial role of the target cortex in the selection of afferents of cortical and subcortical origin as well as the importance of transient juvenile connectivity as a substrate for deviations from normal development. The decision of which of the juvenile afferents to a cortical site should be maintained and which should be eliminated seems to involve specific positive and negative interactions between each type of afferent and different aspects of the target. The decoding of these interactions will be a challenging task for developmental neurobiology and developmental pathology in the coming years.

Acknowledgments

The 2 DG experiment and one of the electrophysiological experiments were executed in collaboration with Dr. P. Melzer. The work was supported by the Swiss National Science Foundation Grant 3.359-0.86 to GMI and by a fellowship from the Spanish Consejo Superior de Investigaciones Scientíficas to PB. Ibotenic acid was kindly provided by Dr. Eugster. We thank Marc Weisskopf for his comments on this manuscript.

7

Functional Brain Asymmetry, Dyslexia, and Immune Disorders

Kenneth Hugdahl

A central idea in most biologically oriented theories and models of dyslexia is that deviations in normal patterns of brain asymmetry are in some way associated with developmental reading disorders (see Hynd and Semrud-Clikeman, 1989). However, the exact nature of these deviations or alterations of normal patterns of brain lateralization remains controversial (e.g., Satz & Soper, 1986) and is the subject of debate (see Hugdahl, Synnevåg, and Satz, 1990; Wofsy, 1984; Salcedo et al., 1985).

The relationship between shifts in brain asymmetry and dyslexia has variously been explained as a result of "bilateralization" (Orton, 1937); "maturational lag in lateralization" (Corballis and Beale, 1976; Satz and Sparrow, 1970); "deficit in interhemispheric integration" (Marcel, Katz, and Smith, 1974); or "intrahemispheric, left hemisphere deficit" (McKeever and Van Deventer, 1975; Beaumont, Thomson, and Rugg, 1981).

A more recent hypothesis was put forward by Geschwind and Galaburda (1985a–c). These authors have argued that there is an association between learning disorders, anomalous brain asymmetry patterns, immune function, and left-handedness (see also Geschwind, 1984; Geschwind and Behan, 1982). Before I turn to a more thorough discussion of the "testosterone hypothesis" as it was formulated by Geschwind (1984) and Geschwind and Behan (1982), a brief historical review of the relation between asymmetry and dyslexia may be of interest. In particular, since Geschwind's more recent views have historical parallels to Orton's work in the late 1920s and early 1930s, a brief review of Orton's model may be in order.

Orton (1937) viewed weakened bilateralization or weakened hemisphere dominance relations as major causes for dyslexia. Orton assumed that visual stimuli (like a text) were stored in both hemispheres, but that the picture stored in the right hemisphere was a mirror reversal of that stored in the left hemisphere. The right-sided, mirror-reversed picture was usually suppressed by the dominant left hemisphere. If the left hemisphere dominance was weak, however, the mirror-reversed storage would interfere with left hemisphere processing, causing confusion and poor reading performance.

Although Orton has been criticized for lacking both theoretical depth and empirical support, there is one clinical observation that supports his view. Some dyslexic children can actually read better with the book turned upside down. Turning the book upside down means that the initial input to the brain is a mirror-reversed image of the text to be read (reading from right to left). According to Orton's (1937) model, this should result in a veridical representation in the right hemisphere. Thus there should be no interference from the "mirror-reversed" right hemisphere image, with the left hemisphere processing the written text. The problem that still remains unsolved is why only a small fraction of all dyslexic children can benefit from reading upside-down texts. It is, on the other hand, difficult to account for mirror-reversed reading without taking Orton's notions of weak hemisphere dominance relations and bilateralization into consideration.

It is also perhaps interesting to note that the testosterone hypothesis of Geschwind (1984) represents an "update" of several central themes in the Orton model. Among other things, both Orton and Geschwind stressed that reading disorders were caused by a failure of development of normal laterality dominance patterns (see, e.g., Geschwind, 1982). What Geschwind did in addition was to provide a plausible explanation for this phenomenon—namely, that testosterone *in utero* inhibited the normal development of the left hemisphere, thus setting the stage for "symmetry" rather than "asymmetry" as a working principle between the cerebral hemispheres.

Hemispheric Asymmetry

In a broad sense, "hemispheric asymmetry" refers to the fact that the two cerebral hemispheres differ in their ability and propensity to process information (see Hellige, 1990, for an updated review). This means that hemispheric asymmetry is an important modulation of cognitive function. From this perspective, it may be surprising to realize that a vast majority of studies of hemispheric asymmetry are restricted to the initial registration and perceptual encoding of a stimulus or to "perceptual asymmetry." Very few studies have been concerned with asymmetry for more integrated cognitive functions, like learning, memory storage, and skill acquisition (see Hugdahl and Brobeck, 1986; Hugdahl, Kvale, Nordby, and Overmier, 1987; Beaumont and Dimond, 1973, for exceptions).

If learning disorders are related to asymmetrical functioning of the brain, it is plausible that the activity of the two sides is also a guiding principle for learning and memory in the *normally* functioning individual. It thus becomes important to understand how learning *in general* is cognitively represented in the two hemispheres of the brain in order to understand what causes *disorders* of learning (cf. Hugdahl and Andersson, 1989).

It has been shown in our laboratory that the learning of simple linguistic associations is differentially represented in the two hemispheres (e.g., Hugdahl and Brobeck, 1986; Hugdahl, Nordby, and Kvale, 1990a). These studies have been conducted within the framework of the dichotic listening procedure (see Hugdahl, 1988, for review), involving the "dichotic learning paradigms" (Hugdahl et al., 1990a).

A basic idea derived from these studies is that the mental representation of a learned association is stored only in one hemisphere, despite both hemispheres having access to the stimuli that are associated during acquisition (or learning phase). The storage of learned associations involves procedural memory processes, which differ from the processes involved in declarative memory (e.g., Mayes, 1989; Squire, 1987). When each hemisphere is probed separately for elicitation of the stored association, the learned response shows up only from one hemisphere. The procedure (described in detail in Hugdahl, 1987; Hugdahl et al., 1990a) involves dichotic presentations of the probe stimulus and a control stimulus. In experiments using event-related potentials (ERPs) as indicators of cortical processing during learning, it has been shown that there is a difference for both early and late components of the ERP when the probe is presented to the left as compared to the right hemisphere (see Hugdahl et al., 1990a).

The development of the dichotic learning paradigm has grown from an interest in basic research related to learning and hemispheric asymmetry. If the results are replicable and if the paradigm can be further validated, the dichotic learning approach could be used in applied settings for studies of learning disorders. One possibility is that children with learning disabilities may demonstrate a lack of asymmetry of associative representations to linguistic stimuli. Such a response pattern may further be a behavioral manifestation of the suggestion by Galaburda, Sherman, Rosen, Aboitz, and Geschwind (1985) of symmetry rather than asymmetry of homologous left and right hemisphere areas in the planum temporale and parieto-occipital cortex (see also the discussion by Hynd and Semrud-Clikeman, 1989). Furthermore, it has been argued that in individuals with learning disorders there is a slowing of the development of the left hemisphere and a compensatory growth of the right hemisphere, resulting in *symmetry* between the hemispheres (e.g., Rosen, Galaburda, and Sherman, 1987). Thus, the net result is a smaller-than-normal left hemisphere and a larger-than-normal right hemisphere.

A possible consequence of brain symmetry is that the normal left-to-right hemisphere balance during a learning task is disrupted, yielding inferior learning performance. This possibility has so far mainly been investigated in learning tests related to overt reading and writing performance. An argument based on the findings by Hugdahl and colleagues (e.g., Hugdahl et al., 1990a) is that if the suggested link between hormones, hemispheric (im)balance, and learning disorders is valid, then a difference should also be observed in studies of *basic* processes of learning (that is, associative learning) involving a procedural memory load. The dichotic learning paradigm (see Hugdahl and Brobeck, 1986) may be a sensitive tool to reveal such disorders at a "low" processing level.

A tentative hypothesis is that dyslexic children will less likely encode a verbal association in the left hemisphere, and in consequence no differences should be detected across the hemispheres in a dichotic learning task. So far this is only a possible development of our current work (e.g., Hugdahl and Brobeck, 1986; Hugdahl, 1987; Hugdahl et al., 1990a), but we certainly hope to be able to pursue the research along these lines.

Brain Asymmetry, Dyslexia, and Immune Disorders

In the following I will report some empirical data from our laboratory that relate more directly to the theorizing of Geschwind (e.g., 1984), Geschwind and Behan (1982), and Geschwind and Galaburda (e.g., 1985a–c) about shifts in hemispheric asymmetry in dyslexic children, possibly accompanied by increased frequency of left-handedness and certain immune disorders. More specifically, Geschwind and Behan (1982) and Geschwind (1984) argued that there might be a common factor causing dyslexia, immune disorders, and left-handedness. The impetus of these speculations were clinical observations that dyslexic children (mostly males) were also left-handers and suffered from various immune disorders (like allergies and migraine). Since more boys than girls were affected by these disorders, it was argued that overproduction or oversensitivity to the male sex hormone testosterone in the fetus might be the "missing link." Geschwind and coworkers proposed that testosterone slows the growth of specific regions of the left hemisphere and that homologous regions of the right hemisphere may develop more raipdly. Thus, there is a shift from left to right hemisphere in the control of language and handedness. This will eventually lead to permanent learning disorders in some cases, particularly in boys. Testosterone also affects the development of the thymus gland, resulting in a faulty cellular-based immune system. Hence, overproduction of or oversensitivity to testosterone may predispose the developing fetus toward increased susceptibility to left-handedness, dyslexia, and immune disorders.

Although theoretically appealing, the Geschwind hypothesis has been hampered by a lack of empirical replications from large-scale samples (see Satz and Soper, 1986, for criticisms) and by conflicting empirical evidence (e.g., Salcedo, Spiegler, Gibson, and Magilavy, 1985; Hansen, Nerup, and Holbek, 1986; Schur, 1986). Given the available empirical evidence, it seems that part of the problem is a failure of conceptual consensus. For example, Rosen et al. (1985) used the phrase "relationship of lefthandedness, (auto)immune disease, and dyslexia" (p. 30), while Satz and Soper (1986) talked about the "association between lefthandedness, and autoimmune-disease, migraine and dyslexia" (p. 453). It is unclear what exactly is expected to interact with what, and in what way. Because of the conceptual confusion, I will use the term *hypothesis* rather than *theory* when describing the work of Geschwind and colleagues (e.g., Geschwind, 1984; Geschwind and Behan, 1982; Geschwind and Galaburda, 1985a–c). Formally, their position may be neither, but rather a set of loosely connected hypotheses. A hypothesis refers to facts that are as yet unexperienced, and it is corrigible in view of new knowledge (Bunge, 1967). The logical sense of the word *hypothesis* is therefore of an assumption, premise, or starting point of an argument. A confirmed hypothesis is called a scientific *law* that is supposed to depict an objective pattern. A *theory,* finally, is a set of related laws characterized by the relation of deducibility. Such theories are called hypothetico-deductive systems (Bunge, 1967). If we consider, for example, the Geschwind and Behan (1982) paper with these definitions in mind, it does not qualify as a theory. Very few, if any, of the suggested relationships are empirically confirmed. The paper did suggest an interesting series of possibly related hypotheses, however, and may prove to be the start of a full-grown theory if the hypotheses are confirmed. It is therefore more appropriate to label the Geschwind and Behan (1982) paper as a set of loosely related hypotheses.

In a broader sense, the more significant effort may be not to develop a theory of dyslexia (or left-handedness) but to generate a new general theory about brain laterality. The work by Geschwind and Galaburda as well as other data related to the issues of immune function, dyslexia, and left-handedness may thus be important steps toward an integrated theory of *brain function and laterality,* rather than contributions only to an explanation of learning disorders.

A three-way interaction

If we look at the testosterone hypothesis in some more detail, it can be argued that comparing the occurrence of dyslexia in samples of left- and right-handed children is not the same as comparing the occurrence of left-handedness in samples of dyslexic and normal children. Since most left-handers probably are normal (as are most right-handers), one might not expect increased rates of dysfunction in a large left-handed group. However, the opposite approach, selecting for a dysfunction and comparing rates of left- and right-handedness, might yield a different outcome (cf. Annett and Turner, 1974).

In the first investigation reported in the Geschwind and Behan (1982) study, the authors selected left-handers and right-handers and compared the frequency of immune dysfunctions and learning disorders in the groups. The results showed an increased occurrence of both disorders in the left-handed as compared with the right-handed group. In the second part of the study, the authors selected patients with immune- and autoimmune-system dysfunctions (migraine and myasthenia gravis) and examined the frequency of left- and right-handedness. The results showed an increased frequency of left-handedness only in the migraine group. In another paper, Geschwind and Behan (1984) selected left-handed and right-handed subjects and examined the frequency of different immune disorders, as well as stuttering and skeletal malformations. Once again, there were elevated occurrence rates in the left-handed group.

Taken together, the original results (Geschwind and Behan, 1982, 1984; Geschwind, 1984; see also Behan and Geschwind, 1985) empirically demonstrated elevated occurrences of immune disorders and learning disorders in left-handers as compared with right-handers, but not in dyslexics.

It should be emphasized that in the original Geschwind and Behan (1982) study there was no comparison of the frequency of left-handedness and immune disorder in groups selected to be *dyslexic* (or having a learning disorder). Only Pennington, Shelley, Smith, Kimberling, Green, and Haith (1987) have to my knowledge directly addressed *the three-way interaction* from the perspective of dyslexia—that is selecting dyslexic and normal children and comparing the frequency of the other two factors in the two groups. The results of this study showed a significant elevation of both autoimmune and allergic disorders in dyslexics (compared with controls). However, they did *not* find an elevated occurrence of left-handedness in the dyslexic group.

Considering the impact that the testosterone hypothesis has had on recent theories of dyslexia (e.g., Rosen et al., 1987), and the criticism that it has aroused (e.g., Satz and Soper, 1986; Wofsy, 1984). We recently performed a study where we selected dyslexic subjects and examined the frequency of immune dysfunction and handedness. We also obtained data on familial sinistrality; hand posture when writing; sightedness (myopia, hypermetropia, and astigmatism); stuttering; and complications during pregnancy and/or at delivery. Information was gathered for these latter variables since they all have been suggested to correlate with hemispheric asymmetry and/or dyslexia (see Bradshaw and Nettleton, 1983; Benbow, 1986). For a more detailed report of this study, see Hugdahl, Synnevåg, and Satz (1990b). The study involved 210 children between 7 and 16 years of age, of whom 105 children were diagnosed as dyslexic. The other 105 children were age- and sex-matched normal controls. There were about twice as many boys as girls in both groups. All of the dyslexic children had been initially referred for special education services by their regular teachers as a result of reading failure. They had all been formally diagnosed by a speech therapist and/or at a special educational unit as being retarded in their reading/writing performance compared with their age peers, according to the standard criteria used in the Norwegian educational system. Thus, all children in the dyslexic group (but none in the control group) had been identified as a "child with a specific disorder of reading and writing" and were in special education training. The diagnostic procedure involved, in addition to a general test of anamnesis, medical examination (for signs of sight and hearing problems), reading and writing tests, IQ tests, teacher evaluations, standardized general ability tests, and tests of concentration and motivation.

A questionnaire (available in English from the author) was developed with items relevant for handedness, sightedness, stuttering, immune diseases, autoimmune diseases, and conditions during pregnancy and at delivery. Handedness was tested with 15 items related to manual preference taken from a modified version of the questionnaire developed by Raczkowski, Kalat, and Nebes (1974). Handedness data were analyzed to determine whether the distribution of hand-preference scores was J-shaped and shifted to the right as in the general population (Annett, 1985). Both groups showed the expected rightward shift with no difference in the shape of the distributions. The questionnaire gives three choices for each item, right, left, or both hands. Thus, no degree of hand preference was measured. Familial handedness was measured by self-report. In order to be classified as a right-hander, each child had to indicate that at least 12 out of the 15 items were preferably performed by the right hand. Children reporting any other combination were considered non-right-handed. The actual distribution of the handedness score for the dyslexic and control children are seen in Figure 7.1. Although the dyslexia questionnaire also taps different subtypes of dyslexia, this issue will not be dealt with in the present chapter.

The questionnaire results show significantly more diseases related to immune-system functioning (eczema, asthma, eye inflammation, migraine, and allergies) in the dyslexic group than in the control group. More than half of the dyslexic children (53 percent) reported one (or more) immune disease, while only 25 percent of the control children reported this. There were, however, no significant differences between the groups in occurrence of left-handedness. Taken

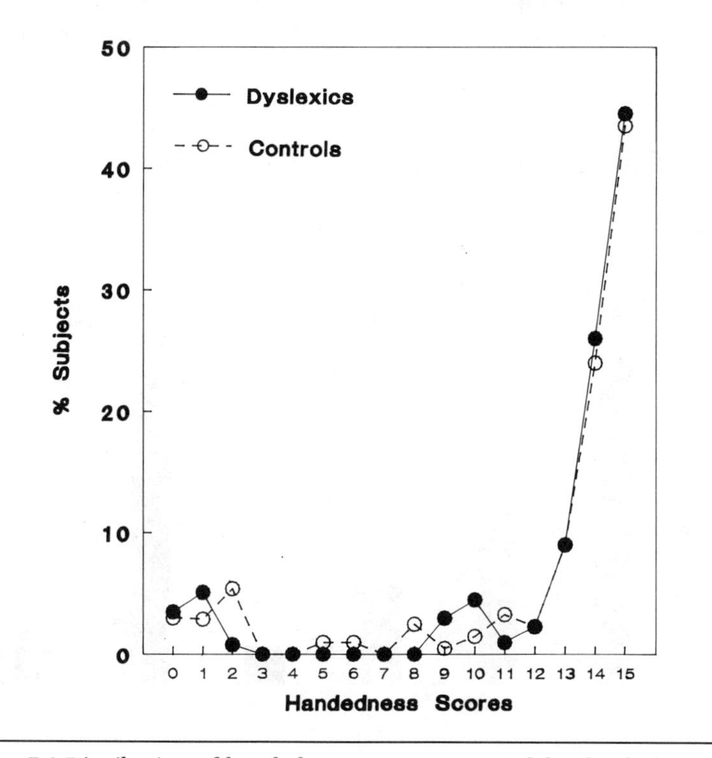

Figure 7.1 Distribution of handedness scores, separated for the dyslexic and control groups. (Data from Hugdahl, Synnevåg, and Satz, 1990.)

together, the results showed 84 percent of the groups were right-handed, 9.5 percent were left-handed, and 6.5 percent could not be classified as either. The immune disease results are seen in Figure 7.2.

The results for reported autoimmune diseases should be read with some caution since only five diseases were registered: gastro-intestinal disorders, myasthenia gravis, rheumatoid disorders (2 cases), and diabetes mellitus. However, all five autoimmune diseases were found in the dyslexic group. In other words, five children with these disorders were identified in the dyslexic group, but none in the control group. The disorders were not verified by a physician, though, so they should be interpreted with caution.

If one looks at the distribution of *diseases* in the two groups (rather than at the distribution of *subjects* within each group that had at least one disease), there were 91 (73.4 percent) immune diseases in the dyslexic group compared to 33 (26.6 percent) in the control group.

In order to analyze further the suggested three-way interaction of

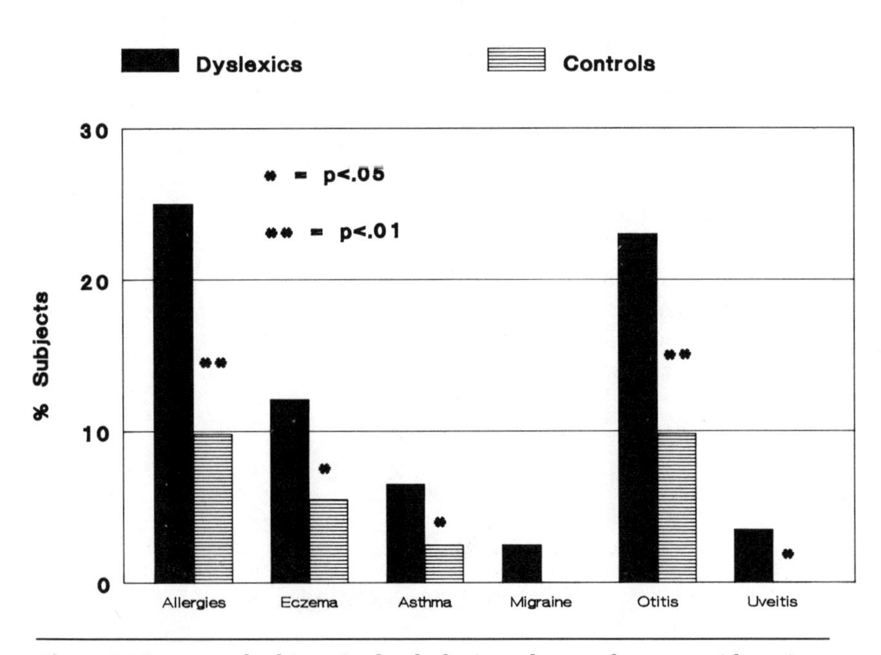

Figure 7.2 Percent of subjects in the dyslexic and control groups with various immune diseases.

dyslexia, non-right-handedness, and immune diseases, we pooled the data of four cohorts: dyslexics *and* non-right-handers; controls *and* non-right-handers; dyslexics *and* children with immune diseases; controls *and* children with immune diseases. Figure 7.3 shows, on the left, the percentage of subjects with at least one immune disease among the dyslexic–non-right-handed (Dysl-NRH) group and the control–non-right-handed (Ctrl-NRH) group; on the right, the percentage of non-right-handed subjects in the dyslexic–immune-disease (Dysl-Immn) group and the control–immune-disease (Ctrl-Immn) group. The most interesting aspect of these analyses was that although there were no differences in frequency of non-right-handedness in the *overall* analysis, there were still more non-right-handers in the *subgroup* of children who were dyslexic *plus* had an immune disease than in the subgroup who were not dyslexic *plus* had an

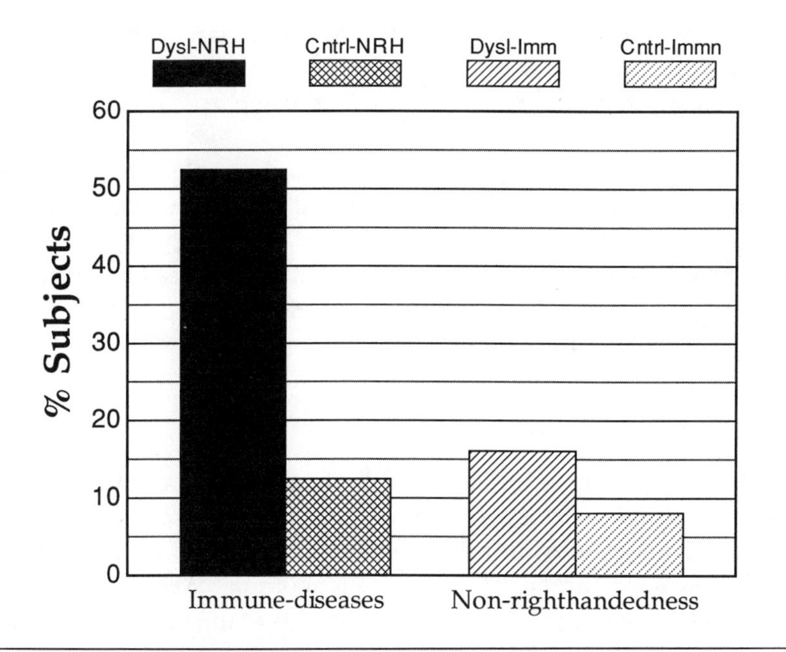

Figure 7.3 Percent of subjects in the dyslexic–non-right-handed (Dysl-NRH) and control–non-right-handed (Cntrl-NRH) groups with immune diseases, and percent of subjects in the dyslexic–immune-disease (Dysl-Immn) and control–immune-disease (Cntrl-Immn) groups who were non-right-handers.

immune disease. Similarly, there were more subjects with immune diseases in the subgroup of dyslexics who were non-right-handers than in the subgroup of controls who were non-right-handers. An interesting speculation is that those dyslexic children with both non-right-handedness and an immune disease may have a unique subtype of dyslexia, a form different from that of the dyslexic children without non-right-handedness or an immune disease. Interestingly, researchers in Bratislava, Czechoslovakia (Katarina Jariabkova, personal communication), have recently replicated several of our findings, using the same questionnaire. They also found more subjects with both left-handedness *and* immune diseases in the dyslexic than in the control group. Thus, children with *both* left-handedness *and* immune diseases and who are also dyslexic may perhaps be the critical subgroup that the Geschwind hypothesis is targeting. This possibility should be further investigated in future research.

Significant differences between the groups were also found for frequency of *hypermetropia* and for *stuttering*. A marginally significant effect was found for *astigmatism*. In both of the significant cases, there were increased occurrences in the dyslexic compared with the control group (see Figure 7.4).

Table 7.1 shows the frequency of right- and non-right-handedness, familiar sinistrality (first- and second-order relatives), and writing hand posture, separated for boys and girls within the dyslexic and control groups. The only significant sex difference was for frequency of non-right-handedness, with proportionally more non-right-handed boys in both groups.

Maternal and Perinatal Health

Table 7.2 gives some measures of the health of the mothers of the children in the dyslexic and control groups during pregnancy and at birth: reports of stress during pregnancy; a record of ultrasound investigations; occurrence of spontaneous abortions; weight and length of newborn; and atypical signs and reactions of the infant at delivery (blue face, oxygen deprivation, etc.) The most striking findings concerned the mothers' experience of pregnancy: almost twice as many mothers of the dyslexic children had experienced major stressful life events and diseases during pregnancy than had mothers of the control children; about five times as many mothers in the dyslexic group

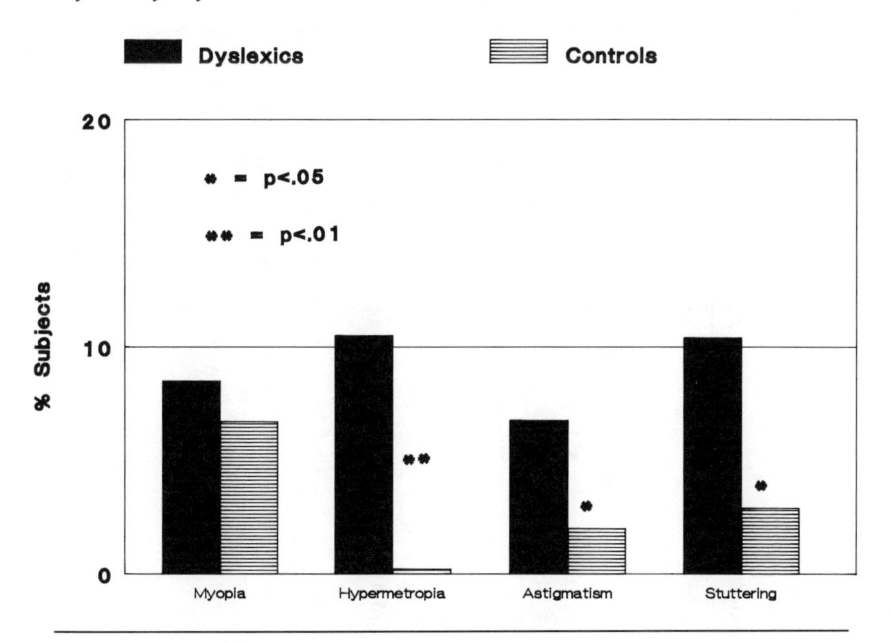

Figure 7.4 Percent of subjects in the dyslexic and control groups who displayed myopia, hypermetropia, astigmatism, and stuttering.

as in the control group reported their pregnancy as "difficult"; mothers of dyslexic children had previously had spontaneous abortions more frequently than had the control mothers.

The single most important result was the increased frequency of immune diseases in the dyslexic group. The study thus provides some empirical support for the hypothesized relationship between immune-system dysfunction and dyslexia. Since there was not an increased frequency of left-handedness among dyslexic children, however, we did not find support for that axis of the hypothesis. Since the absence of more left-handers among dyslexic children has been frequently reported (e.g., Thomson, 1975), one may thus question the triadic association among handedness, dyslexia, and immune dysfunction.

There was, however, a tendency toward more left-handers in the subgroup of children that were both dyslexic *and* had an immune disease than in the subgroup of control subjects with immune diseases. This may be an important qualification of the Geschwind-

Table 7.1. Frequency (and percentages) of boys and girls in the dyslexic and control groups classified by various criteria

Group		RH	NRH	FS+1	FS+2	RI	RNI	LI	LNI
Dyslexics	Boys $n = 69$	54 78.3%	15 21.7%	8 11.6%	32 46.4%	6 8.7%	47 68.12%	2 2.9%	10 14.5%
	Girls $n = 36$	34 94.4%	2** 5.6%	10 27.8%	19 52.8%	3 8.3%	33 91.7%	0 0%	0 0%
Controls	Boys $n = 69$	56 81.1%	13 18.8%	14 20.3%	24 34.8%	3 4.3%	57 82.6%	1 1.5%	8 11.6%
	Girls $n = 36$	33 91.7%	3* 8.3%	12 33.3%	16 44.4%	0 0%	34 94.4%	0 0%	2 5.6%
Total	$n = 210$	177	33	44	91	12	171	3	20

Note: The groups were identified as: RH = right-handed; NRH = non-right-handed; FS+1 = with first-order relatives (mother, father, sister, brother) that included a left-hander; FS+2 = with second-order relatives (grandparent, cousin) that included a left-hander; RI = with writing hand right, inverted; RNI = with writing hand right, not inverted; LI = with writing hand left, inverted; LNI = with writing hand left, not inverted. * = $p < .05$; ** = $p < .01$.

Source: From Hugdahl, Synnevaag, and Satz, 1990b.

Table 7.2. Frequency (and percentages) of mothers of dyslexic and control children who reported stressful experiences during pregnancy, and other indications of maternal and neonatal health

	Dyslexics	Controls	Chi^2 (d.f. = 1)
Mean weight (g) at birth	3364	3493	2.43 n.s.
Mean length (cm) at birth	49.4	50.2	<1 n.s.
Number of mothers who perceived stress during pregnancy (determined by answers to ten questions related to major life events and diseases)	43 (41.5%)	25 (23.8%)	4.76*
Number of mothers reporting pregnancy			
Easy	51 (48.6%)	61 (58.1%)	<1 n.s.
Average	39 (37.1%)	38 (36.1%)	<1 n.s.
Difficult	11 (10.5%)	2 (1.9%)	6.23*
Number of mothers who underwent ultrasound investigation	41 (39.1%)	59 (56.1%)	3.24 n.s.
Number of children displaying atypical signs and reactions at delivery (determined by answers to three questions related to oxygen deprivation, etc.)	14 (13.3%)	8 (7.6%)	1.64 n.s.
Number of mothers who had previous spontaneous abortions	29 (27.6%)	17 (16.2%)	4.13(*)

Note: n.s. = not significant; * = $p < .05$; (*) = $p < .10$.
Source: From Hugdahl, Synnevaag, and Satz, 1990b.

Behan (1986) hypothesis. Satz and Soper (1986) argued that since Geschwind and Behan (1982) had not shown that there was a direct association between left-handedness and dyslexia, the hypothesis could be criticized. As the present results show, the association is probably more complex than Geschwind and Behan (1982) had originally proposed. The critical link in the hypothesis (between left-handedness and dyslexia), which Satz and Soper (1986) concluded was missing, may actually exist, but only in those dyslexics who also have an immune dysfunction. These individuals represent a much smaller subset of the population.

A similar problem may exist with the critique by Pennington et al. (1987). These authors found no evidence of an increased frequency of left-handedness in a dyslexic group when compared with a normal control group. There was, however, a significant increase in both autoimmune and certain immune disorders in the dyslexic group. Their results were almost identical to the present findings: an increase among dyslexics in immune disorders but not in left-handedness in overall comparisons between the groups. Pennington et al. (1987) argued, however, that since there was no association between left-handedness and dyslexia, and since this linkage is one of the cornerstones in the testosterone hypothesis, then Geschwind and Behan (1982) were wrong. But they failed to partition their data for other co-occurrences of the disorders (dyslexia plus immune dysfunction) and handedness, thus actually overlooking a possible source of support for the original hypothesis (see also Urion, 1988).

A first remark about the present results is that *if* there is a three-factor interaction of dyslexia, immune dysfunction, and left-handedness, then it applies probably only to a subset of the population with *either* disorder. It then still remains to be shown in *what* kind of disorders the association holds. This relation may also cast further light on the putative mechanisms involved, an issue still in debate (see, e.g., Satz and Soper, 1986).

One possibility may be that a majority of dyslexic children do not reveal a hormonal imbalance affecting the development of the cellular immune system. However, if there is a deficiency in the preprocessing of the T lymphocytes of the thymus gland shortly before birth, this may also slow the development of the left hemisphere. This developmental difference may, however, cause only dyslexia. If the deficiency in the development of cellular immunity persists, then prolonged suppression of T lymphocytes may in addition cause immune disease, possibly also followed by a shift in control of handedness.

Dichotic Listening and Visual Half-Field Tests of Dyslexic Children

A second study in our laboratory related to the testosterone hypothesis is reported in Hugdahl, Ellertsen, Waaler, and Kløve (1989). The goal of this study was to deduce a set of experimentally testable hypotheses from the testosterone hypothesis. One of the points of departure for the study was the suggestion that as a consequence of testosterone retarding the migration and/or assembly of neurons in the left hemisphere, a right hemisphere dominance will emerge. Right hemisphere superiority has been linked to artistic, musical, or mathematical talent (see Kolata, 1982), all of which emphasize spatial ability. The testosterone-caused right hemisphere superiority has also been linked to the high incidence of immune disorders and dyslexia seen in intellectually precocious children (Benbow, 1986; Benbow and Stanley, 1981). If this is true, one could then suggest that left-handed dyslexics should differ from right-handed dyslexics on tasks designated to tap right hemisphere functions. More specifically, if testosterone inhibits left hemisphere development, which leads to enhancement of the development of the corresponding right hemisphere functioning, and if this causes both left-handedness and dyslexia, then left-handed male dyslexics would be expected to show superior performance compared with their right-handed counterparts on a typical right hemisphere task, like visuo-spatial ability.

Thus left- and right-handed dyslexic boys were compared on their performance of a visuo-spatial task initially presented to either the right or left hemisphere, using the visual half-field technique (VHF) (see McKeever, 1986, for a detailed description of the technique).

The second aim was to compare the subjects on two verbal tasks (which are specific to the left hemisphere). One interpretation of the Geschwind and Behan (1982) hypothesis is that while the control of handedness is shifted to the right hemisphere as an effect of testosterone acting on the fetal brain, language control is not shifted. Thus, according to the Geschwind and Behan hypothesis, if dyslexia results from exposure to increased testosterone levels, then language should not shift to the right hemisphere. A prediction can thus be made concerning the performance of left- and right-handed dyslexics on verbal tasks initially presented to the left hemisphere. If a testosterone effect during fetal life results in left-handedness in some individuals, one could argue that left-handed dyslexics are genetically right-handed and thus the left hemisphere is dominant for language. If so, they should not differ from right-handed dyslexics in language skills, especially not when males are compared. This hypothesis was tested with the VHF and dichotic listening (DL) techniques. See Hugdahl (1988) for a detailed description of the DL technique. Both the VHF and the DL-techniques are considered valid and reliable noninvasive techniques for assessment of hemispheric asymmetry (Geffen and Quinn, 1984; Hugdahl and Franzon, 1985).

The DL technique involved pair-wise presentations of consonant-vowel (CV) syllables. In addition, selective attention was included in the study by having the subjects attend to the right or left ear. We included an attentional task because of an observation of Hugdahl and Andersson (1986, 1987), who reported that while right-handed adults showed a left ear advantage (LEA) during a forced-left attentional recall condition, this was not the case for children, who still reported more items from the right ear. The right ear advantage (REA) in the children during the forced-left condition further correlated with level of reading competence, being more profound the less proficient in reading the child was.

Twenty-six dyslexic boys (13 right-handers, 13 left-handers) of 15–16 years of age were tested by dichotic presentations of CV syllables and a "letter-and-arrow" test with the visual half-field (VHF) technique. Two different kinds of stimuli were used in the VHF test; letters (verbal) and arrows (visuo-spatial). The stimuli were presented in pairs (either two letters or two arrows) on each trial. The stimulus pairs were presented either to the left or the right half-field. The results are illustrated in Figures 7.5 and 7.6.

Figure 7.5 shows essentially two things. First of all, there was a right ear advantage (REA) in nonforced recall in both the right-handed and left-handed groups. Second, the REA increased equally in both groups during forced-right recall.

It is interesting to note that the increase in the magnitude of the REA during forced-right recall was not caused by an increase in correctly recalled right-ear items, but instead by a corresponding decrease in correctly recalled left-ear items, compared with the results for nonforced recall. Furthermore, this effect was almost identical for right- and left-handed subjects. The same drop in left-ear performance during forced-right recall was also observed by Hugdahl and Andersson (1987) studying normal eight- to nine-year-old children. Relating the dichotic listening results to the Geschwind and Behan (1982) hypothesis, we predicted that if dyslexia is caused by a prenatal lesion that shifts handedness to the right hemisphere, then the left-handed dyslexics should still be genetically right-handed and thus have language controlled in the left hemisphere. It was therefore

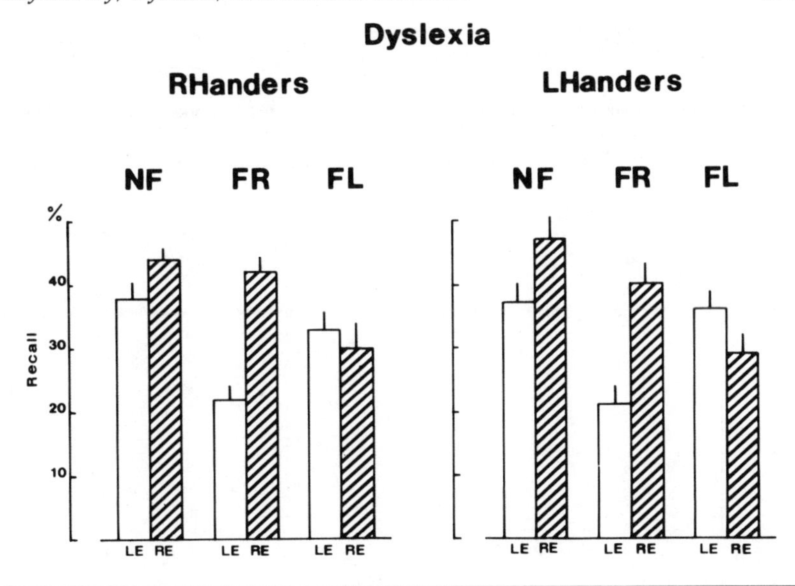

Figure 7.5 Mean percent of correct recall from the left (LE) and right (RE) ears of subjects under nonforced (NF), forced-right (FR), and forced-left (FL) recall conditions. Small bars = S.E. (From Hugdahl et al., 1989.)

hypothesized that the groups should not differ in performance of dichotic listening to CV syllables. As can be seen in Figure 7.5, this was confirmed.

Figure 7.6 shows superior left-field performance for the arrows by the left-handers compared with the right-handers. If identification of arrows in either half-field is acknowledged as an example of processing visuo-spatial information, then the hypothesis of right hemisphere enhancement in left-handed dyslexics was supported. Goldman-Rakic and Rakic (1984) showed that diminution in the fetal monkey of the size of one cortical area may lead to an enlargement in the homologous region of the opposite side. Thus, Geschwind and Behan (1982) hypothesized that *in utero* testosterone exposure may enhance right hemisphere development, in addition to inhibiting left hemisphere development, in humans. Consequently, if increased testosterone levels may cause both left-handedness and dyslexia in males, then left-handed dyslexic boys should show superior right hemisphere performance in response to visuo-spatial stimuli. The left-handers were on the average 12 percent better in

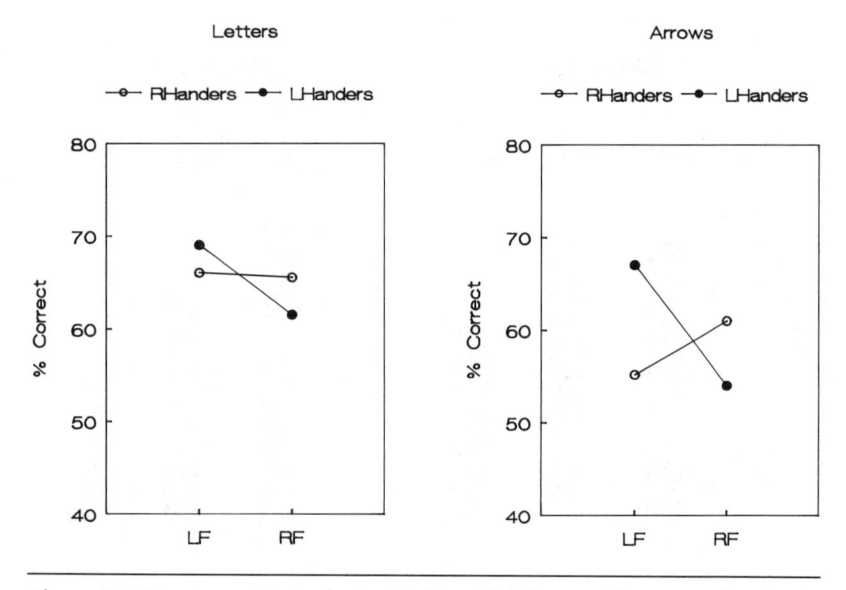

Figure 7.6 Mean percent of correct identifications of letters and arrows, separated for left-field (LF) and right-field (RF) stimulus presentations. Small bars = S.E. (Data from Hugdahl et al., 1989.)

identifying the arrows presented in the left half-field, while the right-handers were on the average 7 percent better in identifying the same arrows presented in the right half-field. However, since the three-way interaction of half-field, stimuli, and groups did not turn out significant, the findings of support for the hypothesis should be interpreted with some caution. The left-handed dyslexics showed a left field superiority for both types of tasks, while the right-handed dyslexics did not. Although this result is in support of the Geschwind-Behan hypothesis, other explanations cannot be ruled out.

Conclusions

Data from two studies related to brain asymmetry and dyslexia have been presented. Although correlations between dyslexia and immune diseases were found, the nature of these covariations remains uncertain. Moreover, using questionnaire data to reveal the nature of the correlations may increase the risk for "type-I" errors (not rejecting the null hypothesis when it in fact is false). Having a dyslexic child may bias parents to be more prone than parents of nondyslexic children to "observe" various diseases and disorders in their children.

Another critical remark concerning the nature of the association between immune diseases, dyslexia, and possibly also left-handedness is that not all studies have found (or even investigated) an effect for all three variables. As in the present study, Pennington et al. (1987) did find increased frequency of immune disorders in cases of familial dyslexia, but not of left-handedness (cf. Urion, 1988). The fact that brain *symmetry* has been found in left- but not in right-handed dyslexics (Parkins, Roberts, Reinarz, and Varney, 1987) complicates matters further, since Galaburda et al. (1985) have argued that brain *symmetry* may be a characteristic feature of dyslexia. If this is true, and if symmetry is associated with left-handed dyslexics (Parkins et al., 1987), one would then expect that left-handedness should be a salient feature among dyslexics (and this would be especially so from the perspective of the testosterone hypothesis). When left-handedness in fact seems to be loosely (at the best) associated with the other variables, one may regard the hypothesized relationships with some skepticism. As also mentioned by Hynd and Semrud-Clikeman (1989), Kinsbourne (1986) and Schachter, Ransil, and Geschwind (1987) have argued that how one operationalizes left-handedness influences conclusions regarding its associations with both dyslexia and immune diseases.

A tentative conclusion is therefore that a hormonal imbalance in uterine life *may* cause a shift in the normal development of hemispheric asymmetry and specialization. The change may be hypodevelopment of one hemisphere and/or hyperdevelopment of the other hemisphere. This in turn *may* cause a shift in control of handedness and deficiencies in language development and of the immune system. One problem with this conclusion, however, is that language-related problems other than dyslexia have rarely (if ever?) been reported to covary with increased frequency of certain immune-system disorders (or with a shift toward left-handedness). Until we have a better understanding of why dyslexia but not, for example, delayed speech development is correlated with left-handedness and immune dysfunction, the testosterone hypothesis remains a hypothesis, not a theory.

Quite another issue, however, is the fact that the research of

Geschwind, Galaburda, and their colleagues has (re)-focused interest on the brain and on biological factors in reading and learning disorders. Seen in this perspective, the testosterone hypothesis has had an important spin-off effect: it has shifted research interests back to neurobiology and brain function (see, e.g., Hynd and Semrud-Clikeman, 1989, for a recent "biological" review). This change in the focus of dyslexia research has also had the consequence that issues related to hemispheric asymmetry and brain specialization (and localization) are brought into view again. As has been shown in the present chapter, studies of the effects of asymmetry on learning in general have, unfortunately, been neglected, with emphasis instead put on *perceptual* asymmetry. I hope the renewed interest in the relationship between learning disorders and functional asymmetry of the cerebral hemispheres may also lead to a renewed interest in the association between *normal* learning and hemisphere asymmetry. An example of the latter is the study by Hugdahl and Andersson (1989) demonstrating a shift in asymmetry patterns in children during the normal developmental transition from preliterate to literate stage. This study thus demonstrated a shift in processing interactions of the hemispheres when the children learned to read (acquired a new skill). It is my hope that researchers in the future will devote more time and interest to how the hemispheres contribute to the process of *normal* skill acquisition in order to gain an understanding of how the same hemispheres contribute to the process of *dysfunctional* skill acquisition. Thus, my plea is for students of dyslexia to go back to the laboratory and rephrase their questions about *disordered* learning and hemisphere functioning to include questions about *ordered* learning and hemisphere functioning.

Acknowledgments

The present research was financially supported by grants from the Nordic Council of Ministers (NOS-S) and the Norwegian Council for Research in the Social Sciences (NAVF-RSF).

8

Fetal Exposure to Maternal Brain Antibodies and Neurological Handicap

Matteo Adinolfi

In 1975, Kirman proposed that maternal antibodies against brain antigens could affect the development of the fetus's brain and cause some forms of learning disability and behavioral disorder in the child. This suggestion was received with a mixture of interest and skepticism. At that time, in fact, the main objection to the hypothesis was based on the widely accepted view that, as in adults, the fetal blood-CSF (cerebrospinal fluid) barrier would not allow the transfer of immunoglobulins or other plasma proteins (Davson, 1967; Dobbing, 1971; Evans et al., 1974; Saunders et al., 1976); consequently, it was believed that maternal brain antibodies, though they could cross the placenta, could not reach the fetal brain.

However, it has now been shown that several plasma components, including IgG immunoglobulins, can cross the blood-CSF barrier during fetal and perinatal life in humans and experimental animals (Adinolfi et al., 1976, 1985; Adinolfi and Haddad, 1977; Statz and Felgenhauer, 1983). Experimental studies have also documented the deleterious effects that maternal brain antibodies or antibodies injected during perinatal life can have on brain development and behavior in laboratory animals (Adinolfi et al., 1985).

In this chapter I will review first the evidence that plasma proteins can cross the fetal blood-CSF barrier; then I will discuss the experimental studies that document the effects of brain antibodies on CNS development and finally analyze the evidence in favor of or against the suggestion that certain forms of mental retardation and behavioral disorder in humans may be due to maternal brain antibodies.

Transfer of Plasma Proteins across the Blood-CSF Barrier during Fetal Life

Early studies of the development of the blood-CSF and blood-brain barriers have been reviewed in detail by Davson (1967, 1976); here, I present only a brief summary. Behnsen (1926, 1927) was the first to observe that trypan blue administered to young mice accumulates in the nervous system and that, in this species, the blood-brain barrier is not fully developed during perinatal life. The staining of the nervous system was extensive in mice only a few days old; not until 7–8 weeks of age was the transfer of the dye similar to that observed in adult animals.

Stern and Peyrot (1927) injected Prussian blue into newborn rabbits, rats, cats, and dogs, as well as mice, and observed that the uptake of the dye was rapid during the first few days after delivery but declined quite swiftly later on in life.

In 1948, Otila first detected higher levels of plasma proteins in CSF from premature neonates than in normal adults; similar conclusions were reached by Arnhold and Zetterstrom (1958), and Nellhaus (1971). In spite of these results, however, for many decades the prevailing view was that in humans and experimental animals the blood-CSF barrier was as effective in preventing the transfer of proteins during fetal life as it was in adult individuals (Grontoft, 1954; Grazer and Clemente, 1957; Millen and Hess, 1958; Evans et al., 1974; Saunders et al., 1976). Grontoft's (1954) investigations were performed using intra-arterial perfusion preparations of dead fetuses, a procedure unlikely to provide information about the permeability of the blood-CSF barrier *in vivo*.

The rediscovery of the permeability of the fetal blood-CSF barrier was made possible by investigation of the causes of the abnormally high levels of alphafetoprotein (AFP) in the amniotic fluid of fetuses with neural tube defects (NTDs) (Brock and Sutcliffe, 1972). These studies indirectly suggested the presence of high levels of AFP in fetal CSF, which was passing into the amniotic cavity *via* an "open" neural tube defect.

To confirm this suggestion, CSF was collected from a group of human fetuses between 14 and 25 weeks of gestation, and the levels of various plasma proteins were estimated, including albumin, AFP, pre-albumin, transferrin, immunoglobulin IgG, α2-macroglobulin, and β-trace (Seller and Adinolfi, 1975; Adinolfi et al., 1976; Adinolfi and Haddad, 1977). High levels of albumin (reaching 3.4 mg/ml) and AFP were detected in the fetal CSF. The highest values of AFP were detected in fetuses 16–19 weeks old, when the levels of these proteins were at their highest in fetal sera. High levels of IgG were also detected in fetal CSF; with genetic markers (Gm groups), it was possible to establish that these molecules were of maternal origin and had crossed the placenta. Transferrins and β-trace were found in fetal CSF in concentrations higher than in CSF from normal adults (Adinolfi and Haddad, 1977). From comparisons of the levels of these proteins in fetal CSF and sera it was concluded that during fetal life the blood-CSF barrier is permeable to plasma proteins with a molecular weight less than 200,000 and that their transfer is passive and thus does not require specific receptors. Only pre-albumin was detected in fetal CSF in concentrations higher than those expected according to the ratios observed for other plasma proteins.

An increased permeability of the choroid plexus to proteins has also been documented during the first months of postnatal life, since elevated values, compared with those detected in CSF from normal adults, have been observed in CSF from neonates and young infants. The levels of albumin, IgG, and α2-macroglobulin decline during the first 2–3 months after delivery and reach "adult" values at about 6 months (Otila, 1948; Arnhold and Zetterstrom, 1958; Widell, 1958; Nellhaus, 1971; Harms, 1975; Wenzel and Felgenhauer, 1976).

Evidence that specific maternal antibodies cross the placenta and can reach the fetal CSF has been obtained by detecting diphtheria and tetanus IgG antibodies in maternal sera and in CSF from young infants (Thorley et al., 1975). Transfer of plasma proteins across the blood-CSF barrier during fetal life has also been documented in experimental animals (Adinolfi and Haddad, 1977). As mentioned, early studies in fetal lambs were interpreted as evidence that the morphological structure of the choroid plexus prevents the transfer of protein (Evans et al., 1974; Saunders et al., 1976) and that, by extrapolation, it would be "unlikely that much protein could penetrate the CSF and brain extracellular space of the human fetus" (Evans et al., 1974).

In rat fetal CSF, however, the levels of albumin, AFP, transferrin,

IgG, and two unidentified α-globulins have been found to be higher than the levels in adult animals (Amtorp and Sorenson, 1974; Adinolfi and Haddad, 1977; Johanson, 1980). The transfer of albumin, AFP, and IgG was confirmed by using ^{125}I-labeled proteins injected into pregnant or newborn rats (Adinolfi and Haddad, 1977; Adinolfi et al., 1985). Furthermore, ^{125}I-labeled IG brain antibodies were found to reach the CNS.

In conclusion, there is clear evidence that several plasma proteins, including IgG immunoglobulin of maternal origin, can readily cross the blood-CSF barrier during fetal life. This is in agreement with results of histological investigations of the human fetal choroid plexus that have been interpreted as evidence of immaturity (Kappers, 1958; Adinolfi, 1976).

Experimental Studies of the Effects of Brain Antibodies on CNS and Behavior

Brain antibodies and maturation of the CNS

The effects that brain antibodies may have on the development of the CNS were first investigated by Gluecksohn-Waelsch (1957), who induced active immunization against brain antigens in female mice and detected morphological abnormalities of the CNS in about 10 percent of the offspring. The progeny of control mice, injected with heart antigens, showed no malformations.

A few years later, Levi-Montalcini (1964, 1966) showed that the injection of antibodies against nerve growth factor (NGF) in newborn mice, rats, and rabbits would induce temporary or long-lasting "immunological sympathectomy" in the absence of adverse effects on other organs. Immunosympathectomy was also observed in the offspring of animals injected with anti-NGF during pregnancy (Klingman and Klingman, 1972). Since the immunological sympathectomy can be induced only in young animals, it seems that anti-NGF can cross the immature perineurium during fetal or perinatal life (Adinolfi et al., 1985).

Studies of the effects of brain antibodies on behavior were pioneered by Auroux and collaborators (Auroux et al., 1967, 1968; Auroux 1968), who observed marked learning deficiencies in the offspring of rats injected toward the end of their gestational period.

Unfortunately, no data were provided about the specificities of the antibodies.

Specific antisera against synaptic-ending antigens were used by Karpiak and Rapport (1975). The antibodies were injected into rats on the nineteenth day of pregnancy; male offspring, tested at about two months of age, manifested behavioral deficits in tests of avoidance conditioning. In fact, increased response latencies, slow acquisition rate, and poor retention of behavioral tasks were observed in the treated animals as compared with untreated controls.

In 1979, a series of experiments was set up to establish whether myelin antibodies, injected into pregnant or newborn rats, could reach the CNS and alter motor functions. Signs of motor deficiency, such as difficulties in walking on a narrow wooden strip suspended 10 cm above the bench surface or in gripping a smooth surface inclined at 30 degrees, were observed in the treated animals. These and other motor disorders disappeared on aging (Adinolfi, 1979).

In another group of experiments, motor-behavioral deficits were observed during the first two weeks of life in the offspring of rats injected during pregnancy with antibodies against brain gangliosides (Rick et al., 1980; Adinolfi et al., 1985). Righting responses, pivoting geotaxic reactions, crawling and walking tests, based on those standardized by Altman and Sudarshan (1975), were performed. Although righting responses and geotaxic reactions were similar in the treated and control groups of neonatal rats, increased time spent pivoting and deficits of walking and crawling were observed in rats born to mothers injected with ganglioside antibodies (Adinolfi et al., 1985). Since the control and treated animals did not differ in righting and geotaxic responses, which depended on the peripheral nervous system being intact (Tang, 1935; Brooks and Peck, 1940), it was concluded that the observed abnormalities were not due to a deficient development of the peripheral circuitry (Adinolfi et al., 1985).

Further investigations have been performed in rats born to mothers injected with antigangliosides and tested at four weeks of age (Rick et al., 1981). The effects of the antibodies were evaluated by comparing the treated animals and controls using an open-field apparatus and a sequence maze. Abnormal behavior was observed in the treated animals (Rick et al., 1981; Adinolfi et al., 1985). The results of the sequence-maze test demonstrated that rats born to treated mothers and investigated at 35–36 days of age needed almost twice as many trials as the control rats.

These and other studies, all performed "blind," since the investigators performing the tests did not know which were the control and which the treated animals, clearly documented that brain antibodies could affect the development of brain and behavior (Rick et al., 1980, 1981; Adinolfi et al., 1985).

Recently De Felipe et al. (1989) have observed that the injection of antibodies against substance P (SP) into two-day-old rats produced long-lasting blockade in the antinociceptive and hypertensive responses to SP. Similar effects had not been observed in previous studies, when the antibodies were injected into adult rats. Treatment of the newborn rats with SP antiserum did not decrease SP receptors but probably reduced SP synthesis from neurons.

Effects of brain antibodies in adult animals

Early studies were performed by injecting brain antibodies directly into the ventricular cavity of the CNS tissue. The results, documenting a variety of abnormalities such as interference with memory and inhibition of passive-avoidance learning, have been summarized in a recent review (Adinolfi et al., 1985). The direct injection into CSF, with consequent possible damage induced by this procedure, was avoided by McPherson and Shek (1970), who immunized adult rats with crude microsomal fractions from brain. The results of one experiment were particularly interesting because they showed that rats who had learned a visual discrimination task following immunization with brain antigens required fewer trials than control animals when they were tested on a reversed discrimination task. This finding is comparable to other results, which show that rats previously trained in a standard two-lever chamber perform better than controls in two learning tasks after the injection of antibodies to brain gangliosides (Rick et al., 1980).

Several more papers have been published on the effects that brain antibodies have on specific CNS functions (Mihailovic and Jankovic, 1961; Heath et al., 1967; Heath and Krupp, 1967; Mihailovic et al., 1969; Bowen et al., 1976; Kobiler et al., 1976; Fariello et al., 1977; Hofstein et al., 1980; Alescio-Lautier et al., 1989). Performed using different techniques, tests, and antibodies with various specificities, these experiments all documented temporary modification of the animals' behavior. However, since in most instances the antibodies employed for the experiments were polyclonal and probably cross-reacted with antigens present in tissues other than the brain, it is not possible to exclude the possibility that some of the observed effects were secondary to functional alterations of organs outside the CNS.

In at least one investigation a monoclonal brain antibody has been used and shown to inhibit the acquisition of a passive-avoidance behavior in rats (Nolan et al., 1987). The use of monoclonal antibodies should reduce the risk of secondary effects.

Maternal Antibodies and Human Brain Development

One of the most intriguing problems of mammalian reproduction is to establish to what extent the pregnant female is immunologically aware of her antigenically alien conceptus. Most studies of the materno-fetal relationship have documented a state of unique "anergy" of both mother and fetus (Adinolfi, 1990). In fact, in many instances, the mother fails to produce antibodies against polymorphic antigens synthesized by her fetus under the control of paternal genes. On the other hand, maternal antibodies are often well tolerated; an example is the successful transfer of allogenic blastocysts to surrogate rabbit mothers previously immunized against "paternal" antigens (Heape, 1891). Also, in breeding experiments between different species, such as the horse and donkey, maternal antibodies against fetal antigens have no deleterious effects on the conceptus. In humans antibodies against class I and II HLA antigens occur in about 30 percent of pregnant women; yet these antibodies, often associated with IgG molecules and thus capable of crossing the placental barrier, do not induce abortions or malformations (see Adinolfi, 1990). In humans maternal antibodies affect the fetus only when they are directed against certain unique antigens, such as the ABO and Rh blood groups (Mollison et al., 1979). Another example is the heart antigens, which may induce fetal cardiac arrest (Scott, 1976).

It seems therefore, that the consequences of maternal immunization against the fetus depend on the nature of the antigen and its distribution, as in the case of the Rh system, where the antigen is exclusively expressed on red cells. These observations suggest that whether or not maternal brain antibodies affect fetal brain development and subsequently behavior may depend on their specificities and the type of tissue against which they are directed. In other words, the detection of brain antibodies in sera of pregnant women is not sufficient evidence to assume that such antibodies can affect fetal brain development. As in other situations, brain antibodies may be present in maternal sera, cross the placenta, reach the fetal circulation—and even the brain—and yet have no deleterious effects.

It is in light of this view that I will now analyze the evidence in favor of or against Kirman's (1975) suggestion that, in humans, some forms of mental retardation and behavioral abnormalities may be due to exposure *in utero* to maternal brain antibodies. When this hypothesis was advanced, the evidence to support it was based on a series of studies by Russian investigators (Semenov et al., 1968, 1972, 1973; see also Sazonova and Yavkin, 1977). In their papers, Semenov and his collaborators claimed to have detected maternal brain antibodies in sera from infants born to mothers with a variety of mental diseases, including "encephalopathies of infectious-allergic, vascular or endocrine origins, schizophrenia and epilepsy"; they concluded from this evidence that "diseases of the central nervous system in postnatal life may sometimes be associated with autoimmune processes of cerebral origin that occurred in the maternal body." According to Sazonova and Yavkin (1977), "on the basis of his and other studies, Semenov (1973) considers it proven that antibrain antibodies produce a neurotropic effect." Semenov et al. (1972) had also investigated a group of children with retarded mental development and detected brain antibodies in 46 percent of them.

In his article, Kirman (1975) quotes studies by Kulbova (1972) and Batmanova (1972) describing the detection of brain antibodies in children with psychological abnormalities. Unfortunately, these studies are difficult to evaluate because a wide range of vaguely described mental disorders are mentioned, the specificity of the antibodies is often not reported, and the tests used for their detection (precipitation and complement consumption tests) are not accurate assays and are affected by subjective evaluation. In fact, low-titer "brain" antibodies are present in most normal individuals (Leibowitz and Hughes, 1983), and the cut-off point between normal and abnormal values is often arbitrary.

It has been suggested that dyslexia may be caused by brain antibodies, presumably reacting against a subpopulation of specialized cells. In fact, two hypotheses have been advanced; that antibodies present in the maternal circulation during pregnancy cause the disorder (Behan and Geschwind, 1985) or that brain autoantibodies are produced by the affected children (see Chapter 7). If dyslexia is associated with maternal brain antibodies, the long-lasting effects in the offspring should be the result of a time-limited action of the antibodies that had reached the fetus and affected brain development during fetal and perinatal life. On the other hand, if the dyslexia is an autoimmune disease, brain function should be affected during childhood. The two claims are somehow mutually exclusive, because there is no evidence that maternal autoantibodies might be responsible for triggering an autoimmune disease in the offspring.

Another problem with any theory based on the "incidence" of allergies and autoimmune diseases in mothers of dyslexic children is that allergic diseases are usually due to IgE antibodies, which do not cross the placenta and therefore cannot affect the fetus. Allergic disorders such as eczema or asthma should be removed from the list of possible causative agents. Equally, autoantibodies belonging to IgM or IgA cannot be held responsible for fetal damage since they do not cross the placenta.

Furthermore, in some studies maternal sera have been tested for the presence of autoantibodies several years after the dyslexic child was born. Samples taken after such a long delay cannot provide useful information about the relationship between maternal autoantibodies and dyslexia. In fact, high levels of brain antibodies are often produced for short periods of time in both humans and in experimental animals.

Any theory based on the hypothesis that maternal brain antibodies may induce dyslexia in the offspring should also come to terms with the problem that this disorder should occur more often in siblings born soon after a dyslexic child. Apparently this pattern has not been observed (Dr. Paula Tallal, personal communication). Furthermore, both monozygotic and dizygotic twins should be equally affected, but this does not seem to be the case (see Chapter 10).

It has been claimed that anti-Ro antibodies are present in about 9 percent of mothers of dyslexic children (Behan and Geschwind, 1985). It is a characteristic of these antibodies to produce fetal heart block (Scott, 1976), yet such a disorder has not been observed in dyslexic children during their perinatal life. Thus this claim needs to be confirmed by further investigations, including eventually those that analyze in more detail the specificity of the autoantibodies.

Equally difficult to judge are the effects that brain antibodies may have on the etiology of infantile autism. In fact, a variety of immune dysfunctions are said to be associated with infantile autism. Abnormalities of T cell responsiveness (Stubbs and Crawford, 1977; Warren et al., 1986), decreased T cell number and altered T helper (T_H) and T suppressor (T_S) cell ratios (Warren et al., 1986), and reduced natural killer (NK) cell activity have apparently been detected in autistic children (Warren et al., 1987). The results of *in vivo* investigations have also been interpreted as evidence that at least some autistic individuals have altered recognition of self-antigens (Weizmann et al., 1982; Todd and Ciaranello, 1985; Todd et al., 1988). However, in a study using a crude dot blot test and extracts from cortex, no differences in total brain antibody titers were detected between groups of autistic patients and normal subjects. The researchers (Todd et al., 1988) were careful to suggest that brain antibodies may have been present in sera of autistic patients at an early stage of their disease. It is equally plausible that maternal brain antibodies might be responsible for inducing autistic behavior, but careful investigations are needed to confirm this claim.

The published literature suggests the possibility that autoimmune procedures are responsible for other neurological disorders in childhood, such as infantile spasm (for references see Plioplys et al., 1989). Using a Western blotting immunological technique for the detection of brain antigens (including antigens expressed in tissues besides the brain), Plioplys et al. (1989) have investigated sera from a group of 121 children with epilepsy and 28 children with a variety of neurological disorders. When sera were tested against extracts from frontal cortex, 27 percent of sera from epileptic patients displayed immunoreactivity, as compared with 10 percent of sera from controls. No differences between the two groups were observed when the sera were analyzed by immunostaining on brain sections.

It would be extremely interesting to perform a similar study on a large number of sera collected from women during pregnancy and relate the presence of brain antibodies (or other autoantibodies) with the development of CNS abnormalities in their children. Such a study would conclusively establish whether maternal antibodies may be held responsible for abnormal brain development in the offspring.

Graft-versus-host (GVH) disease has also been said to impair cerebellar growth in mice, although no lymphocytic invasion of the CNS has been detected (Griffin et al., 1978). Since the human cerebellum acquires 80 percent of its total cell population during fetal development, Griffin et al. (1978) have called the attention of pediatricians to the risks of GVH-derived brain damage in children born after receiving blood transfusions *in utero* or during perinatal life.

In humans, the only clear evidence of a relationship between maternal antibodies and neurological disease in infants is seen in two conditions that are in many ways unique: in one the relationship is indirect and in the other the effect involves the peripheral nervous system. The first is the occurrence of mental retardation of variable degree as a result of fetal exposure to maternal Rh antibodies associated with kernicterus; here the brain damage is only a by-product of the immunological disease. The second condition where maternal antibodies produce neurological disorders in the newborn is myasthenia gravis. In fact, maternal antibodies to acetylcholine receptors (AChR) produce neonatal myasthenia gravis (NMG) in relatively few infants born to mothers with the disease. The neonatal syndrome resembles the adult form but has a transient duration. The clinical improvement generally correlates well with decreasing antibody levels (Keesey et al., 1977; Nakao et al., 1977), and recurrence of the disease does not occur. Since only some infants born to myasthenic mothers are affected, many attempts have been made to establish a relationship between NMG and the IgG titer and specificity of the antibodies. Direct blocking, decamethonium-dependent (DC) blocking, and precipitating antibodies have been detected with different incidence and ratios in the affected infants (Abramsky et al., 1979; Ohta et al., 1981; Donaldson et al., 1981; Lefvert and Osterman, 1983; Bartoccioni et al., 1986; Morel et al., 1988; Vernet–der Garabedian et al., 1989).

The presence of direct and DC blocking antibodies and occasionally of precipitating antibodies in some asymptomatic neonates suggests that they are not inevitably pathogenic. At present it is not possible to define the best predicting indices of NMG from the analysis of the maternal antibodies, although it is clear that the higher the titer, the higher the risk (Keesey et al., 1977; Donaldson et al., 1981; Ohta et al., 1981; Morel et al., 1988).

Conclusion

Many problems about the connection between maternal brain antibodies and abnormal fetal CNS development are still unsolved. Antibodies against brain antigens have been detected in normal individuals and in patients with a variety of neurological diseases (Leibowitz and Hughes, 1983); often only marginal titer differences have been noticed between the controls and the patients. If antibodies are to affect fetal brain development, they should be present in maternal blood in large amounts during a certain time of gestation, and they should be associated with IgG molecules so they can cross the placental and CSF barriers. Furthermore, only a few brain antibodies are expected to be capable of disrupting the biological functions of selected brain cells.

Many immunologists have expressed their reservations about the interaction of the immune system and the brain (Cohen, 1989) and about the role of brain antibodies in the behavior of adult patients. Even more cautious should be our evaluation of the relation between mental retardation and behavioral disorders in children with maternal antibodies against CNS antigens.

There is no doubt that studies in laboratory animals have shown that fetal brain development may be affected by maternal antibodies, but these experimental findings cannot be extrapolated to humans. First, because most investigations have been performed with polyclonal antibodies cross-reacting with fetal cells and tissues other than CNS cells, the effect of the antibodies on the brain may have been secondary to dysfunction of other organs (Adinolfi et al., 1985). Second, it is difficult to correlate behavioral abnormalities in young experimental animals with neurological disorders in humans (Rick et al., 1981).

However attractive the hypothesis may be that certain cases of mental retardation or abnormal behavior in humans result from maternal brain antibodies, the evidence to support this theory is still very scanty. We can only hope that further investigations will be more conclusive. In particular, we await a large prospective study designed to relate the presence of autoantibodies in sera from pregnant women with later development of brain abnormalities in the offspring of these women.

Acknowledgments

The author is grateful to Adrienne Knight for help in preparing the manuscript and to the Spastics Society and the Generation Trust for financial support.

9

Hormones and Cerebral Organization: Implications for the Development and Transmission of Language and Learning Disabilities

Paula Tallal
Roslyn Holly Fitch

Developmental language and learning disabilities in children can take many different forms. Research has focused primarily on specific disabilities characterized by a significant delay or disorder in one aspect of learning against an otherwise normal background of development. Learning disabilities affecting language and/or reading acquisition (developmental dysphasia and dyslexia) have been studied most thoroughly. Verbal learning disabilities have been reported at a considerably higher frequency in males than in females, and there is a higher-than-expected incidence of left-handedness among affected children.

Although there are many possible reasons for delayed or disordered language development, the diagnosis of specific developmental language or reading disorders can be made only after ruling out mental retardation, peripheral auditory or visual dysfunction, autism, gross neurological impairments such as hemiplegia or seizure disorder, and severe social deprivation or lack of educational opportunity. Thus, the typical profile of a developmentally dysphasic or dyslexic is a child who shows a marked discrepancy between nonverbal (performance) IQ and verbal IQ, with a history of delayed or disordered language, speech, and/or reading development. These children usually perform quite normally on visual-spatial tasks while demonstrating severe deficits in tasks of auditory or visual temporal processing, motor sequencing, phonological processing and memory, language, reading, and spelling. This characteristic neuropsychological profile suggests specific left-hemisphere dysfunction and possibly a failure to develop normal cerebral lateralization.

Cerebral Morphology and Lateralization in LLI Subjects

The existence of abnormalities in the cerebral development of language and learning impaired (LLI) individuals is supported by the findings of Galaburda and Kemper (1979). They reported neuronal migration abnormalities, as well as a lack of the expected cerebral asymmetry, the left side larger than the right, in the planum temporale (an area subserving speech) in postmortem morphological examination of the brains of dyslexic individuals. With the recent advent of noninvasive *in vivo* brain imaging, several studies of the neuropathology of developmental reading disability (dyslexia) and developmental language disorder (dysphasia) have been reported (Haslam, 1981; Hier et al., 1978; Jernigan et al., 1987; Larsen et al., 1990; Plante et al., 1989; Rosenberger and Hier, 1980; Rumsey et al., 1986; see Hynd, 1989, for review). These imaging studies support neuropathological findings of reduced or reversed hemispheric asymmetry, especially in posterior temporal cerebral areas. Of these new imaging techniques, magnetic resonance imaging (MRI) offers the most sensitive, noninvasive method for assessing brain morphology during life, and as such it provides a unique opportunity for examining brain structure in children with specific developmental language and learning disabilities.

Normally, the perisylvian and postsylvian regions of the left hemisphere are larger than homologous regions in the right hemisphere in humans. Using MRI, Tallal and colleagues have demonstrated a significant increase in aberrant asymmetry (the right equal to or larger than the left) of postsylvian cerebrum in developmental dysphasics as compared with matched controls (Jernigan et al., 1987, 1991). Plante et al. (1989) also report finding aberrant left-right perisylvian configuration in a young language-impaired boy. MRI results were also presented for this boy's twin sister, who was not language impaired but who also exhibited a larger right side in the perisylvian region.

In a recent study, Larsen et al. (1990) used MRI to examine the size and symmetry of the planum temporale in 19 dyslexic eighth-grade students, as well as in matched controls. They report a 70 percent incidence of planum symmetry among the dyslexics, contrasting with 30 percent symmetry among controls; none of the subjects in either group showed a significant right-greater-than-left pattern. In addition, Larsen and colleagues report that the incidence of pure phonological deficit was highly correlated with the incidence of symmetrical plana. In fact, of three control children found to exhibit pure phonological deficiency, two had symmetrical plana.

These consistent findings of abnormalities in cerebral asymmetry in the brains of LLI children has led to a search for potential models of etiology and to increased interest in the mechanisms of the development of cerebral lateralization in the normal population as well. One potential mechanism that has been addressed is that of hormonal influence on cortical development.

Hormonal Influences in the Development of Cerebral Organization

In 1982, Geschwind and Behan proposed that developmental learning disorders may be linked to both left-handedness and immune disorders through the action of testosterone on brain development in the perinatal period. This proposal was expanded in 1985 in a series of papers by Geschwind and Galaburda. These authors suggested that abnormally high testosterone levels, or sensitivity to the effects of testosterone, might deter maturation of the left hemisphere, resulting in cellular, anatomical, and functional abnormalities associated with the left hemisphere and in abnormalities in the development of cerebral lateralization. The actions of testosterone were proposed to be specific to the left hemisphere on the basis that it was thought to develop slightly later than the right and therefore to be more susceptible to disruption in the latter stages of development. These abnormalities may lead, in turn, to the development of atypical cerebral dominance, such as left-handedness, as well as to a higher incidence of left-hemisphere disorders, including language disorders, among males, since they would be more prone to androgenic deviance. Geschwind also postulated that abnormalities in early androgenic exposure may lead to increased immune disorders in later life. Reported sex ratios for left-handed and LLI individuals and those with immune disorders, as well as evidence for links among these three factors, support this assertion (Geschwind and Galaburda, 1985). The applicability of such a model to the development of language and learning disorders is also consistent with specific left-hemisphere dysfunction and with the cerebral abnormalities and higher incidence of left-handedness among LLI subjects.

This proposal is one of the few global models that has been advanced to explain the actions of hormones on the developing cortex, despite the fact that a vast body of data exists detailing hormonal effects on specific cortical and cognitive variables. At this point it might be useful to review existing data, in order to reiterate the profound impact of early hormonal exposure on normal cortical development.

Sex differences in human lateralization and performance

A number of studies report sex differences in lateralized cognitive abilities, with spatial and mathematical performance higher among males and verbal performance generally higher among females (for reviews, see Baker, 1987; Halpern, 1986; Maccoby and Jacklin, 1974; or Wittig and Peterson, 1979). The sex difference in spatial ability is particularly robust for specific tasks, such as the water-level test, the rod-and-frame test, and mental-rotation tests, all of which show a highly significant and consistent male advantage. The female advantage on verbal tests is somewhat less marked and consistent. However, females have also been reported to exhibit precocial language onset with greater frequency than boys do (Moore, 1967). Finally, a marked ratio of males to females has been found for prepubertal mathematical genius (Benbow and Stanley, 1981, 1983).

These sex differences suggest that gonadal hormones may influence organization of the brain and be expressed as differences in cognitive abilities. But sexually biased cultural pressures (for example, parental and teacher encouragement) are also thought to influence many of the cognitive measures discussed above. Therefore it is important to look at other variables, ones that are not as easily affected by environmental factors.

The fact that sex differences are found for abilities controlled by different hemispheres (with verbal in the left and spatial in the right for the majority of the population) has led some researchers to postulate that the cognitive sex effects may be expressions of underlying differences in cortical organization. For example, Buffery and Gray (1972) postulated that verbal skills would develop better in individuals strongly lateralized for verbal ability, while spatial ability would be facilitated by bilateral representation and interhemispheric communication. They also postulated that females may be more lateralized than males, hence exhibiting better verbal and poorer spatial ability than males, but this hypothesis does not fit with what is known about sex differences in cerebral lateralization.

To some extent the data are better fit by the hypothesis advanced by Levy (1976; Levy and Nagylaki, 1972; Levy-Agresti and Sperry, 1968). Levy posited that spatial ability was facilitated by more lateralization and that the greater lateralization in males was expressed as a spatial advantage. Language, conversely, may be slightly facilitated by representation in both hemispheres. Several major findings to date support this general association. First, males appear to be more lateralized than females, at least for spatial functioning (Kimura, 1987; Kimura and Harshman, 1984; McGlone, 1978, 1980; McGlone and Kertesz, 1973). These measures of lateralization derive largely from subjects with unilateral hemispheric damage. That is, males appear to have more difficulty than females in recovering function after damage to a single hemisphere, leading many to believe that females must have compensatory capacity in the opposite hemisphere while males do not. Measures of lateralization from dichotic listening tests also indicate that males are, in general, more lateralized than females (see Bryden, 1982, for review). Females appear to rely more on bilateral representation for both verbal and visual-spatial abilities (Ray et al., 1979). Finally, some studies have reported a positive correlation between degree of lateralization and spatial functioning (Levy and Reid, 1978; Waber, 1976; Waber et al., 1985). These data support the possibility that increased cerebral lateralization in males may underlie their advantage in performing spatial tasks.

Differences like these are considerably more difficult to ascribe to cultural influences than to the cognitive sex differences themselves. The notion of a link between biological factors and cerebral development has been strengthened by the report of an interaction between sex and handedness in cognitive abilities (Harshman et al., 1983). These researchers found the expected effect of males scoring higher than females on spatial tasks. Among those with high reasoning ability, however, right-handed males were superior to left-handed males, while the reverse was found for females, with left-handers outperforming right-handers. In contrast, while females were better than males on selected tests of verbal ability, right-handed females were better than left-handed, and left-handed males were better than right-handed. These findings support, to some degree, the general notion that sex, handedness, cognitive abilities, and lateralization may all be influenced by a common underlying factor(s) expressed in patterns of cerebral development and organization. This mechanism (or mechanisms) may also relate to the findings that there are greater numbers than expected of both left-handers and males in the LLI population. Geschwind (1985) posited that this critical factor may be early androgenic exposure, a hypothesis that is supported by findings of cognitive and structural effects on the cortex following manipulations of early androgen exposure. These findings will be discussed in a moment.

Some evidence also suggests that maturation rates may contribute to differences in cerebral organization. Several researchers have reported a positive correlation between age at puberty and spatial ability, with later age at puberty correlated to higher spatial ability for both males and females (Ray et al., 1981; Sanders and Soares, 1986; Waber, 1976; Waber et al., 1985). However, the mechanisms underlying these differences in maturation rate are unknown. There are many questions yet to be addressed, including how handedness, which is at least partially influenced by heritable factors, is involved in the development of cerebral lateralization and sex differences.

Sex differences in human cortical structure

The assertion of sex differences in the organization of the human brain is strengthened by reports of sex differences in cortical structure. Some of the most profound findings of this nature are reports of a sex differences in the size of the human corpus callosum, the fiber tract interconnecting the cerebral hemispheres. Many researchers have been unable to replicate the original report by deLacoste-Utamsing and Holloway (1982) that the corpus callosum is larger in women than in men (Bell and Variend, 1985; Bleier et al., 1986; Byne et al., 1986; Demeter et al., 1985; Kertesz et al., 1987), and more recent reports suggest that the callosum may be larger, in general, in men (Clarke et al., 1989; Witelson, 1989, 1990). The relationship between sex and callosal size, however, appears to be rather complex for humans. Witelson (1985, 1989, 1990; Witelson and Kigar, 1988) has reported an interaction between sex and handedness for callosal size, an effect replicated in other laboratories (Denenberg et al., 1991b). Specifically, she has shown that regions of the callosum are larger in nonconsistently right-handed men than in other groups, particularly in the isthmus region (which comprises axons of passage from the parietal and temporal gyri surrounding the posterior segment of the sylvian fissure). The possible association between this finding and the sex and handedness interactions in cognitive ability reported above is quite intriguing. Interestingly, Habib (1989) has shown a negative correlation between the size of the corpus callosum and the size of the left planum temporale for a normal sample of human adults. This finding suggests a significant role for the callosum in cerebral organization, highlighting the importance of sexual dimorphism in this structure.

Performance in hormonally abnormal human populations

Another area of research in which information on the role of hormones in cortical development can be gained is the study of hormonally abnormal human populations. Reports have shown that females with adrenal hyperplasia (marked by increased output of adrenal androgens) have elevated spatial ability relative to control sisters (Resnick and Berenbaum, 1982), and also that males with testicular feminizing syndrome have a female-like performance pattern of verbal ability greater than spatial ability (Masica et al., 1969). Studies have also focused on the cognitive performance of individuals with sex chromosomal abnormalities. Chromosomal aberrations are often associated with abnormalities in gonadal hormones, although genetic effects on cognition are also possible. Females with Turner's syndrome (one X chromosome), for example, have been found to exhibit normal general IQ and verbal abilities but impaired spatial abilities (Garron, 1977; Rovet and Netley, 1982). Conversely, subjects with Klinefelter's syndrome (males with an additional X chromosome) exhibit speech and language impairment (Graham et al., 1982, 1990; Neilson et al., 1981; Puck et al., 1975). Interestingly, these individuals exhibit specific deficiencies in processing rapid non-verbal tonal sequences, a phenomenon also observed in language and learning impaired individuals (Tallal, 1980; Tallal and Newcombe, 1978; Tallal et al., 1985a,b).

Sex differences in cortical structure in animal studies

Data regarding the role of hormones in cortical organization has also accumulated in the field of animal research. One of the most profound findings in this regard is that male but not female rats exhibit a right-greater-than-left pattern of cortical thickness (Diamond et al., 1981; Stewart and Kolb, 1988). Specifically, regions of the cortex measured in homologous areas of the right and left hemispheres show an asymmetry in thickness for males, with specific areas in the right being thicker than their counterparts on the left. Female rats, in contrast, show a trend toward a left-thicker-than-right pattern (Diamond et al., 1981). The inference that this sexual dimorphism is at least partially influenced by neonatal androgen exposure stems from findings that both neonatal castration and prenatal stress demasculinize (i.e., eliminate) the asymmetrical cortical thickness pattern in the male, thereby making the pattern more female-like (Diamond, 1984; Fleming et al., 1986; Stewart and Kolb, 1988).

Neonatal androgens also appear to influence sexual dimorphism in the corpus callosum, which has been shown to be larger in male rats than in females (Berrebi et al., 1988; Denenberg et al., 1989; Fitch et al., 1990a; Zimmerberg and Mickus, 1990). These data are derived from studies in which a mid-sagittal tracing of the corpus callosum, histologically obtained from experimental subjects, is measured for area, length, and width (or thickness). The finding of a callosal sex difference is particularly striking in light of the recent reports that the corpus callosum is generally larger in men than in women (Clarke et al., 1989; Witelson, 1991). Fitch and colleagues (1990a) have found that exposure of neonatal female rats to testosterone (via injection) results in an enlargement or masculinization of their callosa by adulthood. These testosterone-treated females have callosa the size of male callosa. Likewise, prenatal exposure to the anti-androgen flutamide followed by castration and prenatal exposure to alcohol have both been shown to reduce callosal size in male rats (hence eliminating callosal sex differences) (Fitch et al., 1990b;

Zimmerberg and Scalzi, 1989). Such findings support the assertion that neonatal exposure to testosterone influences cortical development.

Sex differences in cognitive functioning in animal studies

These reports of an androgenic influence on cortical development provide a background for studies showing that neonatal manipulations of androgens alter performance on sexually dimorphic cognitive tasks in rats. For example, sex differences have consistently been reported for active and passive avoidance learning in rats (e.g., Beatty et al., 1983; Denti and Epstein, 1972). These sex differences emerge at puberty (Bauer, 1978) and can be reversed by neonatal manipulations of gonadal hormones (Denti and Negroni, 1975). Sex differences in taste-aversion learning are also influenced by androgenic exposure (Chambers, 1985; Earley and Leonard, 1978). Other studies have demonstrated that male rats make fewer errors than females in learning a spatial maze (see Beatty, 1984, 1979, for review) and that this sex difference also appears at puberty (Krasnoff and Weston, 1976). Neonatal castration of males or exposure of females to androgen reverses the sex-typical performance of maze learning (Dawson et al., 1975; Joseph et al., 1978; Stewart et al., 1975). The parallels between these findings and reports that human males score higher than females on spatial tasks and that human sex differences in spatial performance also appear strongest after puberty (Halpern, 1986) are quite intriguing.

Clark and Goldman-Rakic (1989) have also demonstrated sex differences in the organization of the orbital prefrontal cortex (ORB) of monkeys, as measured by alterations in the ability to perform an object discrimination reversal task following ORB lesions. Specifically, intact males made fewer errors than intact females in learning the task, and lesions to the area disrupted the ability of males but not females to perform the task. Furthermore, females given androgen in early perinatal life performed similarly to normal males, and their performance was similarly disrupted by the lesion.

Mechanisms of Hormone Action

In sum, the vast body of data implicates neonatal androgens as a potent influence in establishing patterns of cortical organization, as measured by both structure and cognitive function, but the specific mechanisms of the hormonal action remain unclear. The model proposed by Geschwind and Galaburda (1985), that testosterone may specifically affect the left hemisphere during development as a consequence of differences in hemispheric rates of development, provides a possible mechanism. A recent study by Galaburda (Galaburda et al., 1987), however, suggests that cerebral asymmetry, at least in the planum temporale, is the consequence of decreased size in the *right* planum. Galaburda found the degree of planum asymmetry to be inversely correlated with the size of the right planum, rather than positively correlated with the size of the left. This correlation has been replicated by Habib (1989). Habib has also reported a negative correlation between laterality quotient for handedness and asymmetry in the planum temporale.

Galaburda postulates that testosterone may act to prevent normally occurring cell death in the right planum during development, with high levels of testosterone promoting symmetry by preventing normal cell death in this region. Thus he postulates that testosterone has a *facilitative* effect on cortical growth. This report is partially consistent with the findings of Stewart and Kolb (1988), who showed that neonatal gonadectomy of male rats led to an elimination of cortical-thickness asymmetry by promoting a thicker cortex in the left hemisphere. However, while Stewart and Kolb found that testosterone was acting to induce cortical asymmetry in rats, they found that this effect was mediated by *inhibiting* effects of testosterone on cortical growth in the left hemisphere. Witelson (1990) also suggests that testosterone may act on the development of the callosal isthmus region in humans by promoting cell death (i.e., *inhibiting* growth), and therefore that abnormally *low* levels of testosterone may, because of a decrease in normal cell death, lead to the larger isthmus region seen in nonconsistently right-handed males.

Although these interpretations appear inconsistent, they are united in the assertion that exposure to androgen influences development of the cerebral cortex. It is possible that the mechanism of hormonal action differs in different cortical regions. For example, there is evidence to suggest that androgens can inhibit cell death in sexually dimorphic nuclei of the spinal cord (Nordeen et al., 1985), while other findings, such as the fact that female rats have many more axons than males in the corpus callosum, suggest that androgens promote cell death (Juraska and Kopcik, 1988). Even these findings are confounded by the fact that androgens may influence other components of neural growth. For example, despite having fewer callosal axons, enriched male rats have myelinated callosal axons of greater diameter than those found in enriched females (Juraska and Kopcik, 1988). Steroids have also been shown to influence dendritic sprouting and the growth of dendritic spines *in vitro* (Toran-Allerand, 1984) and *in vivo* (Frankfurt et al., 1990; Gould et al., 1990), as well as synapse formation *in vivo* (Matsumoto and Arai, 1976, 1981; Miyakawa and Arai, 1987).

Each of these findings contributes evidence to the general assertion that early exposure to androgens influences the development of the cortical hemispheres and that this action may differ between cortical regions and hemispheres. Although a direct application of these findings to the putative role of early exposure to androgens in the development of language and learning disorders has not been established, the link is strongly supported by: (1) the crucial role of androgens in establishing sexually dimorphic patterns of structural and functional asymmetry, as well as differences in cognitive performance; (2) the abnormalities in cerebral asymmetry and specific left-hemisphere deficits observed among LLI subjects; (3) the higher incidence of left-handers among LLI subjects; and (4) the higher incidence of LLI subjects reported among boys than among girls.

Development of Cerebral Organization and LLI in Females

One important point of consideration has been bypassed in the current discussion. We have focused on the role of androgens in the development of both cerebral lateralization and cognitive abilities. However, most models and discussions have failed to consider the mechanisms underlying the development of language and learning impairments in female subjects. Most of the findings discussed above assume that early exposure to androgens influences neural development in a male typical fashion and that the absence of testosterone leads to a female pattern of cerebral organization. Thus, under these assumptions, abnormalities in testosterone production or response to testosterone in male subjects may lead to abnormalities in the development of cerebral laterality and hence to language and learning disorders as well. This would explain the higher incidence of LLI among boys, since abnormalities in androgen exposure are presumably more likely to occur in males. And yet, language and learning impairment does affect girls, and the etiology for the development of this disorder in females has barely been broached. Geschwind and Galaburda (1985) suggest that exposure to androgens *in utero* may influence cortical development in females. The presumed source of these androgens would be the maternal ovaries or possibly the adrenals (e.g., via stress to the mother). However, evidence for a mechanism of maternal androgenic effects on the female fetus is scant. Indeed, evidence suggests that although prenatal stress and prenatal alcohol exposure exert demasculinizing effects on male rodent offspring, female offspring are unaffected by these manipulations (e.g., Ward, 1972, 1983; Zimmerberg and Scalzi, 1989).

Adrenal androgens

Another possibility is that endogenous production of abnormal levels of adrenal androgens in girls may influence the development of cerebral laterality and LLI. Some researchers, for example, have examined girls with adrenal hyperplasia and have found increased measures of spatial ability and decreased verbal expressiveness as compared with their control sisters (Resnick and Berenbaum, 1982). However, it is not known whether these girls exhibit an increased incidence of LLI, though the possibility presents a potential area for future research.

Ovarian hormones and cortical development

We might also consider the possibility that a different mechanism may influence the development of cerebral organization in females. In a series of papers, Diamond and her colleagues (Diamond et al., 1979; Pappas et al., 1978, 1979) reported that neonatal ovariectomy (Ovx) increased the cortical thickness of adult female rats. Furthermore, they reported that the increased thickness was a consequence of an increase in the size of neural perikarya in Ovx females. Interestingly, Ovx females had fewer and less densely packed cortical neurons in the areas examined than were found in control females. In a later paper, Diamond and her colleagues (1981) found that neonatal ovariectomy also acted to reverse the left-right cortical thickness pattern typical of females to a more male-like pattern.

Stewart and Kolb (1988) were unable to replicate this latter finding. However, they did report a main effect of gonadectomy in increasing overall cortical thickness; the cortex of gonadectomized females and males was significantly thicker than that of their respective controls.

Gould et al. (1990) have reported that ovariectomy decreases dendritic spine density on CA1 pyrimidal cells in the hippocampus of adult female rats and that this effect is blocked by the concurrent administration of estrogen and progesterone. Miyakawa and Arai (1987) also found that post-pubertal estrogen treatment increased the number of axodendritic synapses in the lateral septum of intact female but not intact male rats.

Recent data also show that ovariectomy influences sexual dimorphism in the rat corpus callosum. Specifically, female rats ovariectomized in the neonatal period had callosa as large as those of males by adulthood (Fitch et al., 1991a). Furthermore, the sensitive period for this effect extends considerably later than the sensitive period for the effects of androgens on the cortex, which ends in the early neonatal period. Also, the effects of ovarian hormones are expressed considerably later in development (Fitch et al., 1992). Effects of sex and testosterone on callosal size are observed in rats as young as 3 days (Zimmerberg and Scalzi, 1989), whereas the effects of ovariectomy on the callosum are not seen until after puberty.

Corticosteroids and cortical development

One possibility is that these ovarian effects on the cortex are mediated in part by secondary effects on the glucocorticoid system. This hypothesis derives from the observation that ovarian and testicular feedback lead to significantly different corticosterone levels in adult male and female rats, with higher levels among females (Kime et al., 1980). Also, neonatal ovariectomy significantly reduces corticosterone levels of adult female rats to the male level (Fitch et al., 1991b) and reduces cortisol levels in female monkeys (Heisler et al., 1990). Finally, neonatal ovariectomy depresses and prolongs the response-recovery curve for ACTH in female rats (Fitch 1992).

Glucocorticoids have been found to exert profound effects on cortical development as assessed by adrenalectomy (Meyer, 1983a,b; Meyer et al., 1985). Meyer has reported that adrenalectomy leads to increased myelination of the rat cortex, and this effect is reversed by corticosterone. Neonatal handling, which elevates levels of corticosterone in the neonate (Denenberg et al., 1967) and produces more sensitive corticosterone feedback in adult rats (Denenberg and Zarrow, 1971; Meaney et al., 1988), also influences cortical development. Meaney has reported increased levels of hippocampal and cortical glucocorticoid receptors and improved memory for rats that have been handled in infancy. Handling also enhances lateralization of function in male rats (Denenberg and Yutzey, 1985). Finally, handling has been found to exert sexually dimorphic effects on both cortical structure (Berrebi et al., 1988; Denenberg et al., 1991a) and behavior (Camp et al., 1984; Weinberg and Levine, 1977).

However, the extent to which sexually dimorphic gonadal hormones modulate the corticoid system, and in turn cortical development, has barely been addressed. Most researchers in this area have used only male rats. Therefore it will be of critical importance to focus in the future not *only* on the role of testicular androgens in cortical development but also on the role of other gonadal and adrenal hormones and the interactions between these hormonal systems.

Summary

Accumulated findings that hormones other than testicular androgens are involved in the mediation of cortical development take on new relevance in light of recent assertions made by Witelson (1991). On the basis of structural-functional correlations obtained between callosal and behavioral measures in human males but *not* females, Witelson suggests that males and females are "dichotomous" with respect to cerebral organization; that is, rather than representing overlapping populations distributed along a continuum, males and females may be categorically different with respect to cerebral organization. Furthermore, she suggests that different mechanisms may influence the development of cerebral organization in males and females. That ovarian hormones have been shown to influence the development of sexually dimorphic structure in the cortex of female rats adds support to this notion; it also provides a possible mechanism whereby hormonal abnormalities in females could lead to abnormalities in patterns of cerebral development.

It will be important in the future to examine more closely the etiology of language disorders in males and females—particularly in light of new data suggesting that many females may be afflicted but not detected (Shaywitz et al., 1990). It is possible that the expression of LLI may differ in subtle and as yet undefined ways as a function of sex.

Familial History and Transmission of Language and Learning Impairments

Another approach to investigating genetic and hormonal bases for language and learning disabilities focuses on the evaluation of family history data of children with specific developmental language disorders. Following this approach, Tallal et al., (1989a,b) reported the results of a family history study of a large, well-defined cohort of four-year-old children with specific developmental language impairment. The majority of language-impaired subjects evaluated in this study had both severe expressive and receptive language deficits, although some were specifically more impaired in the receptive or expressive mode. Over 60 percent of this sample also exhibited speech articulation impairment, although children with articulation deficits alone were not included in this study. Over 75 percent of these children also showed evidence of severe reading and learning disabilities by the age of eight.

Familial history information was obtained from questionnaires completed by the biological parents of each subject. Classification of parents as "affected" was based on self-reports of: (1) a history of language problems; (2) a history of below-average school achievement as far as the eighth grade in reading, writing or both; or (3) a history of having been kept back in grade school. These questions were used for classification because of the difficulty in obtaining specific diagnostic histories for the parents. Siblings of the subject, also evaluated in the questionnaire, were considered "affected" if the parent reported a history of difficulties in reading, writing, language development, or other learning disabilities. The analysis of these reports yielded extremely interesting results.

The 62 language-impaired subjects in this study consisted of 44 males and 18 females, a sex ratio of 2.4:1. This is consistent with prior reports of higher LLI incidence among males (Shaywitz et al.,

1988). However, this sex ratio was significantly (and unexpectedly) influenced by the familial aggregation for the disorder. Analysis of subjects *without* an affected parent failed to show a significant difference between the number of boys and the number of girls with LLI. Analysis of subjects *with* an affected parent (mother or father) showed a 3:1 ratio of males to females. When the affected parent was the father the ratio was 1.8:1. When the affected parent was the mother, however, the ratio was 4:1. This ratio increased to 5:1 if both parents were affected. These findings strongly suggest that the high ratio of males to females with language impairment reported in the literature derives specifically from children with an affected parent, particularly an affected mother.

Unexpected sex ratios were also found among the siblings of the language-impaired probands. Specifically, these children had twice as many brothers as sisters (a 1.9:1 ratio). Further analysis again pointed to parental affliction as the source of this sex-ratio difference. Affected fathers had only slightly more sons than daughters, not including the proband (1.4:1). However, affected mothers had 2.5 times as many sons as daughters.

In an effort to clarify the contribution of the sex of the affected parent, families with both parents affected were excluded and families of affected mothers and fathers were analyzed separately. In families with an affected father and nonaffected mother, the sex ratio of offspring was found to approach 1:1. When only affected offspring (excluding the proband) were examined, the sex ratio was still about 1:1. However, among families with an affected mother and nonaffected father a highly aberrant sex ratio was found. A boys-to-girls ratio of 3.5:1 was found if probands were included. This sex ratio increased to 5.3:1 if only affected offspring (including the proband) were counted. This ratio rose to 8:1 affected sons to daughters when the proband was excluded, to avoid possible ascertainment bias.

Analysis of the sex ratio of affected sons and daughters as a function of absolute numbers of sons and daughters revealed an interesting result. For families with affected parents, male and female offspring were equally likely to be affected. However, families with an affected mother had a disproportionate number of male births, leading in turn to a disproportionate number of affected boys in the families of an affected mother. Consequently it can be stated that the higher frequency of language-impaired males ascertained through this study was directly or indirectly the consequence of a markedly higher number of sons born to affected mothers, approximately half of whom were found to develop LLI. These affected sons born to affected mothers account almost entirely for the skewed ratio of affected male to female children in the study.

These findings have profound consequences for the study of language and learning impairment, because they strongly support a biological link between the parental incidence of LLI, the sex ratio of offspring, and the ratio of males to females affected by LLI in the general population. While Shaywitz et al. (1988) have posited that both environmental and genetic factors might contribute to the sex-ratio differences in LLI children in the general population, environmental factors cannot explain the dramatic and consistent pattern of deviant sex ratios in the families of affected mothers. This pattern, combined with the high risk of maternal transmission of LLI (about 50 percent), results in a significantly higher number of boys than girls affected by LLI. Indeed, these findings cannot be addressed without consideration of a biological mechanism.

As one considers possible mechanisms to explain these results, several important questions also arise. First, what is the mode of familial transmission, and does having an affected father versus an affected mother, or having no affected parent, result in subtle differences in the etiology of the disorder? Second, how does the higher ratio of sons to daughters born to affected mothers relate to the transmission of LLI, if at all?

Hormonal factors affecting sex ratio in humans

One potential mechanism to be considered is that of a hormonal factor influencing both the development of LLI and the sex ratio for the offspring of affected mothers. As a hypothetical example, consider that abnormal perinatal hormonal exposure in a female may lead both to the development of atypical cerebral organization and LLI in early life and to abnormalities in the sex ratio of offspring in later life. The transmission of LLI to offspring may involve abnormal hormonal exposure of the offspring *in utero* of the affected mother, or it may be a consequence of a genetic factor producing an abnormal endogenous hormonal environment in the offspring, which may in turn result in LLI. The notion of a genetic factor influencing the endogenous hormonal environment appears to be more consistent with the fact that paternal affliction increases the risk of LLI, which cannot be explained by *in utero* exposure. Alternately, both factors may combine, contributing to the increased risk of LLI if the mother rather than the father is affected.

These notions are highly speculative, and yet they are substantiated by several factors. First, there is evidence that hormonal factors can influence sex ratio. For example, increased male births have been associated with maternal stress and abnormal levels of gonadal hormones, particularly testosterone (James 1986, 1988). James (1980, 1983) has also suggested that maternal gonadotropin levels at the time of conception may be related to the sex of the fertilized zygote, high levels of hormone being associated with the conception of females. Sas and Szollosi (1980) reported that men with low sperm counts who were given methotestosterone therapy sired 45 boys and 17 girls during treatment and up to three months later. Bernstein (1954) reported variations in sex ratio as a function of the "masculinity" of the occupation of each parent (masculinity being defined as the relative number of males in that occupation). Kreuz et al. (1972) also reported that stress reduces testosterone levels in men and that men in stressful occupations had more male than female children. Although these data are scanty, and the specific mechanisms influencing sex ratio are not clearly defined, the studies reviewed here support the general notion of hormonal involvement in the occurrence of deviant sex ratios. Finally, Hugdahl et al. (1990) have reported an increased incidence of spontaneous abortion and pregnancy complications among dyslexic mothers, which may relate both to the higher incidence of surviving male offspring and to the transmission of LLI to the offspring.

Conclusions

From recent studies of language and learning impairment a number of observations have emerged: (1) atypical cerebral asymmetry in LLI subjects; (2) an increased incidence of left-handedness among LLI subjects; (3) an increased frequency of LLI among males, specifically those with affected mothers; (4) a deviant sex ratio in the offspring of mothers affected by LLI; (5) mounting evidence for hormonal influences on cerebral development; and (6) evidence for hormonal mechanisms influencing sex ratio. In combination these present an intriguing scenario for the creation of a model addressing the developmental mechanisms of LLI. Although we recognize that the implications for hormonal involvement in the development of LLI are strong, it must also be noted that the evidence is circumstantial; the hypothesis of abnormal hormonal influences on the development of LLI has not been directly tested in humans.

As stated earlier, a number of questions remain to be addressed. These include an assessment of the behavioral and neuroanatomical etiology of LLI among female subjects, to determine whether sex differences in the expression of the disorder exist; an examination of the mode(s) of transmission of LLI from affected parents to offspring; a determination of the mechanism(s) of deviant sex ratios in the offspring of affected mothers; and an assessment of the exogenous and endogenous hormonal exposures of LLI subjects. It is hoped that further studies and the continued integration of data obtained from human and animal research will contribute to the development of a model for the role of hormonal factors in cerebral organization and the ontogeny of language and learning disorders.

10

Genes and Genders: A Twin Study of Reading Disability

J. C. DeFries
Jacquelyn J. Gillis
Sally J. Wadsworth

Although results of twin and family studies (e.g., DeFries, Fulker, and LaBuda, 1987; Smith and Goldgar, 1986) suggest that deficits in reading performance have a heritable basis, the etiology of reading disability may differ as a function of group membership (Finucci and Childs, 1983; LaBuda, DeFries, and Pennington, 1990). For example, in a commentary for the proceedings of a conference on "Sex Differences in Dyslexia," Geschwind (1981) speculated that girls may be "less affected by certain environmental influences, such as the quality of teaching, social class differences, or outside pressures within society," (p. xiv). Thus, the cause or causes of reading disability in females may differ from those in males.

In a review entitled "The Contribution of Twin Research to the Study of the Etiology of Reading Disability," Harris (1986) concluded that twin studies should play a larger role in etiological research. More specifically, she suggested that twin studies could be employed to assess the possibility of differential etiology in males and females as follows: "In addition to the comparison of the concordance rates of the MZ and DZ twins, separation of the sample by sex may provide information about the causes of sex differences in population prevalence. Given that an individual has a 'vulnerable' genotype, as evidenced by an affected MZ co-twin, does the risk of being affected vary between males and females? Does the pattern of concordance for MZ and DZ same-sex twins differ by sex? . . . Further research in this area is warranted" (p. 16).

Although it has not been previously noted, results from an early twin study of reading disability provide some evidence for differential etiology in males and females. Bakwin (1973) ascertained 338 pairs of same-sex twins, ranging in age from 8 to 18 years, through clubs for mothers of twins. Data were obtained through interviews with the parents (usually the mothers), supplemented by telephone calls and mail questionnaires. Although Bakwin's (1973) diagnostic criteria were not specified explicitly, he defined reading disability "as a reading level below the expectation derived from the child's performance in other school subjects" (p. 184). A positive history of reading disability was obtained for 97 of 676 children, a prevalence of 14.3 percent. This relatively high prevalence may be due to a higher risk for reading disability among twins than among singletons (Hay and O'Brien, 1982). Alternatively, it may be indicative of ascertainment bias or a tendency of parents in this study to over-report problems for their children. Nevertheless, prevalence was highly similar for members of identical and fraternal twin pairs—14.0 and 14.9 percent, respectively.

Comparison of the identical and fraternal twin concordances reported by Bakwin (1973) suggests a substantial genetic etiology for reading disability. Pairwise concordance was 84 percent for identical twins versus only 29 percent for fraternal twins. As may be seen in Table 10.1, pairwise concordances were essentially equal for male and female identical twin pairs—84 and 83 percent, respectively.

Table 10.1. Differential concordance for reading disability in male and female twin pairs reported by Bakwin (1973)

	Male		Female	
	Identical	Fraternal	Identical	Fraternal
Number of pairs concordant	16	8	10	1
Number of pairs discordant	3	11	2	11
Pairwise concordance (%)	84	42	83	8
Probandwise concordance (%)	91	59	91	15

However, male fraternal twin pairs were more concordant than female fraternal twins—42 and 8 percent, respectively. Although not reported by Bakwin (1973), this difference in concordance for male and female fraternal twin pairs is significant ($\chi^2 = 3.97$, d.f. = 1, $p < 0.05$, corrected for continuity). Consequently, the difference between female identical and fraternal twin concordances (83 versus 8 percent) exceeds that between male identical and fraternal twin pairs (84 versus 42 percent). This pattern of results suggests that genetic factors may be more important as a cause of reading disability in females than in males.

In order to test this hypothesis of differential etiology as a function of gender more explicitly, we subjected Bakwin's twin data to a 2 × 2 × 2 (gender, zygosity, and concordance) factorial loglinear analysis (SPSS-X HILOGLINEAR; SPSS-X, 1990). The two-way interaction between zygosity and concordance provides a test for genetic etiology in the total sample, whereas the three-way interaction tests for differential etiology in males and females. Results of this analysis revealed a highly significant ($p < 0.001$) interaction between zygosity and concordance, substantiating Bakwin's (1973) report of a highly significant difference between the MZ and DZ pairwise concordance rates. In addition, the three-way interaction is marginally significant ($p = 0.09$, one-tailed test). (Because we hypothesized that genetic factors are more important as a cause of reading disability in girls than in boys, a one-tailed test of significance is appropriate.) Thus, results of the loglinear analysis of data from Bakwin's small twin study provide at least some evidence for differential etiology of reading disability in males and females.

Because of a paucity of well-designed twin studies of reading disability, a twin study was initiated in 1982 as part of the ongoing Colorado Reading Project (Decker and Vandenberg, 1985; DeFries, 1985). An extensive psychometric test battery is currently being administered to identical (monozygotic, or MZ) and fraternal (dizygotic, or DZ) twin pairs in which at least one member of each pair is reading disabled and to a comparison group of twins with no history of reading problems. Following a brief review of methodological considerations in twin research, we shall utilize twin data from the Colorado Reading Project to test the hypothesis of differential etiology in males and females.

Twin Concordance

Previous twin studies of reading disability (e.g., Bakwin, 1973; Stevenson, Graham, Fredman, and McLoughlin, 1984, 1987; Zerbin-Rüdin, 1967) employed a comparison of concordances in identical and fraternal twin pairs as a test for genetic etiology. A pair is concordant if both members are affected with the same disorder, but discordant if only one member of the pair is affected. Although the concept of concordance is conceptually very simple, its estimation depends upon the method of sample ascertainment (DeFries and Gillis, 1991). Both members of a pair may be affected (++, i.e., Twin 1 and Twin 2 are both affected), only one may be affected (+− or −+), or neither may be affected (−−). If Twin 1 is arbitrarily specified (e.g., the first-born member of a twin pair), there are two types of discordant twin pairs (+− and −+) and such pairs are expected to occur with equal frequency in the population. Thus, the frequencies of ++, +−, −+, and −− twin pairs may be symbolized as A, B, B, and C. If ascertainment is "complete" (e.g., all school-age twin pairs in Barcelona, Spain, are tested), the sample size is $A + 2B + C$. However, if twins are ascertained only because at least one member of each pair is affected, two different types of "incomplete" ascertainment may be employed (Thompson and Thompson, 1986): (1) "single selection," in which only one member of each pair (e.g., Twin 1) is selected as a proband; or (2) "truncate selection," in which each affected member of a pair is ascertained as a proband.

If, for example, only first-born twins were screened for reading deficits, the number of twin pairs with an affected proband (++ and +−) would be $A + B$. Thus, in a sample ascertained by single selection, the "pairwise" concordance (Plomin, DeFries, and McClearn, 1990) would be computed as the ratio of the number of concordant pairs to the total number of pairs, or $A/(A + B)$. However, if both members of the pair were screened and each could be identified as a proband, the number of pairs with an affected proband (++, +−, and −+) would be $A + 2B$. To compensate for this increased number of discordant pairs, "probandwise" concordance (Plomin et al., 1990) should be computed using a procedure in which each member of a concordant pair is counted twice, once as a proband and once as a co-twin. This "double-entry" of concordant pairs effectively increases the sample size to $2A + 2B$ and results in a probandwise concordance comparable to that of pairwise concordance with single selection, as $2A/(2A + 2B) = A(A + B)$.

Previous twin studies of reading disability (e.g., Bakwin, 1973; Stevenson et al., 1984, 1987) employed truncate selection to ascertain samples of affected twins. However, with the exception of the later publication by Stevenson et al. (1987), pairwise concordances were reported for these studies. When truncate selection is employed but pairwise concordance is computed, twin concordance will tend to be underestimated, because $A/(A + 2B) < A/(A + B)$. For example, it may be seen in Table 10.1 that probandwise concordances are higher than the pairwise rates reported by Bakwin (1973) for each of the four zygosity-by-gender comparisons. The difference is greatest for DZ males in which pairwise concordance is intermediate. Thus, a comparison of MZ and DZ probandwise concordances provides a somewhat stronger case for differential etiology as a function of gender than does a comparison of the pairwise rates reported by Bakwin (1973).

Multiple Regression Models

Although a comparison of concordances in MZ and DZ twin pairs is an appropriate test of genetic etiology for dichotomous variables, reading disability is operationally defined (Wong, 1986) and its diagnosis is based upon a continuous measure (e.g., reading performance) with arbitrary cutoff points (Stevenson et al., 1987). An alternative test that facilitates the analysis of both between-group and within-group variation was suggested by DeFries and Fulker (1985). A multiple regression analysis of twin data was proposed in which each co-twin's score is predicted from that of a proband (the member of the pair with a deviant score on a continuous measure such as reading performance) and the coefficient of relationship (0.5 and 1.0 for DZ and MZ twin pairs, respectively). Two regression models were formulated: (1) a basic model in which the partial regression of a co-twin's score on the coefficient of relationship provides a test of significance for genetic etiology; and (2) an augmented model that also contains an interaction term between a proband's score and relationship. Partial regression coefficients included in the augmented model yield direct estimates of heritability (h^2), a measure of the extent to which individual differences within the selected group are due to heritable influences, and the proportion of variance due to environmental influences shared by members of twin pairs (c^2).

DeFries and Fulker (1985) noted that the results of fitting the basic model to such twin data could be used to estimate h_g^2, a measure of the extent to which the proband deficit is due to heritable influences. They also suggested that a comparison of h_g^2 and h^2 could be used to test the hypothesis that the etiology of deviant scores differs from that of variation within the normal range. For example, unexpectedly low reading performance could be due to a major gene effect or to some gross environmental insult, whereas individual differences within the proband group may be due to polygenic or multifactorial influences. If such were the case, h_g^2 and h^2 should differ in magnitude. However, if probands merely represent the lower tail of a normal distribution of individual differences, estimates of h_g^2 and h^2 should be similar in magnitude.

The basic model in which a co-twin's score *(C)* is predicted from the proband's score *(P)* and the coefficient of relationship *(R)* is as follows:

$$C = B_1P + B_2R + A \tag{10.1}$$

where B_1 is the partial regression of co-twin's score on proband's score, B_2 is the partial regression of co-twin's score on relationship, and A is the regression constant.

The augmented model is similar to the basic model but also contains an interaction term:

$$C = B_3P + B_4R + B_5PR + A \tag{10.2}$$

where PR is the product of the proband's score and the coefficient of relationship. Inclusion of the interaction term in the augmented model changes the expectations for the partial regression coefficients estimated from the basic model; thus, the coefficients of P and R are symbolized B_3 and B_4 in equation 10.2.

DeFries and Fulker (1985) asserted that B_1 is a measure of average MZ and DZ twin resemblance and that B_2 estimates twice the difference between the means of the MZ and DZ co-twins after covariance adjustment for any difference between the MZ and DZ probands. To the extent that the deficit of probands is due to genetic influences, scores of DZ co-twins should regress more than those of MZ co-twins toward the population mean. Therefore, B_2 was advocated as a test of statistical significance for genetic etiology. DeFries and Fulker also demonstrated that B_3 and B_5 provide direct estimates of c^2 and h^2 respectively. Subsequently, LaBuda, DeFries, and Fulker (1986) derived the expected partial regression coefficients for the basic and augmented models. Given the standard assumptions of quantitative genetic analyses of twin data (e.g., a linear polygenic model, little or no assortative mating, and equal shared environmental influences for MZ and DZ twin pairs), they showed that B_1 is a weighted average of MZ and DZ twin resemblance and that B_2 equals twice the difference between the means for the MZ and DZ co-twins, adjusted for any difference between the means of the MZ and DZ probands. Moreover, they showed that B_3 and B_5 provide unbiased estimates of c^2 and h^2, respectively. DeFries and Fulker (1988) subsequently noted that B_2 directly estimates h_g^2 and B_4 estimates $h_g^2 - h^2$ when the twin data are suitably transformed (viz., each score is expressed as a deviation from the mean of the unselected population and then divided by the difference between the proband and control means) prior to multiple regression analysis.

The basic and augmented regression models can easily be extended to include other main effects and interactions. For example, in order to test for differential etiology as a function of gender, the following extended basic model can be fitted to data from male and female twin pairs simultaneously:

$$C = B_1P + B_2R + B_6G + B_7PG + B_8RG + A \qquad (10.3)$$

where G symbolizes gender (a dummy variable coded $+0.5$ and -0.5 for males and females, respectively). B_7, the partial regression of co-twin's score on the product of the proband's score and gender, is a measure of differential twin resemblance in male and female twin pairs. Of greater interest for the purpose of the present analysis, B_8 provides a test of statistical significance for differential genetic etiology as a function of gender.

When truncate selection has been employed to ascertain a sample of twins, data from concordant pairs should be double-entered for regression analysis. Thus, as was the case in the estimation of probandwise concordance rates, each member of a concordant pair should be considered as both a proband and a co-twin. Of course, with double entry, standard errors and tests of significance provided by conventional computer regression programs must be adjusted accordingly.

The Colorado Reading Project

In order to minimize the possibility of referral bias, twin pairs in the Colorado Reading Project are systematically ascertained through co-operating school districts. Without regard to reading status, all twin pairs within each district are identified by school administrators and permission is then sought from parents to review the school records of both members of each pair for evidence of reading problems. If either member of a twin pair manifests a positive history of reading problems (e.g., low reading achievement test scores, referral to a reading therapist because of poor reading performance, reports by classroom teachers or school psychologists), both members of the pair are invited to complete an extensive battery of tests in laboratories at the University of Colorado. At the Institute for Behavioral Genetics, a psychometric test battery is administered that includes the Wechsler Intelligence Scale for Children–Revised (WISC-R; Wechsler, 1974) or the Wechsler Adult Intelligence Scale–Revised (WAIS-R; Wechsler, 1981) and the Peabody Individual Achievement Test (PIAT; Dunn and Markwardt, 1970). A discriminant function score is then computed for each member of the pair using the PIAT Reading Recognition, Reading Comprehension, and Spelling subtest data. The discriminant weights employed in this computation were estimated from an analysis of PIAT data obtained from an independent sample of 140 reading-disabled and 140 control non-twin children tested during an earlier phase of the Colorado Reading Project. In order to be diagnosed as reading disabled in the present study, an individual must have a positive school history for reading problems, be classified as affected by the discriminant score, and have an IQ score of at least 90 on either the Verbal or Performance Scale of the WISC or WAIS. Standard exclusionary criteria (e.g., no evidence of neurological, emotional, or behavioral problems; and no uncorrected visual or auditory acuity deficits) are also employed.

Control twins are matched to probands on the basis of age, gender, and school district. In order for a twin pair to be included in the control sample, both members of the pair must have a negative school history for reading problems and at least one member of the pair must be classified as unaffected by the discriminant analysis.

Selected items from the Nichols and Bilbro (1966) questionnaire are administered to determine zygosity of same-sex twin pairs. If the results of this determination are ambiguous, zygosity of the pairs is confirmed by analysis of blood samples. As of December 31, 1989, 99 pairs of MZ twins and 73 pairs of same-sex DZ twins met our criteria for inclusion in the proband sample (i.e., at least one member of the pair is reading disabled). In addition, a total of 99 pairs of MZ twins and 68 pairs of same-sex DZ twins constituted the control sample. These twins ranged in age from 8 to 20 years at the time of testing and all had been reared in English-speaking, middle-class homes.

In contrast to referred samples of non-twin reading-disabled children in which the gender ratio is often three males to each female or higher (Finucci and Childs, 1981), the numbers of reading-disabled males and females in the sample of same-sex twin pairs are 121 and 127, respectively. Because female MZ pairs tend to be over-represented in twin studies (Lykken, Tellegen, and DeRubeis, 1978), this lower gender ratio for reading-disabled members of twin pairs (0.96 to one) may be due in part to a differential volunteer rate of male and female twin pairs. In fact, the sample of twins screened for reading problems in the Colorado Reading Project contained 542 male and 652 female same-sex twin pairs (a gender ratio of 0.83 to one). Moreover, the ratio of males to females in the sample of MZ probands is somewhat lower than that in DZ probands: 68 MZ males and 84 MZ females (a gender ratio of 0.81 to one) versus 53 DZ males and 43 DZ females (a gender ratio of 1.23 to one). However, neither gender ratio approaches that typically reported in referred samples. Thus, the great excess of males invariably found in system-identified populations of reading-disabled children is almost certainly due at least in part to a referral bias (Shaywitz, Shaywitz, Fletcher, and Escobar, 1990; Vogel, 1990).

Concordance

A comparison of MZ and DZ probandwise concordances estimated from data of the Colorado Reading Project suggests a substantial genetic etiology for reading disability. Of the 99 pairs of MZ twins in the proband sample, 53 are concordant for reading disability. In contrast, only 23 of 73 DZ pairs are concordant. The corresponding estimated probandwise concordances for the MZ and DZ twin pairs are 70 and 48 percent, respectively, a highly significant difference ($p < 0.001$).

As may be seen in Table 10.2, the concordances of MZ and DZ female twin pairs are both lower than those for male pairs. However, the difference between the probandwise concordance rates of MZ and DZ female twin pairs is somewhat larger than that for male twin pairs, 27 versus 19 percent, respectively. Although this result is also consistent with the hypothesis that genetic influences are more important as a cause of reading disability in females than in males, the three-way interaction (gender-by-zygosity-by-concordance) estimated from a loglinear analysis of these data is not significant ($p = 0.38$).

It is also of interest to compare the concordance of DZ twins to that of non-twin siblings. In an earlier family study (DeFries, Vogler, and LaBuda, 1986; Lewitter, DeFries, and Elston, 1980), 133 reading-disabled children, their parents and siblings, and members of 125 control families were administered an extensive psychometric test battery that also included the PIAT subtests. In 102 families of probands, one or more siblings were tested for a total of 166 possible sib pairings. By the discriminant function score criterion, 70 of these 166 sibling pairs were concordant for reading disability, a pairwise concordance rate of 42 percent. (Because single selection was employed to ascertain the probands in the Colorado Family Reading Study, pairwise concordance is reported.) This concordance rate for

Table 10.2. Differential concordance for reading disability in male and female twin pairs tested in the Colorado Reading Project

	Male		Female	
	Identical	Fraternal	Identical	Fraternal
Number of pairs concordant	26	15	27	8
Number of pairs discordant	16	23	30	27
Probandwise concordance (%)	76	57	64	37

non-twin siblings is only slightly lower than that estimated for fraternal twins in the Colorado Reading Project (48 percent).

Multiple regression analyses

To allow us to assess the heritable nature of reading disability, as well as individual differences among probands, the basic and augmented regression models (equations 10.1 and 10.2) were fitted to discriminant function data from the total sample of probands and co-twins. Then, the basic model was fitted to data from male and female twin pairs separately in order to assess the genetic etiology of reading disability in each group. Finally, the extended model (equation 10.3) was fitted to data from male and female twin pairs simultaneously to test the hypothesis of differential etiology of reading disability as a function of gender.

The average discriminant scores of the MZ and DZ probands and co-twins, expressed as standard deviations from the mean score of 316 control twins, are presented in Table 10.3. From this table it may be seen that the average scores of the MZ and DZ probands are highly similar and deviate substantially (over three standard deviations) from the mean of the controls. It may also be seen from this table that the MZ and DZ co-twins have regressed differentially toward the control mean: the difference between the means of the MZ probands and co-twins is only 0.25 standard deviations, whereas the difference between the DZ proband and co-twin means is 0.90 standard deviations. In order to estimate h_g^2, the basic model was fitted to discriminant function data transformed by expressing each subject's score as a deviation from the control mean and then divid-

Table 10.3. Mean discriminant scores of 99 pairs of identical twins and 73 pairs of same-sex fraternal twins in which at least one member of each pair (the proband) is reading disabled

Zygosity	Probands	Co-twins
Identical	−3.28	−3.03
Fraternal	−3.04	−2.14

Note: Discriminant scores are expressed as standard deviations from the mean score of 316 control twins.

ing by the difference between the proband and control means. Because of small mean differences among the average scores of the male MZ, female MZ, male DZ, and female DZ probands, different transformations were employed for the four gender-by-zygosity groups. The resulting estimate of $B_2 = h_g^2 = 0.44 \pm 0.11$ is highly significant ($p < 0.001$) and suggests that almost one-half of the reading performance deficit of the probands, on average, is due to heritable influences.

When the augmented model was fitted to the transformed discriminant function data, $B_5 = h^2 = 0.94 \pm 0.41$ ($p < 0.0125$, one-tailed) and $B_3 = c^2 = -0.14 \pm 0.33$ ($p > 0.25$, one-tailed). These results suggest that individual differences in reading performance within the proband sample are highly heritable, whereas environmental influences shared by members of twin pairs are not an important source of variation. Moreover, although the estimates for h^2 and h_g^2 are rather different (0.94 and 0.44, respectively), suggesting that probands may not merely represent the lower tail of a normal distribution of individual differences (DeFries and Fulker, 1988), the difference between these two parameter estimates is not significant ($B_4 = -0.50 \pm 0.42$, $p > 0.50$, two-tailed).

Estimates of h_g^2 obtained by fitting the basic regression model (equation 10.1) separately to transformed discriminant function data from male and female twin pairs are presented in Table 10.4. The resulting estimate of h_g^2 for males (0.42 ± 0.17, $p < 0.025$) is somewhat smaller than that for females (0.48 ± 0.15, $p < 0.01$), but the difference between these two estimates of h_g^2 is not significant. When the extended model (equation 10.3) was fitted to the transformed discriminant function data of male and female twin pairs

Table 10.4. Genetic etiology of reading disability in male and female twin pairs

	Number of twin pairs			
Gender	MZ	DZ	$B_2 = h_g^2$	p
Males	42	38	.42 ± .17	<.025
Females	57	35	.48 ± .15	<.01

simultaneously, $B_8 = -0.06 \pm 0.22$ ($p > 0.50$). This difference between the estimates of h_g^2 obtained for male and female same-sex twin pairs is somewhat larger than that previously reported by Olson, Wise, Conners, Rack, and Fulker (1989). In their analysis of PIAT Reading Recognition data from a smaller sample of twins tested in the Colorado Reading Project, estimates of h_g^2 for males and females were 0.40 and 0.41, respectively. In addition to the difference in sample size between the two studies, Olson et al. (1989) did not double-enter data from concordant pairs for their analysis.

Discussion

Although the topic of gender differences in reading disability has been widely discussed (e.g., Ansara, Geschwind, Galaburda, Albert, and Gartrell, 1981; Vogel, 1990), little definitive evidence is currently available regarding the magnitude or etiology of any such difference. Unfortunately, the possibility of sample bias is a major problem with research in this area (Shaywitz et al., 1990). In a recent review of gender differences in children with learning disabilities, Vogel (1990) concluded that "findings from research thus far must be interpreted with caution, since the majority of samples comprised system-identified samples that may be biased. A growing body of research suggests that females experiencing learning difficulties are not identified as frequently as males" (p. 50). Several possible causes for the underidentification of females with learning disabilities were suggested: (1) a mismatch between children's problems and the screening agent's expectations due to a scarcity of research involving females with learning disabilities; (2) a higher prevalence of attentional deficits and behavioral problems in boys that results in their more frequent referral for special education services; and (3) a differential attitude of teachers that favors the referral of more males than females even when they have identical problems. Because most previous research regarding reading disability has been based upon referred samples, we are still unable to answer definitively the rhetorical question posed over a decade ago by Finucci and Childs (1981): "Are there really more dyslexic boys than girls?" (p. 1).

This dearth of information pertaining to the issue of gender differences in reading disability, as well as the possible underidentification of females with learning problems, is lamentable. Geschwind

(1981) clearly articulated the importance of researching gender differences in dyslexia as follows: "the arguments for studying the problem are powerful ones. In the first place, this type of investigation can have a useful impact on research. For example, the strikingly uneven proportion of males and females raises the possibility that the nature of dyslexia may not be uniform and that different retraining techniques may be necessary. Furthermore, the unequal sex ratio can suggest research approaches that may lead to a better knowledge of the biological substrates of dyslexia and therefore lead to better prevention or treatment" (p. xiv).

Differential etiology

Although the gender ratio for reading disability may not be as disproportionate as Geschwind (1981) believed, the possibility of a differential etiology in males and females is still tenable. In Bakwin's (1973) small twin study of reading disability, the difference between the MZ and DZ concordances in female twin pairs was larger than that in males, suggesting that genetic factors are more important as a cause of reading disability in females. This pattern of results, although less marked, was also obtained in the larger Colorado Reading Project. Moreover, the estimate of h_g^2 (a measure of the extent to which the deficit of probands is due to heritable influences) was somewhat larger in female than in male twin pairs (0.48 versus 0.42, respectively). Although this difference is consistent with Geschwind's hypothesis that environmental factors may be less important as a cause of reading disability in girls, the difference between these two parameter estimates is not significant. Thus, a larger sample of twins will be required to test more rigorously the hypothesis of differential etiology as a function of gender.

Genetic models

Although a number of different single-locus models have been proposed to account for the familial transmission of reading disability, there is no consensus at this time regarding a particular mode of inheritance. In order to account for the apparent difference in prevalence between males and females, Symmes and Rapoport (1972) proposed that reading disability may be caused by a sex-linked recessive gene. Females have two X chromosomes (in addition to 22 pairs of autosomes), whereas males have one X and one Y chromosome. Thus, the presence of a recessive allele on the one X chromosome of a male might be sufficient to cause a reading disability, whereas recessive alleles on both X chromosomes of females would be necessary to cause the condition. Consequently, if the frequency of a sex-linked recessive allele were q in a random mating population, then q of the males and q^2 of the females would be expected to be affected. For example, if q were 0.1, then the prevalence rates in males and females would be 0.1 and 0.01, respectively. In addition, parent-child resemblance for sex-linked characters differs as a function of the gender of the family members as follows: father-daughter resemblance should approximate mother-son resemblance, both of which should exceed mother-daughter resemblance, and father-son resemblance should be near zero. DeFries and Decker (1982) subjected this model of sex-linked recessive inheritance for reading disability to several tests using data from the Colorado Family Reading Study. In brief, little or no evidence was obtained to support the hypothesis that reading disability is caused by a sex-linked recessive gene.

In order to test for possible autosomal (not sex-linked), major-gene influence, Lewitter, DeFries, and Elston (1980) subjected data from the Colorado Family Reading Study to complex segregation analysis (Elston and Yelverton, 1975). Preliminary analysis of a composite reading-performance measure indicated that a mixture of two normal distributions fit the data significantly better than did a single normal distribution. Segregation analysis was thus deemed appropriate and the adequacies of various single-locus models were tested. When reading-performance data from the families of all probands were analyzed, chi-square tests of goodness of fit indicated that each of these models must be rejected. When these models were fitted to data from families of male probands only, similar results were obtained. However, when data from families of female probands only were analyzed, a hypothesis of autosomal recessive inheritance could not be rejected. On the basis of a comparison of chi-square estimates, Lewitter et al. (1980) argued that the inability to reject the recessive-gene model was not due to the smaller number of female probands included in the Colorado Family Reading Study. Thus, results of this segregation analysis also suggest that the etiology of reading disability may differ in males and females.

The best evidence for single-locus transmission of reading disability was provided by Smith, Kimberling, Pennington, and Lubs (1983). Linkage analysis in families with apparent autosomal dominant transmission for reading disability tentatively suggested localization of a hypothesized gene to chromosome 15. Families were selected for testing if a history of reading disability occurred for three successive generations. Each family member was administered a series of standardized achievement tests to detect affected cases and to ensure that the disability was limited to reading and spelling. Children were diagnosed as being reading disabled if they had a full-scale IQ greater than 90 and a reading level at least 2 years below expected grade level. Diagnosis of adults was based upon self-reports if there was a discrepancy between test results and a history of reading disability. Eighty-four individuals in nine extended families were tested. It is of interest to note that although all 9 probands were male, the gender ratio of their 41 affected relatives was 1.16 males to one female.

Smith et al. (1983) subjected pedigree data for 21 routine genotyping markers, as well as chromosomal heteromorphisms, to linkage analysis. Co-transmission of reading disability and a heteromorphism on chromosome 15 yielded a LOD score of 3.24. (LOD is an abbreviation for logarithm to the base 10 of the odds, where the "odds" is a ratio of the probability of linkage given the pedigree to the probability of no linkage.) However, about 70 percent of the LOD score was accounted for by transmission in only one family, and another family had a large negative score indicative of nonlinkage. Although a LOD score above 3.0 (equivalent to a ratio of 1,000 to 1 in favor of linkage) was considered sufficient to establish linkage, the authors noted that confirmation by a second study will be required before linkage can be accepted with confidence.

Subsequently, this group has extended its findings by adding to the original sample of families with apparent autosomal dominant transmission and by testing for genetic heterogeneity. The augmented sample contains 250 individuals in 21 informative kindreds (Smith, Pennington, Kimberling, and Ing, 1990). In addition to the traditional genotyping markers and chromosomal heteromorphisms, family members are also being typed for DNA restriction fragment length polymorphisms. With the addition of data from 12 new kindreds, the maximum LOD score for linkage to chromosome 15 was reduced to 1.328 (Smith, personal communication). In this larger sample, only about 20 percent of the families of probands manifest apparent linkage to chromosome 15. Thus, reading disability in the majority of these families must have some other genetic and/or environmental etiology. If it is genetic, it could be due to the effect of a major gene at another chromosomal location or it could be due to the combined effects of two or more genes.

Bisgaard, Eiberg, Moller, Niebuhr, and Mohr (1987) recently attempted to replicate the evidence for linkage of a major gene for reading disability to chromosome 15. Only five nuclear families were included in this small Danish study, and the diagnosis of dyslexia was based upon questionnaire data rather than test results. A negative LOD score for linkage between dyslexia and chromosome 15 heteromorphisms was obtained, thereby confirming the recent evidence of Smith et al. (1990) that a majority of families of children with reading disability do not manifest linkage to chromosome 15.

Current studies by Smith (personal communication) are also employing three markers for loci on chromosome 6 to test for other possible linkages. To date, results of these very recent studies suggest that some families manifest linkage of reading disability to chromosome 15 but not 6, whereas a greater proportion show linkage to chromosome 6 but not 15. In addition to increasing her sample size and the number of informative markers on chromosomes 15 and 6, Smith is beginning to employ the method of interval mapping to search for quantitative trait loci (Lander and Botstein, 1989) that may influence reading disability. With this approach, the relative contributions of several loci to the reading deficit of probands can be simultaneously assessed.

Concluding Remarks

Results of twin and family studies reviewed in this chapter suggest that the etiology of reading disability may differ in males and females. Although linkage analyses have yet to address this specific issue, recent evidence obtained by Smith and her colleagues indicates that reading disability is etiologically heterogeneous. Future linkage analyses of data from fraternal twin pairs tested in the Colorado Reading Project will assess the extent to which the reading-performance deficit of probands is due to quantitative trait loci. In this manner, the methods of molecular genetics, medical genetics, and quantitative genetics will be combined to attain a better understanding of the cause or causes of reading disability in both genders.

Acknowledgments

This work was supported in part by a program project grant from NICHD (HD-11681), and the report was prepared while J. Gillis was supported by NIMH training grant MH-16880. The invaluable contributions of staff members of the many Colorado school districts and of the families who participated in this study are gratefully acknowledged. We also thank Robin P. Corley for aid with the statistical analyses and the late Rebecca G. Miles for expert editorial assistance.

11

Neurological Arguments for a Joint Developmental Dysphasia-Dyslexia Syndrome

Charles Njiokiktjien

This survey deals with several nosological aspects of developmental dysphasia and dyslexia (some of which are discussed in Njiokiktjien, 1990) that are relevant to the argument that both these disorders have common biological causes but different clinical manifestations, depending on age. The pathophysiology of developmental dysphasia is complex and age related. In the preverbal and early verbal stage, the severity of the clinical picture of abnormal speech production is primarily determined by concomitant motor pathology (motor dysfunction, dysarthria, general and oral dyspraxia) and by receptive pathology (hearing, auditory perception and processing). During the toddler years, linguistic problems start to play a role (word-finding, agrammatism, verbal memory, auditory discrimination, and verbal comprehension), often in combination with oral motor symptoms. The various language syndromes become clear some time later. After the kindergarten stage, the oral motor and perceptual problems decrease while the language disorders continue to influence the child's conversation, inner speech, and learning at school. In a relatively small number of children there can be a basic syndrome of "pure dysphasia" without any other neurological signs. In approximately half the children this basic syndrome is accompanied by other neurological signs, most of which are indicative of functional disorders of the left hemisphere. There can also be functional disorders of the right hemisphere, the corpus callosum, and the afferent pathway systems for auditory perception.

The nature and causes of these anomalies can be multifarious, so that it is unfeasible to speak of *the* substrate, *the* pathogenesis, or *the* etiology. Nearly all the children in our study with developmental dysphasia have some form of dyslexia in later years, and most of the dyslexic children were found to have been dysphasic in early childhood. The developmental dyslexias, especially the auditory-dysphonemic, and linguistic subtypes seem to be superimposed on and the result of dysphasia. Although developmental dysphasia seems the premorbid condition of dyslexia, it is under dispute which characteristics of dysphasia determine the reading disorders of dyslexia. Dyslexia not only seems to be the result of linguistic deficits *per se* but might also be the consequence of an altered neural substrate with a reduced synaptic space for the reading process induced by dysphasia. The view that dysphasia is somehow causal to dyslexia, however, is invalid when dyslexia is considered only as a pure visual-auditory (cross-modal) disconnection or as a disturbance in phonological awareness that hinders word-recognition in a very early stage.

Behavioral Neurology

The term *behavioral neurology* refers to the field of neurology that focuses on the relation between behavior (disorders) and neurological functions and substrates. Pediatric behavioral neurology brings this new field into the clinical context in the treatment of children with disorders in learning, social behavior, contact, attention, speech, and praxis. Since it is concerned with the development of higher brain functions, this branch of neurology is strongly oriented toward developmental neurology and neuropsychology (Njiokiktjien, 1988). Along with theoretical knowledge, clinical thinking has always been a mainstay of decisions on appropriate action. Changing insights make the continuing reformulation of the framework for clinical thinking a great challenge. This is particularly the case in behavioral neurology, where clinical thinking, until recently, was more often supported by hypotheses than by verifiable facts.

To understand the place that behavioral neurology occupies within the field of pediatric neurology, one has only to glance at the contents of a pediatric neurology textbook (see Table 11.1). The disorders

Table 11.1. Disorders often seen in pediatric neurology clinics

Disorders and diseases leading to behavioral neurology problems
Vertigo and migraine (−)
Neuromuscular diseases: sometimes with cognitive disorders (+)
CNS malformations, agenesis of corpus callosum, etc. (+++)
Cerebrovascular diseases: Moya-Moya disease, etc. (++)
Metabolic diseases leading to regression (+++)
Chromosome anomalies, such as Down syndrome or fragile-X (+++)
Heredodegenerative diseases, such as Huntington's disease (+++)
Demyelinating diseases, the leucodystrophies (+++)
Intracranial tumors and effects of treatment (++)
Extrapyramidal motor disorders, such as Tourette disorder (++)
Postnatal head injuries with concussion (++)
CNS infections, as with hydrocephalus (++)
Cerebral palsies: major and minor motor handicaps (+++)
Spinal dysrhaphism with CNS involvement, such as hydrocephalus (+++)
The epilepsies, especially partial epilepsy (++)
Perinatal disorders: premature birth, asphyxia, etc. (+++)
Specific and nonspecific behavioral developmental disorders
Spectrum of unclassified psychomotor retardation, including not well delineated cognition problems
Developmental dysphasias
Developmental dyslexias and dyscalculias
Developmental attention deficit and memory disorders (AD-HD, developmental amnesias, chronagnosia)
Developmental dyspraxias (ideomotor, ideational, constructive, gestural, graphomotor)
Perception disorders and agnosias (visual, auditory)
Paroxysmal behavior disorders
Autistic spectrum disorders and psychoses
Nonverbal communication disorders (prosopagnosia, dysprosodia, some dyspraxias)
Atypical pervasive developmental disorders (mixed group)

Note: Items in the upper column are listed by estimated increasing frequency from above to below; the diseases and disorders may lead to mild (+), moderate (++), or severe (+++) behavioral neurology problems.

mentioned in the lower column in the table sometimes occur separately, but in many children they occur in syndromic combinations, which makes it difficult to select homogeneous subgroups for research. The developmental dysphasias and dyslexias are among the most frequently seen disorders in the behavioral neurology clinic. Although the clinical picture is age dependent, these disorders belong to the so-called chronic and sometimes pervasive childhood disorders. Developmental dysphasia and dyslexia can also occur after early-postnatal insults to the CNS. Sometimes they come into being during a regression caused by heredodegenerative or other CNS diseases (upper column of table), in which case they are called symptomatic learning disorders. When they are developmental (or congenital) in nature they are called primary learning disorders. In other words, the underlying neurological cause of primary learning disorders is already present before or around birth. Reading and spelling problems caused by low intelligence, attention deficits, lack of motivation, and psychological problems are called secondary learning problems. By definition, developmental dysphasia and dyslexia are not symptomatic nor are they secondary speech and reading problems.

Our approach to the problem of connected speech and reading problems is both theoretical and clinical. It is argued that many of the functional and neuropsychological characteristics of dyslexia appear to be characteristics of dysphasia (see also Tallal and Katz, 1989; Chapter 9, this volume) and that both disorders have common neurobiological characteristics (see Table 11.2). We refer also to data collected from dysphasic and dyslexic children. Although it is suggested here that one might speak of a developmental dysphasia-dyslexia syndrome, we do not advocate a new terminology or a change in the terms *verbal* or *linguistic dyslexia.*

If one approaches dyslexia from a neurological, nosological viewpoint, it becomes clear that a child's developmental history plays an enormous role, especially in his or her cognitive competency. This nosological approach and its consequences pertains not only to dyslexia; they have a great influence on the outcome of a number of other developmental conditions. For instance, if a newborn suffers from asphyxia, it is a decisive factor for the outcome whether that child is a term neonate, a premature baby, or a very-small-for-date baby, because the vulnerability of the brain as a whole as well as

Table 11.2. Comparison of developmental dysphasias and developmental dyslexias

Disorders of developmental dysphasia	Frequency	Same disorders found in subtype of developmental dyslexia	Frequency
Sequencing in words or stories	++	Linguistic dyslexia	++
Syntax	++	Linguistic dyslexia	++
Morphology	++	Linguistic dyslexia	++
Word-finding	+++	Linguistic dyslexia	+++
Verbal short-term memory	+++	Linguistic dyslexia	+++
Low verbal IQ, higher performance IQ	++	Linguistic dyslexia	++
Language comprehension	++	Inner-language dyslexia	+++
Phonology (speech)	++	Auditory or dysphonemic dyslexia	+
Phonology (auditory processing)	++	Auditory or dysphonemic dyslexia	+++
Neurological signs of left hemisphere	++	Verbal and dysphonemic dyslexia	+
Abnormal hemisphere asymmetry	++	Verbal and dysphonemic dyslexia	++
Cortical migration disorders	++	Verbal and dysphonemic dyslexia	++
Visual recognition	–	Dyseidetic or visual dyslexia	+++
Cross-modal problems	–	Cross-modal dyslexia	+++

Note: Frequency is classified as never (–) seen, sometimes (+), many times (++), or nearly always (+++) seen. It is presumed that developmental dysphasia precedes dyslexia. Consequently, some features of dysphasia have already disappeared by the time dyslexia emerges.

that of various separate brain areas is not the same in these children. Leviton (1987) warns against single-cause attribution in developmental disorders. Developmental dysphasia seems the premorbid condition of most children with developmental dyslexia and will therefore be clarified in the next section.

What Is Developmental Dysphasia?

Although disorders in language development are frequently diagnosed, the concept of developmental dysphasia is relatively unknown. In adults and children, we speak of aphasia or childhood aphasia in cases where cerebral damage has been responsible for speech or language-comprehension disorders and where the neuromuscular system is otherwise still intact (Van Dongen et al., 1985). Developmental dysphasia is a neurodevelopmental disorder that can even begin prenatally (Njiokiktjien, 1988, 1990). In such cases development in expressive speech skills lag, sometimes considerably, behind those of language comprehension. There is also a discrepancy between language competence and the rest of the cognitive development, which can sometimes be rather highly developed. Most, but not all, children with developmental dysphasia do show a discrepancy in their verbal and performance IQ scores, especially those with language-comprehension deficits. When referring to young children, we must be careful about using the term *language,* since it refers to a mature cognitive system of concepts not yet mastered in childhood. Moreover, in the case of the very young child it is frequently still uncertain whether there is indeed any language disorder involved. Some authors refer to "developmental dysphasia" (Rapin and Allen, 1986; Woods, 1985; Wyke, 1978), some to "developmental speech disorders" (Ingram, 1971), "developmental language disorders" (Aram and Nation, 1975; Ludlow, 1980; Rapin and Allen, 1982), "specific or developmental language impairment or delay" (Bishop and Edmundson, 1987; Wolfus et al., 1980), and others to "enfants dysphasiques" and "troubles du langage" (Aimard, 1982; Bruner, 1983; De Ajuriaguerra et al., 1976), "Störungen der Sprachentwicklung" and "Entwicklungsdysphasie" (Grohnfeldt, 1986; Holtz, 1987). In the older literature, one also comes across the terms "developmental aphasia" or "dyslogia" (Benton, 1964; Eisenson, 1968) or "primary childhood aphasias" (Cohen et al., 1976). There are also authors who refer to "retarded language development" and "delayed speech" (Hall, 1989) but we take "language retardation" to mean a lag whereby the function manifests itself expressively and receptively much as it would in a younger child. Many children with developmental dysphasia, however, exhibit features never displayed by younger children. In the Netherlands, the term "dysphatische ontwikkeling" (which literally means dysphasic development) was first used by the child psychiatrist X. S. T. Than, who is the founder of the Developmental Dysphasia Foundation (1989).

In a very young child, it is impossible to separate affective development and its expression from motor development and speech. The age-related changes and the relations between affective, motor, speech, and language development are thought to constitute a neural change and lateralization process that can be disturbed at various areas in the brain. This disturbance can occur prenatally, perinatally, or early postnatally; consequently, developmental dysphasia is not confined to any single pathogenetic concept.

The clinical core symptom of developmental dysphasia

The core symptom is a disturbance in the transition from the observed object, the memory, the feeling, the thought, or the idea to the spoken word, to story-telling, and to verbal expression in general, but also to inner speech. Word-finding or fluency problems are core symptoms and so striking that one tends to view the entire clinical picture of developmental dysphasia as a verbal memory-retrieval disorder. The skills required for speech—articulation, sentence construction, verbal short-term memory, auditory perception, and language comprehension—can exhibit great variation and can influence the entire picture. The combination of these symptoms along with the core symptom is what determines the linguistic subtyping of the developmental dysphasias.

Hearing

The notion that abnormal speech development is generally caused by poor hearing is erroneous. In the older child a loss of more than 40 dB does, however, certainly lead to a speech distortion. In younger children at the stage of acquiring speech, an even slighter loss of 20 to 40 dB, especially high-tone losses, can lead to slow and deficient speech, but not to actual linguistic and memory disorders. Temporary conduction losses do not have any permanent effect on the acquisition of language. It happens even more often that there are central auditory perception and discrimination disorders. On this topic the reader is referred to the discussion between Gordon and Bishop in 1988.

Auditory memory problems

A substantial number of dysphasic children have poor short-term memory skills. They have an extremely poor verbal auditory retention and reproduction skill, which manifests itself when they repeat something they have heard. This phenomenon causes severe problems when the people around them use sentences that are too long, and it later causes difficulties in reading long sentences as well. The short-term memory problems are confined to a problem with phonological memory preventing the acquisition of a vocabulary (Gathercole and Baddeley, 1989). Although some people ascribe the memory problems to a disturbed phonological processing (and too-rapid speech), it is our impression that many children have not only true verbal imprinting difficulties but also nonverbal auditory imprinting and reproduction difficulties (see below). The memory-retrieval mechanism is also frequently disturbed in the form of word-finding disorders (Murphy et al., 1988). In repetition, spontaneous speech, and reading, these disorders can cause problems for the child. In addition to the cortical sites in the Wernicke area (Selnes et al., 1985), the limbic and diencephalic memory circuits might also be involved in these disorders.

Oral motor problems

Oral motor coordination is frequently clinically normal even among children who make phonological pronunciation errors, a symptom named "dyslalia." Other children exhibit evidence of oral (or buccofacial) dyspraxia, which makes speech more complicated. The only way oral dyspraxia and dyslalia due to perception disorders can be distinguished from each other is by a praxis examination of the oral region, and they should both be differentiated from the dysarthrias. All these motor problems are nonlinguistic disorders.

Some children have no general oral dyspraxia but exhibit apraxia for word use only. This is "apraxia of speech" or "verbal dyspraxia." These children are nearly mute, though they can learn to read and they have good language comprehension (Rapin and Allen, 1988). It might be questioned whether it is always possible to discriminate between verbal dyspraxia and extreme word-finding problems in these children. Verbal dyspraxia, however, appears to us to be much more rare.

Auditory perception and speech production

Infants are already capable of distinguishing among various categories of speech sounds (Eimas, 1985; Hillebrandt, 1983; Morse, 1972). On the grounds of prosodic differences, they can distinguish between speech in their home language and in a foreign language (Mehler et al., 1988). In young infants, the left hemisphere reacts more than the right one to speech sounds (Molfese et al., 1975; Entus, 1977) and by age five or even earlier the right hemisphere is better in detecting nonverbal sounds (Knox and Kimura, 1970). Dysphasic children, on the contrary, have problems with phoneme identification. In young dysphasic children, there is often also dyslalia and their speech, particularly perceptually based speech is unclear (Tallal, 1980, 1981). At school age such children have clear speech, but they still have difficulties with auditory discrimination and lack phonological awareness; these difficulties are the basis of so-called dysphonemic dyslexia (Bradley and Bryant, 1983; Shankweiler and Liberman, 1989).

A number of authors have erroneously cited the auditory perception problem as the cause of developmental dysphasia (Benton, 1964; Eisenson, 1968; Tallal et al., 1980). They held that abnormal language development was due to deficient speech perception. Many children with abnormal speech development do indeed have more trouble than normal children discriminating a rapid succession of speech sounds and thus have a perception disorder on the phoneme level (Tallal et al., 1980). However, if the speech sounds are presented more slowly, the perceptual problems are reduced, although the expressive element continues to exist. The perception problems are certainly not the cause of linguistic problems, although they do partly account for the articulation problems these children have. Moreover, perceptual deficits are not characteristic for all dysphasic children;

they are also dependent on age (Frunkin and Rapin, 1980). Other authors have similarly failed to confirm the claim that linguistic disorders are caused by auditory perception disorders (Wolfus et al., 1980; Hecox, 1988). In conclusion, auditory processing disorders and the resulting lack of phonological awareness seem important for early speech quality and for the connection of written language to speech at a later age.

Dysgrammatism

In cases of developmental dysphasia, babbling is sometimes absent in the baby and by the end of the second year the young child has still not begun to use words. The older child usually does not speak much, and when he or she does it is with short sentences in telegraphic style. Story-telling takes place chaotically and is accompanied by recurrent hesitation and word-finding problems, which occur more frequently in dialogue than in spontaneous speech. Language usage is frequently characterized by dysgrammatism: morphological errors, pertaining for example to declensions or connecting words; syntax errors; and paraphasic substitutions, replacing a word or part of a word by one that resembles it in the same semantic category. There are also phonologically based substitutions. Words and parts of words are put in each other's place. Speech is not fluent and is sometimes accompanied or replaced by excessive gestures. The quality of speech also depends on the general cognitive level. Although they frequently occur together, all these expressive linguistic characteristics have to be distinguished from the clearness and the speed of speech.

Classification and terminology

Cerebral functioning changes with age, and this is what makes the picture presented by developmental dysphasia at a later age quite different from that of early dysphasia. In general, a classification is used that distinguishes between the expressive groups with or without oral motor problems, the mixed expressive-receptive group, and the markedly receptive group with poor language comprehension. Within this main classification, different authors use different terms for the various types and subtypes (Aram and Nation, 1975; Ingram,

1971; Rapin and Allen, 1986; Wolfus, 1980). If a child does not understand what is said to him, then what he says in response is often not appropriate for the particular situation. His speech is not communicative and can be accompanied by echolalia and repetitions of entire sentences, be it with clear articulation and rather good syntax. This relatively unusual "semantic pragmatic syndrome" (cf. Rapin), however, differs clinically and pathogenetically from the common expressive or mixed expressive-receptive syndromes. As will be noted below, children with this syndrome are likely to be diagnosed as autistic, some of whom are indeed, some not.

The problems in terminology pertain to the professional background of the person using the terms and the concept he or she has of speech development (see also Bishop and Edmundson, 1987). In scientific literature, the further the professional is from the clinical nosology, the greater the tendency is to describe "language disorders" in children as isolated linguistic syndromes. This is even more true for the dyslexias. Dysphasic children are frequently the patients of pediatricians and child neurologists, because their speech problems often suggest mental retardation. With these children, apart from their speech and motor problems, nothing treatable is found and there is no follow-up during the school-age years. Dyslexic children, on the other hand, are more often seen by (school) psychologists than by physicians. The lack of a nosological approach by nonphysicians prevents them from taking medical histories. This means that an important developmental disorder that precedes dyslexia (developmental dysphasia) is missed, although many people say that the majority of dyslexic children exhibit associated defects in speech and language development (Lovett et al., 1989; Tallal and Katz, 1989; Vellutino, 1987) or even that delayed language development may be causative in some cases of dyslexia (Warrington, 1967; Pennington, 1990).

Although we are apt to distinguish among a number of speech or language syndromes (the developmental dysphasias), we do confine them to a developmental neurological and pathogenetic framework. Within this framework there are various neurological symptoms, one of which is dyspraxia. The etiological dimension plays an explanatory role and covers such possibilities as cerebral damage or malformation; prenatal developmental disorders on the cellular level, such as axonal retraction and migration anomalies; genetic and chromosomal aberrations; and metabolic abnormalities. A nosology cannot be based only upon linguistic criteria if one wishes to arrive at a pathogenetic concept of abnormal speech and language development, including reading. The age dimension should also be taken into consideration, as should the total affective, neurological, and neuropsychological functioning of the child and the genetic background. The classification that would seem to bear the greatest relevance to the field of clinical neurology to date is the one drawn up by Rapin and Allen (1986, 1988).

What Is Developmental Dyslexia?

The term *dyslexia* comes from the notion, more than a century old, of "word blindness," a notion that is commonly viewed, as by Critchley (1964), as being obsolete. Critchley has given excellent descriptions of virtually all aspects of (specific) developmental dyslexia. *Developmental dyslexia* or *dyslexia* is used to refer to difficulties in written language (reading and writing) and is indicative of a neurological cause. We do not consider dyslexia one nosological entity, and this is why we prefer to speak of dyslexias or, in other words, of the various subtypes of dyslexia caused by a variety of neuropsychological defects. The combination of these defects with other symptoms accounts for the fact that dyslexia is found in syndromes with a variety of localizations. It is inaccurate to use the term *dyslexia* for all the reading and writing problems that children can have; indeed, spelling problems, collectively referred to as dysorthography, are much more common than severe reading deficits. Not everyone who reads and spells poorly or slowly has dyslexia. The German term *Legasthenie* does include all the groups of poor readers and spellers and is usually used in a broad sense. The distinction is rightly drawn between the reading delay of slow or backward readers (as determined by comparison to the expected level for their mental age) and dyslexia as a disorder that cannot be attributed to any general delay or to factors that cause secondary learning deficits. The terms *specific dyslexia, developmental dyslexia,* or just *dyslexia* refer to a developmental disorder pertaining to the use of written language. In practice, there is no clear border between slight dyslexia and a reading delay, just as it is sometimes difficult to distinguish between developmental dysphasia and a speech-development delay. In literature, a distinction is sometimes made between primary and symptomatic reading deficits.

The Neuropsychological Pathogenesis of Dyslexia in the Process of Learning to Read

When a normal child starts learning to read, she already speaks her "native tongue" and has been familiar with auditory language through speech for years. Since our system of writing is a phonology-based code derived from speech, it is obvious that children whose speech development is already impaired will also have difficulty in reading and writing.

Learning to read has been recognized by Utah Frith (1986) as a three-stage process. The logographic, the alphabetic, and the orthographic strategies are used by the child, in that order. Early reading requires mechanisms other than those for experienced reading. First, the child has to learn to recognize, retain, and remember the shapes of the letters (or phonemes or short words) visually and as a Gestalt or logographically, like somebody seeing Chinese characters for the first time. At this stage some children do not see precisely what is written. For them some words resemble each other as Gestalts; in the deaf child this process is very obvious. The hearing child then has to link these words to familiar sounds or spoken words, a process referred to as grapheme-phoneme linking. For the most part, this is probably a bihemispheric or right-hemispheric process, as is shown by electrophysiological evidence (Licht et al., 1988). Remembering or recognizing what is written is a visual-perceptual and cross-modal memory process, which cannot function if there is a visual impairment (rotations, inversions, inaccuracy, poor Gestalt recognition of small words). This is termed *visual* or *dyseidetic dyslexia* (Myklebust, 1978; Boder, 1973). It concerns perception of the form of the letter, Gestalt perception, and the visual memory of the word, and it occurs much less frequently than the other subtypes of dyslexia (Mattis et al., 1975). The visuo-verbal agnosia aspect (the nonrecognition of the word as a word) is probably more important than the perceptual inaccuracy. A variant of the visual-perception type seems to be the inversion or reversal of letters or words (Orton's strephosymbolia), which is physiological in young readers and which can, in isolation, occur in a severe form. Vellutino (1987) does not agree with the theory that strephosymbolia is a visuo-spatial disorder and argues that the problem is confined to language. Possibly strephosymbolia represents a disturbance in the somatic and extracorporal laterality awareness and is neither purely visual nor linguistic.

The next stage of technical reading is the alphabetic one. The child recognizes phonemes and letters (if he learns the alphabet) as sounds. In this stage the order is very important for the meaning of the word (think of *dog* and *god*). This stage of grapheme-phoneme linking goes wrong, in many instances, because of a lack of phonological awareness, which is called *auditory or dysphonemic dyslexia.* This dyslexia type (the lack of single-word and nonword reading as a phonological processing skill) seems also to load highly in heritability studies (Pennington, 1990). In the third stage of beginning to read the orthographic strategy is used. The child no longer sounds out phonemes but reads whole words. Some children master this stage of reading very early without comprehending what is written. They are called hyperlexic.

Grapheme-phoneme linking (the alphabetic stage) is a cross-modal association and memory-retrieval process. This may be sound oriented in normal children in the beginning, but finally many children read visually without "sounding out." Reading is also the linking between two symbol systems in different modalities, for which parieto-temporal involvement of the left angular gyrus is required. A lesion in this area causes alexia and agraphia in advanced readers. The limbic memory areas for imprinting and retrieval, which are impaired in some children, are also required. In many children, impairment in the memory systems can be most dominant. Research shows that, for example in dyscalculia, these deficits are by far the most dominant (Shalev et al., 1988). In children with impaired reading, short-term memory deficits appear again to be important, especially where visual presentation is concerned (Wood et al., 1989). See also the description of "auditory memory disorders" in the discussion above of developmental dysphasia.

Grapheme-phoneme linking can go wrong in two ways. The first possibility is that it does not establish itself as a result of deficiencies in the association area (left angular gyrus) or in the tracts leading to them. This problem is referred to as a pure visual-auditory disconnection and this type of dyslexia is referred to as *cross-modal dyslexia* (Myklebust, 1978). Since Birch and Belmont (1964) first described this type, not many articles have been published about it. Although cross-modal (intermodal) memory deficits seem to be of less importance than the intramodal memory deficits (Wood et al., 1989) that occur in most children, this pathogenetic aspect of dyslexia deserves more attention.

Cross-modal dyslexia has to be distinguished from the second possibility of incorrect grapheme-phoneme linking, which is *dysphonemic* or *auditory dyslexia*. In this type, although a transmodal connection is established, the auditory process is impaired (also a problem in the very young dysphasic child!) and the child has a problem pronouncing the words. Children with normal speech acquisition need formal instruction to learn the phonological structure of words (auditory analysis). This talent is still unknown to the young reader. The children who suffer from this type of dyslexia are also those who have auditory processing problems in early childhood, allegedly and logically by a cortical temporal dysfunction, and they should be distinguished from those children whose hearing impairment is severe. Deficient phonological perceptual processing, especially in the case that speech stimuli are offered with the normal speed, occurs often in children with a developmental dysphasia (Frumkin and Rapin, 1980, Tallal et al., 1980, 1981). These children also have disordered phonological speech production, but this problem disappears earlier in development than the auditory problems. Consequently, speech production disorders do not always relate to phonological processing deficits, though difficulties in the temporal control necessary for rapid speech production are still present at a later age. The lack of rapid speech production is also a motor problem that may be pathophysiologically analogous to deficient hand motor function—for example, bimanual supination/pronation, as mentioned below.

At school age the lack of phonological awareness (the understanding that words can be broken down into their constituent sounds) is a cause of spelling and reading problems. Some authors attribute a dominant role to the lack of phonological awareness in the pathogenesis of dyslexia. Deaf children also have great trouble in learning to read for this very reason (Bradley and Bryant, 1983; Shankweiler and Liberman, 1983; Lundberg, 1989). The importance of normal phonological processing for learning to read has been illustrated by studies showing that successful deaf readers use nonauditory phonological strategies, such as the use of oral movements (Hanson, 1989). A disturbed phonological processing influences not only speech production but also speech comprehension (Crain, 1989), and verbal short-term memory suffers as well (Shankweiler and Liberman, 1989). In the light of these studies, one can arrive at the opinion that developmental dysphasia and dyslexia exist under the common causal denominator of disturbed phonological processing. However, not everyone shares this opinion.

An alternative explanation is that signs of developmental dysphasia (dyslexia) occur simultaneously because they stem from a broader adjacent neural substrate, the left parieto-temporal area, which has been identified by aphasiological and morphometry studies. For instance, the right hemisphere does not seem to have the ability to translate orthography into sound (Zaidel and Peters, 1981). Consequently, a dysfunction of the left hemisphere will inhibit grapheme-phoneme linking. If the left hemisphere is dysfunctional, all the symptoms will occur more or less simultaneously without a mutual causal relation (e.g., word-finding and syntactic problems, unclear speech, verbal memory problems, and abnormal auditory processing). This of course does not exclude mutual influencing. In this model, developmental dysphasia and simultaneously disturbed phonological processing are precursors of dyslexia. It is thus claimed that the underlying disturbance in phonological or temporal processing can occur in both developmental disorders, as shown by Tallal and Katz (1989).

There are children (including those whose language comprehension and memory are impaired) who are normal and sometimes even extremely talented or *hyperlexic* as regards reading and writing. As toddlers, hyperlexics are already good at the technique of reading, though they do not necessarily comprehend what is read. This phenomenon may also occur in mentally retarded and autistic children (Cossu and Marshall, 1986; Smith and Bryson, 1988). Furthermore, there are severely dysphasic, nonspeaking children with normal language comprehension who are hyperlexic, and there are also deaf children with normal intelligence who are hyperlexic. All these children are able to carry out written instructions (Elliott and Needleman, 1976). With respect to language comprehension there are two types of hyperlexic children, those who are able to ascribe meaning to what they read and those who are unable to understand, or who understand very little of, what they read. If one considers dyslexia as a language disorder, then the latter type of hyperlexics are dyslexic. Myklebust calls this *inner language dyslexia,* and it occurs also of course in children who are not hyperlexic.

Boder (1971, 1973) distinguishes the types dyseidetic and dysphonemic dyslexia (and combined types). Electrophysiological studies by Flynn and Deering (1989) do confirm the existence of these subtypes but, as far as dyseidetic dyslexia is concerned, the results are more indicative of a hyperfunction of the left hemisphere than the dysfunction of the right hemisphere that is frequently assumed. Flynn and Deering's interpretation (1989) of the result is that a hyperfunction of the left hemisphere (too quick a pace in auditory reading) causes hypofunction of the right hemisphere, which results in visual reading impairments. Flynn and Deering's findings (1989) are also corroborated by Bakker (1987). Children with Bakker's L-type dyslexia read rapidly but with visual inaccuracy and are thought to shift to the left-hemisphere process too quickly. Despite the fact that this process, termed linguistic-type dyslexia, is named after the linguistic strategy that these children use, the errors are perceptual in nature and, as far as terminology is concerned, Bakker (1987) deviates from the classification of dyseidetic (visual-perceptual) and linguistic dyslexia (in terms of language disorders). Boder's classification (1973), which is usually applied, is based on the defect and not on the reading strategy.

Finally, while the process of learning to read is advancing, a connection is established with the semantic network (reading comprehension), which is predominantly a left-hemisphere process. This is why alexia in adult readers occurs almost exclusively when the left hemisphere is damaged. The shift to the left hemisphere for advanced reading by normal subjects, which has been demonstrated by electrophysiological studies, can sometimes be impaired during development. This is usually the case in children with abnormal development of spoken language or a developmental dysphasia. This type of dyslexia (termed *verbal* or *linguistic dyslexia,* Boder, 1971, 1973), which comprises more than three-quarters of the dyslexics, is often combined with the dysphonemic dyslexia type. The shift to advanced reading can go wrong in two ways. The technique of reading may be very good (no dyseidetic and dysphonemic dyslexia), but there is a lack of comprehension of the language (which may indicate an impairment in Wernicke's area in the left temporal cortex). The child does not comprehend the text and, for that very reason, can repeat it only poorly (inner language dyslexia). Second (and in the majority of cases), there are children (usually with developmental dysphasia) who do comprehend the spoken language well, or at least reasonably well, but whose speech is abnormal or at least not fluent. These children have a language-production and fluency problem, which can be explained, for instance, by impaired memory-retrieval mechanisms (word-finding and naming defects) or by a speech motor problem. In dyslexics, word-retrieval deficits are of great importance and they seem to be mainly restricted to oral language functioning (Murphy et al., 1988; Wolf, 1986; Wolf and Goodglass, 1986). Word-finding defects often accompany errors, which are either semantic (paraphasias) or phonological in nature (Katz, 1986). It is probable that there is often a problem in the "syntax generator" (which is likely to include mainly Broca's and/or Wernicke's area) as well as dysgrammatism. The children with this problem are those who are not able to reproduce what they have read as a result of disturbed fluency and dysgrammatism, which, incidentally, can also result in language comprehension defects.

The shift to advanced reading can also go wrong in the process of imprinting the meaning of a sentence, since the child (reading too slowly) is unable to remember enough of what she has read. In a child with verbal dyslexia, verbal imprinting defects almost always occur and are persistent until adolescence, after which they decrease. Cross-modal imprinting-retrieval deficits (visual-verbal and verbal-visual) tend to occur in young children with verbal dyslexia and tend to disappear by the age of 12 (Lindgren et al., 1984). An alternative explanation for short-term memory disorders is an overloaded working memory during deficient phonological processing (Shankweiler and Liberman, 1990). Many dyslexics also have short-term memory or imprinting problems for nonverbal auditory stimuli, as can be shown by Stambak's rhythm-reproduction test (personal observations). This means that there could be general, not only linguistic, auditory short-term memory deficits. There is evidence, however, that serial-order deficits in dyslexics are restricted to the auditory domain (Gould and Glencross, 1990).

Table 11.2 summarizes the common characteristics of developmental dysphasia and dyslexia. So far in this discussion, the steps the child follows in the process of learning to read have been presented in a neat order. It is unlikely, however, that every child will follow this order. Some children, for example, connect script directly to the semantic network without necessarily making the grapheme-phoneme linking aloud. They master the visual language rapidly and do not need to hear or say the words. This is also the case with some deaf people and with the hyperlexics described above. This phenomenon has two consequences: because the technique of reading and reading comprehension are two completely different processes, at least two types of dyslexia can be expected; and since a normal semantic network can exist, even though the child cannot speak (intact passive language comprehension), access to and output from this network can be obtained through reading and writing. In some children, visual reading is predominant and problems occur in reading aloud (called "re-auditorization"). These are the children with a spatial cognitive strategy who think in pictures; they belong to the high-performance and lower-verbal IQ group and seem to have a strong bihemispheric representation of their spatial competencies (Witelson, 1977). Writing and script are not always a visual realization of speech, and reading is not always necessarily "re-auditorization." This has consequences for the treatment of reading problems (see below, and Söderbergh, 1984).

As far as a well-defined typology is concerned, the conclusions so far are disappointing: we often see combined types of dyslexia, and the number of possibilities of how and where things can go wrong is so large. Dyslexia is a continuum of disorders of which a few main points can be identified: memory aspects, auditory and visual perception, and use and comprehension of language. In addition, there are often other neurodevelopmental disorders, such as dyspraxia and attention deficits, which can already be found in toddlers with developmental dysphasia.

When reading is seen as information processing, it becomes obvious that divided attention is an important factor (the process of the technique of reading and remembering and understanding the contents), and it is one which is already difficult for children (De Sonneville and Njiokiktjien, 1988). When the technique of reading has become an automatic process, the child can concentrate better on reading comprehension and the "working memory" is not occupied with two tasks at the same time. Working-memory disorders and problems with sustaining and dividing attention, imprinting, storage, and retrieval can all play an important role in early reading and arithmetic dysfunction (Tunmer, 1989). These factors should be distinguished from the problems of, for example, language comprehension and phonological processing deficits, even though mutual influences are always a point to be considered.

Common Neurological and Brain Anomalies in Dysphasia and Dyslexia

It is important to distinguish between the pathogenesis and pathophysiology of developmental dysphasia and dyslexia (functional disorders in one or both of the cerebral hemispheres) and the etiology of these disorders (gene defects; fragile-X and other chromosome aberrations; infectious, toxic, hormonal, and other damaging influences, such as hypoxic ischemia). In this section pathogenesis is the main topic.

Hemisphere differences noted during neurological examination of dysphasic and subsequently dyslexic children

The results summarized here come from a study we made of neurological anomalies in 220 children, a third of whom had been diagnosed at the Developmental Dysphasia Foundation (Njiokiktjien, 1988, 1990). By definition, this group did not include any children who were hard of hearing (impairment more severe than 30 dB on both sides), nor did it include children who, after a thorough examination, could be classified as mentally retarded. It did include dysphasic children who, with the exception of verbal auditory agnosia, exhibited the language patterns described by Rapin and Allen (1986). The children were examined clinically and with EEGs and CT scans.

Marked hemisyndromes occurred in less than 15 percent of the cases and the symptoms indicated a predominance of left-hemisphere functional disorders (six times more common than right-hemisphere disorders). In the entire group (with the exception of the hemisyndromes), ninety-one children (41 percent) exhibited unmistakable neurological abnormalities: fifty-nine children had abnormalities related to the left hemisphere, eighteen had abnormalities related to the right hemisphere, and fourteen had abnormalities relating to both hemispheres. Neurological abnormalities were indicated by motor symptoms, EEG aberrations, and abnormalities on the CT scans. These are conservative estimates: the abnormal language picture itself, asymmetry of motor coordination in the face, oral and ideomotor dyspraxia, disturbed finger localization, and pathological left-handedness were not interpreted as definitive lateralizing symptoms. If these symptoms, with the exception of the abnormal language, are taken into consideration as neurological symptoms, then there were indeed very few children who did not exhibit any organic signs of abnormality.

Children with semantic-pragmatic syndrome, which is characterized by disturbed language comprehension and superficially intact language usage, showed strikingly few signs of hemisphere anomalies, and some of these children had no technical reading problem. For the rest, the neurological abnormalities were still not statistically related to speech and reading patterns. The overall picture showed that although there was a relatively high incidence of symptoms lateralizing to the left hemisphere, there was usually no focal failure. We conclude that the developmental dysphasias are connected to pathology in both hemispheres, although predominantly in the left hemisphere. In 1990 nearly all these children were at school age. The majority (85 percent) attended special schools and hardly no child, even among those attending normal schools, was free from reading problems.

Dysphasic children with high performance and lower verbal IQ scores

The forty-two children with "pure" dysphasia, dyslexia, and a high performance IQ constituted a special subgroup. This combination is not rare in children with relatively high visuo-spatial intelligence, who might for example become artists, or in those with a high mathematical intelligence (Patten, 1973). In our research group, the children with a discrepancy between verbal and performance IQ scores of at least thirty points and a performance IQ of 100 or higher were selected. These children come twice as frequently from families with developmental dysphasia and dyslexia as compared with the rest of the group (nearly 50 percent), and they exhibit strikingly few neurological anomalies and little left-handedness (Njiokiktjien, 1988, 1990b). This would tend to suggest the absence of brain damage and the presence of a genetic factor. Pennington (1990) mentions a familial aggregation of dyslexia and high performance in mathematics and visuo-spatial tasks. There is now some information available as to whether the children in this group also exhibit anatomical anomalies of the hemispheres, and this will be discussed in the following section.

Abnormal cerebral dominance pattern

As has been noted above, approximately half the children with developmental dysphasia do not exhibit any clear classical neurological, electroencephalographic, or neuro-imaging anomalies. In the case of these children, the explanations for their dyslexia are not related to brain damage. There are various hypotheses pertaining to the dominance of the left hemisphere and its disorders, resulting in developmental dysphasia and dyslexia. The most important genetic theory is Annett's right-shift theory (Annett, 1985), which pertains to disturbances in the genetic expression of the localization of language and manual preference in the left hemisphere and which interacts with gender differences. Annett held that the absence of the RS gene makes for a good chance of a deviation in the usual left-hemispheric lateralization pattern for language and the accompanying manual preference.

Geschwind and his colleagues approached disturbances in dominance from a different angle, one based on the knowledge that in normal adults and neonates there is a brain asymmetry in the classical speech areas (Galaburda et al., 1978). Their theories were focused on the embryological genesis of dyslexia. Starting in the thirtieth week or even earlier, it is possible to see in fetuses that the left temporal plane, which is related to language functions, is usually larger than the right one. Few brains are symmetrical (15 percent), and even fewer brains (5 percent) are strongly asymmetrical to the right (see review by Galaburda, 1988). Galaburda et al. (1979, 1985, 1988) and Cohen et al. (1989) focused on neuropathological brain symmetry data found in deceased dyslexics. When the history of the adult was known, the subject in all cases was shown to have suffered from developmental language disorders. Galaburda et al. (1985) mention a "speech delay" in two patients, and Cohen et al. (1989) describe a severely dysphasic girl. A number of radiological studies have also focused on the absence of brain asymmetry in children with developmental dyslexia (Hier et al., 1978; Haslam et al., 1981; Rosenberger and Hier, 1980, Rumsey et al., 1986). Hier et al. (1978) found an abnormal asymmetry especially in children with a low verbal IQ and "delayed speech." Rosenberger and Hier (1980) explicitly stated that abnormal symmetry concurs with "language delay." Another study (Haslam et al., 1981), however, did not confirm a relation between symmetry and low verbal IQ or language delay, while Rumsey et al., (1986) did not mention early-childhood language development at all.

At our clinic, a study was carried out that focused on hemisphere asymmetry anomalies in children with developmental dysphasia. In 1989, we studied a group of twenty-three children with developmental dysphasia using MRI scans (de Grauw and Njiokiktjien, 1989). These children did not have classic hemisyndromes, nor did they have any neurological anomalies other than dyspraxia. They were not viewed as brain-damaged and their MRI scans were evaluated by classical standards as being normal. The purpose of this study was to measure hemisphere asymmetry on various levels as specified by Kertesz et al. (1986).

According to our hypothesis, if one could assume that dyslexic children, known from the literature to have abnormal hemisphere asymmetry, were often dysphasic as well, there might be this kind of abnormal pattern in dysphasic children. Differences were indeed found to support this assumption. The patients had a larger right hemisphere in a longitudinal direction significantly more frequently than the control group (in whom no subject showed this). Width measurements illustrated that, compared with the control group, the patients' left temporal lobe tended to be no larger than their right one. The children who were at school age (in 1990) all attended a special school and all had severe reading problems. Recently we enlarged the study group and added also four clinical control groups. A preliminary analysis showed that there is no sharp border between these groups and the developmental dysphasia group regarding brain asymmetries. Only one-third or less of the children in all groups have a normal asymmetry pattern, regardless of age, handedness, and sex. Six children with normal speech development but with general learning disorders, reading included, did not differ from the developmental dysphasia group; four children with a semantic-pragmatic syndrome, who had no reading problems, had abnormal asymmetry and did not differ from the developmental dysphasia group. There were no convincing overall differences between the developmental dysphasia group and the children with mental retardation (twelve). These children also had developmental dysphasia characteristics, but they had other cognition problems and dyspraxias as well. Labeling children (for instance, "mentally retarded") according to the DSM-IV classification is done for practical reasons but does not seem to make neurobiological sense. In many children with mild, unclassified retardation, developmental dysphasia and/or dyslexia seem to us embedded in a more general syndrome with dyspraxia and cognition deficits. The question now is how one can subdivide the various clinical pictures presented on the grounds of the anatomical findings. At the moment, it is still unclear just exactly why symmetry anomalies come into being, but some hypotheses have already been forwarded.

It has been known for some time that the incidence of dyslexia is higher in boys than in girls and that dyslexic boys are often left-handed. Geschwind and Behan (1982) first noted a correlation between left-handedness, learning difficulties (dyslexia), and autoimmunity diseases in boys and in members of their families. The hypothesis was then formulated that dyslexia and left-handedness were based upon a left-hemisphere disorder that expressed itself in abnormal asymmetry and cortical migration disorders (Galaburda and Kemper 1979). Finally, reasoning from the example of thymus anomalies and immunity diseases, they proposed the notion that a hormonal factor, increased prenatal testosterone, might be the cause of both immune and cerebral dominance disorders. The correlation between dyslexia and immune disorders has been confirmed by Hugdahl (see Chapter 7), although it had earlier been disconfirmed (Urion, 1988). The correlation between handedness and immune disorders has not been confirmed (Chavance et al., 1990; McKeefer and Rich, 1990; Chapter 7). The role of testosterone, although it acts on nervous system development and hemisphere differentiation and protects against cell death in certain brain areas (see Kelley, 1991; Chapter 9), therefore, is not yet clear. However, there is known to be a relation between dyslexia and abnormal hemisphere asymmetry, and this is now being interpreted in a different way. After subjecting previous results to new calculations, Galaburda et al. (1987) concluded that the disorder might be related to an absence of physiological embryological cell death in the right hemisphere, with the result that the two hemispheres remain symmetrical. Upon close reading, the articles on dyslexics do often appear to concern children with developmental dysphasia. One and the same cerebral dominance disorder might therefore be assumed to be responsible for developmental dysphasia as well as for verbal dyslexia.

Electrophysiological studies

Electrophysiological studies have also confirmed the dysfunction of the language areas of the left hemisphere in children with developmental dysphasia (Dawson et al., 1989). Partly because of the selection problems referred to above, however, hemisphere dominance disorders of the kind that characterize developmental dysphasia have still not been convincingly illustrated in autistics (Leboyer et al., 1988). A special subgroup consists of children with acquired aphasia with verbal auditory agnosia and clinical epilepsy, the so-called Landau-Kleffner syndrome. The pathologic nature of this syndrome is not known, but developmental forms of this language syndrome with epileptic discharges, mostly on the left side, are known (Maccario et al., 1982).

From a developmental and pathophysiological viewpoint the experiments by Ojemann et al. (1989) are interesting, for they help us to understand the connection between developmental dysphasia and dyslexia. These authors have found, in electrophysiological studies, that the essential areas used for sentence reading are located in one site in the inferior posterior frontal cortex and in one or more sites in the temporo-parietal cortex. These sites, however, occupy a much wider area than that used for naming. In the nonessential areas as well, neurological activity was more widespread for reading than for naming. Zones with electrophysiological changes for both reading and naming are the Broca and Wernicke areas. In subjects with low verbal IQs, the areas essential for naming alone were larger and extended into areas such as the superior temporal cortex, which is essential for phoneme identification. As a consequence, an insufficient number of neuronal circuits are available to subserve later-acquired language skills such as reading. These observed phenomena, which are supposed to happen in certain cases of developmental dysphasia, could be explained by neuronal plasticity: if the available neural substrate has been reduced because of neuronal changes, the chances that dyslexia will follow dysphasia may increase.

Language and the corpus callosum

Sometimes one encounters symptoms that bring to mind the possibility of interhemispheric disconnection. Various authors have devoted ample attention to interhemispheric disconnection in older children with developmental dysphasia and dyslexia symptoms (Denckla, 1985; Klicpera, 1984). Most of the children we have examined with corpus callosum anomalies exhibit abnormal speech and reading development (Njiokiktjien, 1988; Njiokiktjien et al., 1990). We believe that abnormal speech perception is mainly a result of a right auditory pathway system disorder projecting to the left temporal lobe, in combination with an inadequate callosal transfer from right to left. This may cause speech perception anomalies in the left temporal lobe, and there is evidence that this is indeed the case (Njiokiktjien, 1984, 1988). In normal children and adults, there is a light asymmetry of the brainstem auditory evoked potential in favor of the right, which would tend to suggest a right ear dominance of auditory perception (Pinkerton et al., 1989). In dysphasic children, there is thought to be a dysfunction of this pathway system. In children with abnormal speech development, the corpus callosum also plays a role with respect to hand motor coordination (Amorosa, 1981). This pathway might be indirectly connected to the production of speech. It is not uncommon for bimanual coordination to be disturbed in dysphasic children (Njiokiktjien, 1983; Njiokiktjien et al., 1991). Disturbed interhemispheric connections may contribute on various levels to the emergence of developmental dysphasia, either in an isolated state or as a component of cortical anomalies. Later in development, the effect on the patient's speech is less marked: split-brain patients, whose corpus callosum has been surgically severed, speak mostly "normally."

A Joint Dysphasia-Dyslexia Syndrome

The aim of this study is to show that the developmental (longitudinal) rather than the cross-sectional approach can bring us to the conclusion that developmental dysphasia and dyslexia are in many instances two sides of the same coin. This two-sided coin is developmental in nature and shows us in a fixed order first its dysphasic and subsequently its dyslexic side. The dyslexic child, when properly examined, nearly always also reveals dysphasic characteristics. Both the nature and the degree of these characteristics varies, however, and the child has usually been able to improve his language problems either by years of treatment or simply by maturing over time. Ontogenetically and phylogenetically, the spoken language has a biological advantage, because it is based on a network in the brain that matches speaker and listener from birth onward and that has existed since the beginning of mankind. This network is highly specialized and lateralized and is probably vulnerable to pre- and perinatal adverse influences, especially in the development of white matter (Gilles et al., 1983). Reading, which commences much later than the spoken language, begins in most children at the same age. This might be too late, however, for the children who have already partly lost neuronal plasticity. This is especially true in dysphasic children, because neural space in the language areas of the left hemisphere of these children may have been occupied to compensate for early linguistic dysfunction (see Ojemann, 1989). The supposed loss of plasticity is one of the reasons to treat dysphasic children with early reading exposure. We have chosen the teaching method employed for deaf children (Söderbergh, 1984). This method may also initially bypass the lack of phonemic awareness, discussed above.

Reading, which belongs to culture, involves attempting to connect arbitrarily chosen visual symbols to the natural phonetic module in the brain. The connection of written material to an already disordered language (phonetic) network, whether the problems are related to perception, retention, processing, or production of speech, must intuitively lead to a reading problem.

The ideas presented here, concerning the common nature of the developmental disorders under study, depend upon how one conceptualizes brain function. What are thoughts, memories, recognitions, inner speech, reading—in short, the so-called mental functions—in neurophysiological and neuronal terms? Although we have some knowledge of the functioning of neurons, we have less knowledge of how neuron populations interact (reciprocal inhibition and excitation, oscillating systems, gating, convergent and divergent activation, and temporo-spatial reciprocal interactions). Still less information is available about the integrated functioning of neuron ensembles within and between the hemispheres, referred to as the "mental object" by Changeux or the "mental hologram" by Pribram. Pribram compares the memories we have stored away to (mental) holograms. Holograms remain intact even if parts of them are removed, because the information is stored in every point. Pribram argues that if part of the brain is damaged, this does not mean that part of the memory is lost. Moreover, the memory or the concept obeys the rule of perceptual and motor constancy: an individual recognizes the outside world no matter what angle it is perceived from (or reads upside down or distorted text or by Braille), and acts that have been learned by one hand can be performed by either hand or even by the foot. Pribram views the structure of mental holograms as wave formations of massive neuronal activity generated by neuron populations at numerous sites. By way of resonance, stimulation from outside or inside activates these wave formations, which leads to recognition. We suppose that the brain code (or the mental hologram) in children is quite different from that in adults. It is, for instance, characterized by concreteness, field-related perception, affect-relatedness, absence of intermodal connections, and internal instability.

The substrate of the brain for recognition has recently been more clearly elucidated by Damasio (1989, 1990), working with visually agnostic patients. According to Damasio, there are no neural systems dedicated to the representation of particular conceptual categories (e.g., animals, people, furniture). Rather, different neural systems are dedicated to the processing of certain characteristics and conditions of entities and events in knowledge domains. Entities and events, such as an object or a person, generate multiple separate representations (of all sorts of characteristics) in all or more than one sensory modalities and cortices (that are linked to each other) and mostly in more than one way within a modality (e.g., form, direction, and color within the visual modality). All these representations, "laid down" in widely distributed loci, are connected in a network that can be activated or triggered by one stimulus (one representation or a combination of representations). Success or failure of a recognition must depend on access to certain representations. A deficient progress in the build-up of a recognition—i.e., the reconstruction of multiple representations into an entity—can be due to the damage of a certain area or a disconnection. In the view of Damasio, the direct representations and concrete knowledge of the external world are mapped in the posterior association cortices near the primary cortices that are aware of the concrete physical structure. Neuron units that are farther removed from these areas (higher-order units) have a less-concrete representation of reality and a character more like abstract knowledge. Activation by a gustatory, olfactory, visual, or auditory complex of stimuli or by a word or a gesture (an event or an entity) can also arouse a whole world of experiences, memories, and images, thus activating large parts of the brain. In his novel *A la recherche du temps perdu* Marcel Proust describes, for instance, the famous *madeleine,* a French biscuit, which elicits numerous memories of his youth.

I would like to refer to this process of multimodal association as divergent (receptive) activation, since one complex of stimuli (smells, objects, gestures, behaviors, and also words), in which numerous meanings have been condensed, can lead to a chain reaction of associations, a receptive process, which can sometimes lead to an inner language. This is image-processing (Pribram) or recognition and concept reconstruction (Damasio). Seen the other way round, this process allows a person to reduce a whole world of experiences to a word, a sentence, or a gesture, as a condensed symbolization or abstraction and an expressive process. I would like to call this process which leads to the use of symbols convergent (expressive) activation, since a whole world of experiences is channeled into one symbol or an abstraction.

The child's development changes from pure auditory-perceptual-motor functioning (simple echolalic speech) to complex meaningful speech (symbolization). The child first recognizes, has a passive understanding, and later expresses this understanding by use of words. The most widespread problem among children lies within this field: defective word-finding or verbalization—or disturbed convergence to speech—in the course of developmental dysphasia. In the young child, this problem is physiological, and in the child with developmental dysphasia it is deviant and continues to be so for a long time. For the child with developmental dysphasia, recognition and receptive reconstruction of concepts, even by the auditory-verbal way, are much easier than the opposite process, when concepts have to be channeled into words, sentences, or inner speech or when the word has to be reconstructed as a symbolic concept from multiple inner representations. These children are unable to connect experiences and ideas and put them into words, and this deficit mainly manifests itself in dialogue and increases as the abstraction level becomes higher. Gestalts or mental objects cannot be converted into the right words (word-finding disorders and paraphasias) or into speech sequences (chaotic narration and dysgrammatism). The child also has difficulty putting feelings into words (alexithymia). If the word has to be reconstructed from multiple representations, located in several sites, word-finding difficulties may result from damage to widespread loci, even in the right hemisphere.

Learning to read entails saying out loud a recognized auditory structure of a visual form sequence with semantic meaning. Here the recognition can go wrong because vital information required for the building up (or reconstruction) of that recognition can be lacking. This information can pertain to physical visual characteristics of the words but also to the phonemic structure, which may be unclear if a child has no phonological awareness. It is the cross-modal nature of recognition that makes the reconstruction more difficult. Finally, there is meaning assignment, a link to the semantic network, that during word-reading might also be a lexical referential process mediated by the right hemisphere (Bever et al., 1989). Lexical processing does not provide access to stored word codes. Spelling and pronunciation and meanings of words are not kept in separate stores. Rather, lexical processing has to be viewed as an activation of different types of information (Seidenberg and McClelland, 1989). It is, however, usually more difficult than the processing of natural speech. Advanced reading (of sentences) requires more processing, because the child must extract meaning from grammatical structures, thus relying more on left-hemisphere functioning. It does not, however, exclude lexical referential processes. The more advanced reading is, the more it resembles natural speech with the usual grammar, fluency, and prosody (convergent activation). One or more of these elements are lacking in developmental dysphasia. The more the child misunderstands spoken language, the more words become nonwords for the child, which implies nonreading or nonrecognition (or inner language dyslexia, according to Myklebust).

The importance of developmental dysphasia for the interpretation of the clinical picture of dyslexia is summarized in the first section of this paper. It is shown that many characteristics of dysphasia are found some years later in dyslexic children, thus explaining their reading problems. In the second section, the common underlying neurobiological causes of dysphasia and dyslexia are discussed. In our studies, we found clinical indices indicative of anomalies or functional disorders in both hemispheres, particularly in the left one; this has also been noted by other authors (Johnston et al., 1981; Rentz et al., 1986; Ludlov, 1980). There is, however, no evidence of clinical neurological anomalies in approximately half of the children with developmental dysphasia, and this is even more common among children with "pure" dysphasia. EEG studies tend to lead to the same conclusion. Radiological studies indicate damage to the left hemisphere (Dalby, 1977) or, in the absence of classical radiological anomalies, an unusual asymmetry.

There are also arguments supporting the notion that, at least in young children, some of the aspects of disturbed language development are due to a maturation lag (Bishop and Edmundson, 1987). This latter notion, however, does not lead to any conclusion as to the abnormal development of speech or language comprehension. What is the pathophysiological mechanism? The development of speech on the sentence level has yet to be neurophysiologically understood. Is there any indication of a specified neuroanatomical correlate? In order to understand this, we should bear the following in mind: although it is true that in the young child the hemispheres are not equipotential (functionally equivalent) at birth (Molfese et al., 1975; Mehler et al., 1988; Bertoncini et al., 1989), they have not yet become specialized, in other words "irreversibly inhibited," for other functions. This does not occur until between the ages of four and twelve, by which time there is then a reduction in the chance of recovery from aphasia. In the initial stage postnatally, expression is strongly linked to simultaneously occurring body and oral movements (Trevarthen, 1986). Speech development shortly thereafter is an affective senso-motoric process that comes into being by way of the interaction with the speaking mother (Tan, 1990).

Although infants are able to perceive speech sounds in the left hemisphere, there is still no sign of what might be referred to as formal language. In the preverbal stage, speech development begins as a Gestalt-like and emotional prosodic interaction on the basis of possibly predominantly subcortical and right-hemisphere mechanisms, which also dominate nonverbal contact or body language, such as facial recognition and emotional expression. At this early stage speech and expressive gestural motor mechanisms are still strongly linked. It has become clear, mainly as a result of split-brain experiments, that the right hemisphere has a large lexicon and a reasonable capacity for passive language comprehension (Sperry, 1985). The left hemisphere probably plays a greater role as soon as speech becomes symbolic and syntactic constructions appear (the "birth" of language between the ages of one and three). At this age there are indications of left-sided lateralization (Piazza, 1977). In addition, there is a specific lengthy physiological immaturity (delayed myelination) of the auditory thalamocortical pathways and the interhemispheric pathways between the secondary auditory cortices (Yakovlev and Lecours, 1967; Gilles et al., 1983). The target neurons of these pathways also mature late. The destruction of neuronal centers that are already stabilized in adults, and in whom there is no longer the plasticity still evident in children, is different in effect from the damage to the maturation processes in children, in whom the cerebral functions are still plastic. Acquired aphasia in the older child is virtually exclusively caused by left-hemisphere lesions (Van Dongen et al., 1985). This is why the clinical picture of developmental dysphasia is age-related and why one cannot refer to *the* substrate: the brain undergoes changes in the course of the lateralization process of language to the left hemisphere. This also explains why the detection of neurological anomalies depends on the age of the child and on the occurrence of concomitant cerebral dysfunctions, such as motor problems.

We would thus disagree with the notion that "developmental aphasia," as Eisenson (1968) called it, can be explained by a unitary concept. Despite the increase in our knowledge on the neuroanatomy of speech and language functions, it must be admitted that exact information on the correlation between the various aspects of spoken language and reading and how it is acquired is still lacking.

12

Parallel Processing in the Visual System and the Brain: Is One Subsystem Selectively Affected in Dyslexia?

Margaret Livingstone

The primate visual system comprises two, possibly three, major processing pathways that remain largely segregated and independent throughout the visual system, possibly even up through higher cortical association areas. This subdivision is most apparent, and was first discovered, in the lateral geniculate nucleus, where cells in the ventral, or magnocellular, layers are larger than cells in the dorsal, or parvocellular, sets of layers. Anatomical, physiological, and perceptual studies indicate that the magnocellular and parvocellular subdivisions of the primate visual system are largely independent and are responsible for different aspects of visual perception: the magno system is fast, has high contrast sensitivity, is colorblind, and has slightly lower acuity than the parvo division, which is slower, responsive to color contrast, and much lower in contrast sensitivity. Physiological studies at higher levels indicate that the magno system carries information about motion and stereopsis, and perceptual studies suggest that it may be responsible for spatial localization, depth perception from many kinds of depth cues, figure-ground segregation, and hyperacuity. The parvo system seems to be concerned with color perception and object recognition. This chapter reviews the evidence for the different functions of the magno and parvo systems.

Different aspects of human visual perception, such as motion detection, stereopsis and other forms of depth perception, color discrimination, and high-resolution form discrimination, differ markedly in the same four characteristics that distinguish the geniculate subdivisions, suggesting that at even higher levels the magno and parvo systems maintain their segregation, functionally as well as anatomically.

At the end of the chapter I turn to the implications of these differences for dyslexics. It has been suggested that dyslexic subjects are defective in fast, or phasic, visual information processing. In the primate, fast, transient visual information is carried by the magno subdivision of the visual pathway. The observations that dyslexics often have poor stereoacuity, visual instability and problems in visual localization, and poor double-flash order discrimination are all consistent with a selective deficit in the magnocellular pathway. Other sensory and motor systems probably are similarly divided into a fast and a slow subdivision, and I hypothesize that dyslexics are specifically affected in the fast subdivision of many cortical systems.

Anatomical Evidence for Parallel Channels

Though it is most obvious in the geniculate, the subdivision of the primate visual system begins in the retina and is perpetuated into the primary visual cortex. The large type A retinal ganglion cells project to the magnocellular geniculate layers, which in turn project to layer 4Cα of V1; smaller type B ganglion cells project to the parvocellular layers, which project to layers 4Cβ and 4A of V1 (Hubel and Wiesel, 1972; Lund et al., 1975; Leventhal et al., 1981). From Golgi and horseradish-peroxidase studies Lund (1973, 1987), Lund and Boothe (1975), and Fitzpatrick et al. (1985) determined that magno-recipient layer 4Cα projects to layer 4B, which in turn projects to visual area 2 and to area MT, and parvo-recipient layer 4Cβ projects to layers 2 and 3 and from there to visual area 2; both 4Cα and 4Cβ also send less-dense projections to the deeper layers. There is some evidence that there is a third subdivision, specifically concerned with color information, which may originate in the W cells of the retina (Casagrande et al., 1990). In the geniculate this system may be represented by the color-opponent center-only (type 2) cells, which are found in or near the parvocellular interlaminar leaflets of the geniculate and project directly to the blobs of the upper layers of V1 (Fitzpatrick et al., 1983; Livingstone and Hubel, 1982, 1984a, and unpublished observations). This tangle of connections is diagrammed in Figure 12.1.

Though the magnocellular and parvocellular inputs to the cortex seem to remain separate in layer 4C, the separation at subsequent stages may not be as complete. Nevertheless, the specificity of intrinsic connections within layers 2 and 3 of V1 (Livingstone and Hubel, 1984b; Ts'o and Gilbert, 1988) and the specificity of the connections between V1 and V2 (Livingstone and Hubel, 1984a, 1987a; Cusick and Kaas, 1988) suggest that a significant degree of segregation is perpetuated.

Functional Differences between the Channels

In the geniculate, the magno and parvo subdivisions differ in their color selectivity, temporal characteristics, contrast sensitivity, and spatial resolution. Most parvocellular cells are color-opponent cells whereas magnocellular cells are not (Figure 12.2) (Wiesel and Hubel, 1966; De Valois, Abramov, and Jacobs, 1966; Gouras, 1968, 1969; deMonasterio and Gouras, 1975; De Valois, Snodderly, Yund, and Hepler, 1977; Schiller and Malpeli, 1978; Derrington, Krauskopf, and Lennie, 1984).

Magno cells are much more sensitive than parvo cells to luminance contrast (Shapley, Kaplan, and Soodak, 1981; Kaplan and Shapley, 1982; Derrington and Lennie, 1984; Kaplan and Shapley, 1986). The third difference between magno and parvo cells in their field-center sizes. For both systems the average size of the receptive-field center increases with distance from the fovea, in a pattern that is consistent with the differences in acuity between foveal and peripheral vision. Yet at any given eccentricity, magno cells have larger receptive-field centers than parvo cells, by a factor of 2 or 3. Last, magno cells respond faster and more transiently than parvo cells. This sensitivity to the temporal aspects of a visual stimulus suggests that the magno system may play a special role in detecting movement. As we will see, many cells at higher levels in this pathway are indeed selective for direction of movement (Dubner and Zeki, 1971; Maunsell and Van Essen, 1983).

The response properties of cells at stages beyond visual area 2 suggest that the segregation of functions begun at the earliest stages

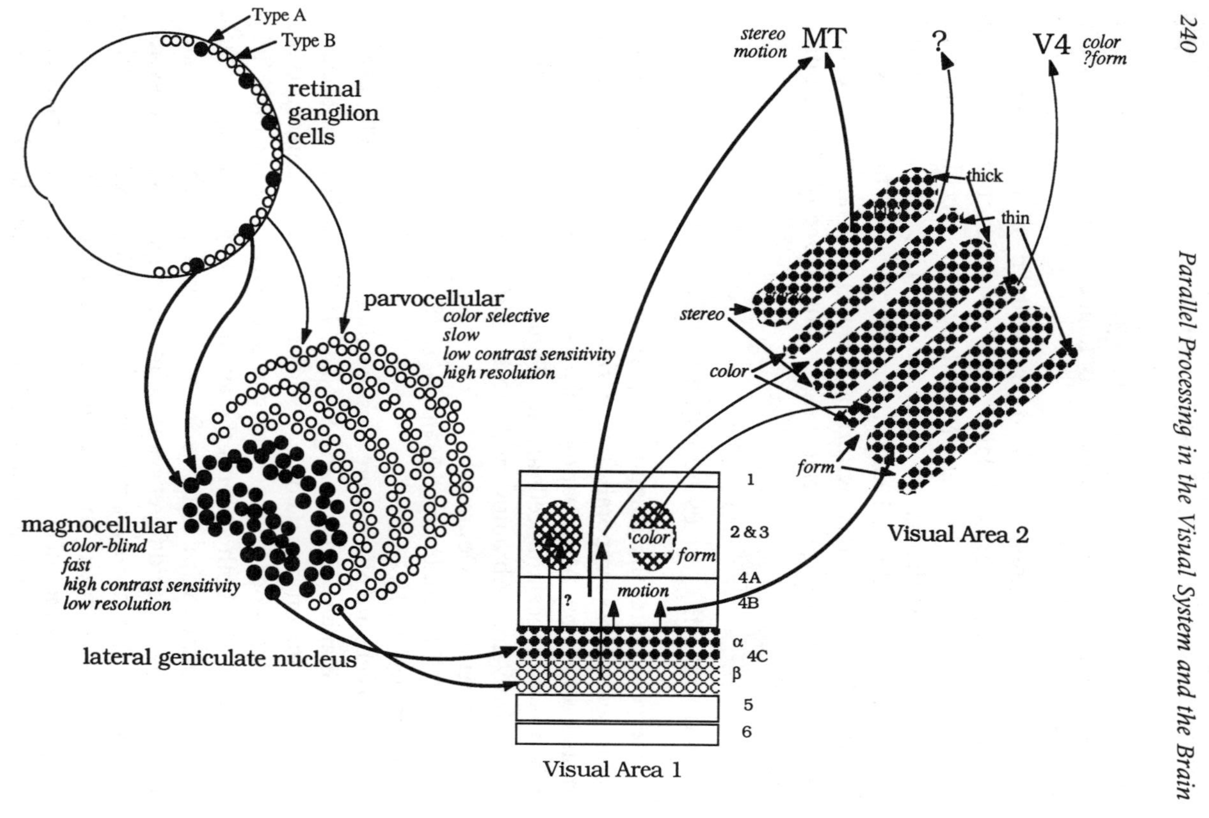

Figure 12.1 Functional segregation in the primate visual system.

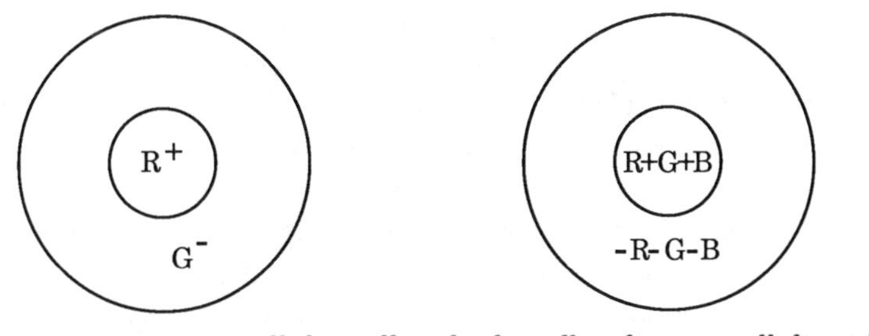

Figure 12.2 Receptive fields for *(a)* typical color-opponent parvocellular geniculate neuron, which is excited over a small region by red light and inhibited over a larger region by green light, and *(b)* typical broad-band magnocellular neuron, which is excited by all wavelengths in the center and inhibited by all wavelengths in its surround.

is perpetuated at the highest levels so far studied. Indeed, the segregation seems to become more and more pronounced at each successive level, so that subdivisions that are interdigitated in visual areas 1 and 2 become segregated into entirely separate areas at still higher levels. As indicated in Figure 12.1, cells in cortical areas V2 and MT receiving input from the magno system show direction selectivity or binocular disparity selectivity, or both, whereas cells receiving input mainly from the parvo system seem to be involved in color or form perception. Movement perception and stereoscopic depth perception should therefore be magno-like when tested for the four variables listed above, whereas form and color perception should be parvo-like. Tests of human perception bear out these predictions to a remarkable degree (for references see Livingstone and Hubel, 1987b). For example, stereopsis and motion perception both deteriorate with stimuli that have color contrast but no luminance contrast (equiluminant stimuli), and they both show high contrast sensitivity, whereas orientation discrimination and grating resolution have lower contrast sensitivity and are only slightly diminished for equiluminant images.

The idea that different aspects of vision are processed separately in humans as well is supported by clinical observations. People with strokes can suffer surprisingly specific visual losses—for example, loss of color discrimination without impairment of form or motion perception, loss of motion perception without loss of color or form perception, or the selective loss of face recognition (Bodamer, 1947; Damasio et al., 1980; Pearlman et al., 1979; Zihl et al., 1983). Moreover, patients with open-angle glaucoma, which selectively damages the large-diameter fibers of the optic nerve and should therefore selectively damage the magno system, have been found to have greater impairments in stereoacuity and sensitivity at low contrasts than in color perception or Snellen acuity (Bassi and Galanis, 1990; Liebergall et al., 1990; Pfeiffer et al., 1990). Though the degree to which the subdivisions remain separate is a subject of some controversy in the fields of both human psychophysics and monkey physiology and anatomy (De Yoe and Van Essen, 1985; Burkhalter and Van Essen, 1986; Cavanagh, 1988; Schiller et al., 1990; Logosthetis et al., 1990; Levitt and Movshon, 1990), it is nevertheless clear that parallel processing must be a fundamental strategy of the primate visual system.

Despite many gaps, the picture beginning to emerge from the anatomy and electrophysiology just outlined is that the segregation apparent at very early stages of the visual system gives rise to separate and independent parallel pathways. At early stages, where there are two major subdivisions, the cells in these two subdivisions exhibit at least four very basic differences: in their color selectivity, speed, acuity, and contrast sensitivity. At higher stages the continuations of these pathways are selective for quite different aspects of vision—form, color, movement, and stereopsis—thus generating the counterintuitive prediction that different kinds of visual tasks should differ in their color, temporal, acuity, and contrast characteristics. To test this prediction we asked whether the differences seen at the early stages of the visual system can be detected in conscious human visual perception by comparing the color, temporal, spatial, and contrast sensitivities of different visual functions. Many of these questions have, not surprisingly, already been asked, and the answers are strikingly consistent with what is known about anatomy and physiology. For several decades psychologists have accumulated evidence for two channels in human vision, one chromatic and the other achromatic, by showing that different tasks can have very different sensitivities to color and brightness contrast. Given what we know now about the electrophysiology and the anatomy of the subdivisions of the primate visual system, we can begin to try to correlate the perceptual observations with these subdivisions. Though at higher cortical levels there seem to be three subdivisions, possibly with some mixing of magno and parvo inputs to the blob system, the most important distinction is probably between the magno system (magno–4Cα–4B–MT) and the parvo-derived subdivisions (parvo–4Cβ–interblobs–pale stripes–?V4 and parvo+?magno–4Cβ–blobs–thin stripes–V4). In my discussion of human perception I will therefore stress the distinctions between functions that seem to be carried exclusively by the magno system and those that seem to be carried by the parvo-derived pathways.

Human Perceptual Correlates

Temporal differences in color and brightness discrimination

Knowing that the magno system is colorblind and is faster than the parvo system, we can predict that discrimination of color and discrimination of brightness should have different temporal properties. This is indeed so: Ives (1923) showed that brightness alternations can be followed at much faster rates than can pure color alternations.

Movement perception

The high incidence of movement and direction selectivity in area MT suggests that this area may be particularly concerned with movement perception. Because anatomically MT receives its major inputs from layer 4B of the primary visual cortex and from the thick stripes of visual area 2, both part of the magno pathway, one would predict that human movement perception should somehow reflect magno characteristics: colorblindness, quickness, high contrast sensitivity, and low acuity. Perceptual experiments indicate that movement perception does indeed have these characteristics. First, it is impaired for patterns made up of equiluminant colors: Cavanagh, Boeglin and Favreau (1985) found that if they generated moving red and green sine-wave stripes, "the perceived velocity of equiluminous gratings is substantially slowed . . . the gratings often appear to stop even though their bars are clearly resolved . . . the motion is appreciated only because it is occasionally noticed that the bars are at some new position." Second, movement perception is impaired at high spatial frequencies, an outcome that is consistent with the lower acuity of the magno system. Fergus Campbell and Lamberto Maffei (1981) viewed slowly rotating gratings and found a loss of motion perception at the highest resolvable frequencies (Figure 12.3): "At a spatial frequency of 16 and 32 cycles/deg[ree] a strange phenomenon was experienced, the grating was perceived as rotating extremely slowly and most of the time it actually appeared stationary. Of course, the subject could call upon his memory and deduce that the grating must be moving for he was aware that some seconds before the grating had been at a particular 'clock-face position'. Even with this additional information that the grating must be rotating the illusion of 'stopped motion' persisted." What is most surprising about the perception of both the equiluminant stripes and the very fine stripes is that even though the sensation of movement is entirely, or almost entirely, lost, the stripes themselves are still clearly visible—they are clear enough that changes in their position can be seen, even though they do not seem to be *moving*.

Finally, movement can be vividly perceived with very rapidly alternating or very low-contrast images (Cambell and Maffei, 1981;

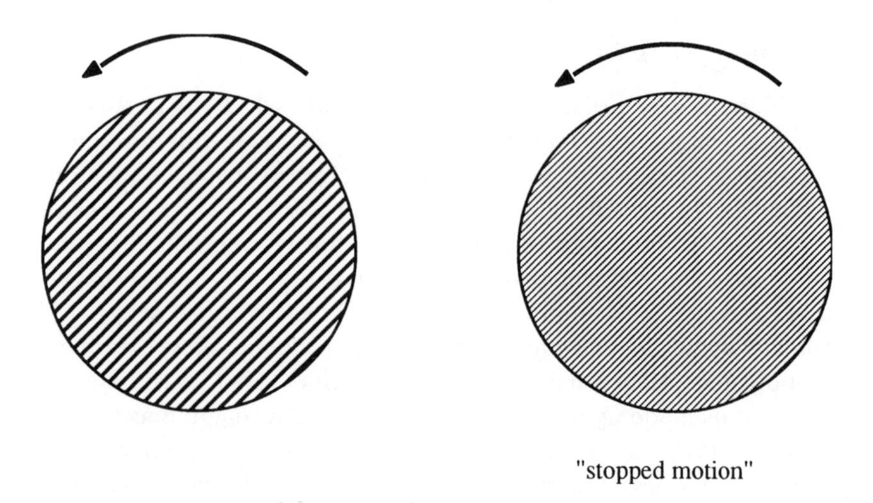

Figure 12.3 Loss of movement perception with high spatial frequency patterns: stimulus designed by Campbell and Maffei (1981) in which a disk with a high spatial-frequency pattern (16 cycles/degree) appears to rotate more slowly than the same disk with a coarser pattern.

Livingstone and Hubel, 1987b). Thus, as summarized in Table 12.1, the properties of human movement perception are remarkably consistent with the properties of the magno system.

Stereopsis

Finding retinal-disparity-tuned cells in the thick stripes of visual area 2 and in arca MT suggests that the magno system is also involved in stereoscopic depth perception. Cary Lu and Derek Fender (1972) found that subjects could not see depth in equiluminant color-contrast random-dot stereograms even though the dots making up the stereogram remained perfectly clear. This finding has been disputed, but we found that differences in results can arise from variations in an individual's equiluminance point with eccentricity, which make it difficult to achieve equiluminance across the visual field. Like movement perception, stereopsis fails for stereograms containing only high, but resolvable, spatial frequencies, but it is not diminished for rapidly alternating or very low-contrast stereograms (Livingstone and Hubel, 1987b). (See Table 12.1.)

Deduction of further magno or parvo functions from perceptual tests

Since the functions that electrophysiological studies had suggested should be carried by the magno system did indeed show all four distinguishing characteristics of that system, David Hubel and I decided to ask whether other visual functions—ones not predicted by single-cell response properties—might also manifest some or all of these properties.

If a particular magno cell sums red and green inputs, there will be a red-to-green ratio at which the red and green will be equally effective in stimulating the cell. This need not imply that every magno cell has the same ratio of red to green inputs and therefore necessarily the same equiluminance point. Nevertheless, the fact that movement and stereopsis fail at equiluminance implies that, for a given observer, the null ratio must be very similar for the majority of the individual's cells responsible for that function. Krüge (1979) and, later, David Hubel and I (Hubel and Livingstone, 1990) found that mag-

Table 12.1. Comparison of physiological findings and human perceptual tests

	Color selectivity	Contrast sensitivity	Temporal resolution	Spatial resolution
Physiological system				
Geniculate subdivision				
magnocellular	no	high	fast	2× lower
parvocellular	yes	low	slow	high
Cortical subdivision				
magno–4Cα–4B–thick	no	high	?	
parvo–4Cβ–interblob	yes	low	?	low
parvo+magno–4C–				high
blob	yes	high	?	
Psychophysical test				
Movement perception				
Movement detection	no	high	fast	low
Phi movement detection	no	high	fast	
Depth cues				
Stereopsis	no	high	fast	low
Binocular rivalry	no			low
Parallax	no			
Depth from motion	no	high		
Shading	no	high		
Perspective	no	high	fast	low
Contour lines	no			
Occlusion	no			
Linking properties				
Linking by movement	no	high		
Linking by collinearity (illusory borders)	no	high	fast	
Figure-ground discrimination	no			
Shape discrimination				
Orientation discrimination	yes	low	slow	high
Shape discrimination	yes	low	slow	high
Color perception				
Color determination	(yes)		slow	low

nocellular geniculate cells are unresponsive to a moving color-contrast border at a particular relative brightness—a brightness ratio that was very close to a human observer's equiluminant point. Cells in the parvocellular layers also show response decrements at some color ratio, but this ratio varies from cell to cell, depending on the cone inputs to the receptive-field center. In the cortex, cells in magno-derived divisions also show loss of responsiveness at equiluminance, and many cells in parvo-derived regions do not. Thus, not only do individual cells in the magno system seem to be colorblind, but the properties of stereopsis and movement perception indicate that the magno system as a whole is colorblind. (There is, however, currently some disagreement about whether the magno system is inactive at equiluminance; Logothetis et al., 1990.)

Depth from motion

Since both motion perception and stereoscopic depth perception are lost at equiluminance, David Hubel and I suspected that the ability to use relative motion as a depth cue might also. Relative motion is a very powerful depth cue: when an observer moves her head back and forth or moves around in his environment, the relative motion of objects provides information about their distance. In the experiment shown in Figure 12.4, the position of the middle bar was

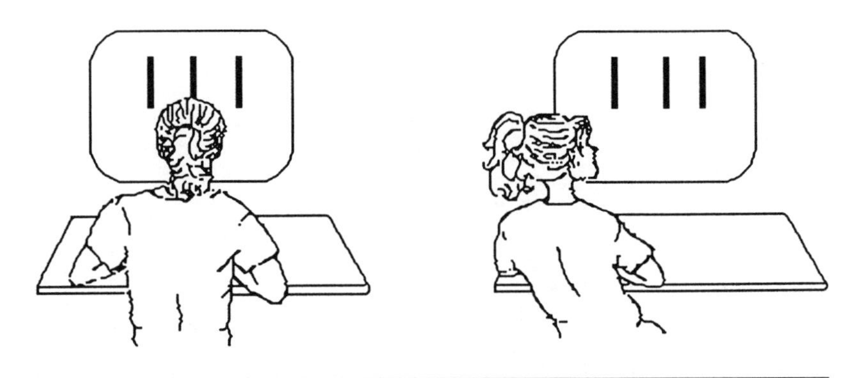

Figure 12.4 Loss of depth from parallax at equiluminance. The position of the middle bar is correlated with the observer's head position. In this case, the center bar appears to lie in front of the reference bars, except when the bars are made equiluminant with the background.

coupled to head movement, and the middle bar appeared to be either behind or in front of the reference bars, depending on whether its movement was the same as, or contrary to, the head movement. When the bars were made equiluminant with the background, all sensation of depth disappeared.

Relative movement in the flat projection of a three-dimensional object is also a powerful depth cue. Figure 12.5 shows two frames of a movie in which random dots move, some to the right and some to the left, as if they were pasted on a rotating spherical surface. The movie gives a powerful sensation of a rotating spherical surface—except when the dots are equiluminant with the background, and then all sensation of depth is lost and the dots seem to dance aimlessly.

Thus depth from motion, both from viewer parallax and from object motion, seems also to depend on luminance contrast and could well be a function of the magno systems. Consistent with this idea, we could see depth from motion at very low levels of luminance contrast.

Depth from shading

The retinal image is two-dimensional, and to capture the three-dimensional relationships of objects the visual system uses many

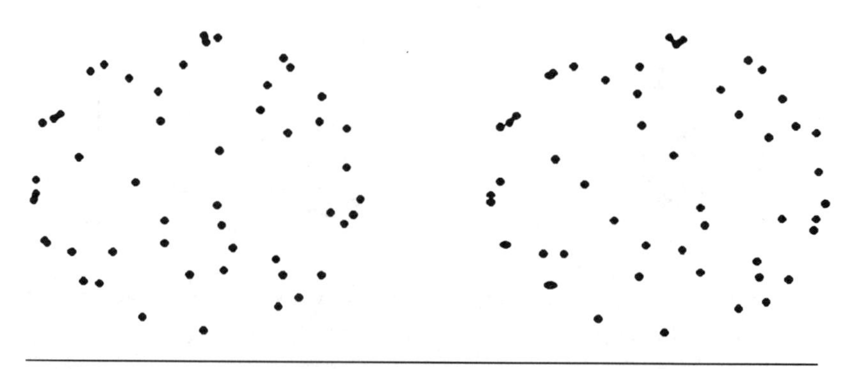

Figure 12.5 Two frames of a movie in which the movement of dots generates the sensation of a three-dimensional object. The dots appear to lie on the surface of a sphere (you can see it by stereo-viewing these two frames). All sensation of depth is lost when the dots are equiluminant with the background.

kinds of cues besides stereopsis and relative motion—perspective, gradients of texture, shading, occlusion, and relative position in the image. We wondered whether the sensation of depth from any of these other cues might also exhibit magno characteristics.

It seemed especially likely that the ability to perceive depth from shading might be carried by an achromatic system, because shading is almost by definition purely luminance-contrast information; that is, under natural lighting conditions a shaded region of an object has the same hue as the unshaded parts, simply darker. But in biology just because something could, or seemingly even should, be done in a certain way does not mean that it will be. Nevertheless, Cavanagh and Leclerc (1985) found that the perception of three-dimensional shape from shading indeed depends solely on luminance contrast. That is, in order to produce a sensation of depth and three-dimensionality, shadows can be any hue as long as they are darker than the unshaded regions of the same surface. Many artists seem to be aware of this—for example, in some of the self-portraits of Van Gogh and Matisse the shadows on their faces are green or blue, but they still convey a normally shaped face. Black-and-white photographs of these paintings (made with film that has approximately the same spectral sensitivity as humans have) confirm that the shadows are actually darker than the unshaded parts.

Depth from perspective

As was well known to artists by the time of the Renaissance, perspective is a powerful indicator of depth. Converging lines or gradients of texture are automatically interpreted by the visual system as an indicator of increasing distance from the observer. We found that when images with strong perspective are rendered in equiluminant colors instead of black and white, the depth sensation is lost or greatly diminished, as are illusions of perspective (Figure 12.6).

As with movement and stereopsis, the most startling aspect of this phenomenon is that even though the sensation of depth and the illusory distortions due to inappropriate size scaling all disappear at equiluminance, the lines defining the perspective and the individual elements in the image are nevertheless still clearly visible. This seems to us to rule out high-level cognitive explanations for the perception of depth from perspective and the illusions of perspective—if you

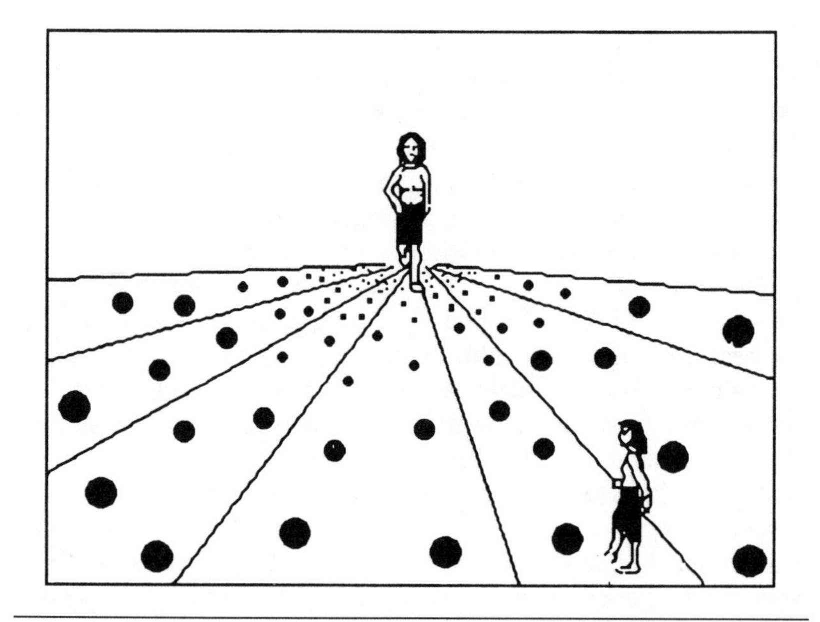

Figure 12.6 Depth from perspective and illusory size difference due to size scaling. If this image had been rendered in two equiluminant colors instead of in black and white, both the appearance of distance and the illusory size difference of the two figures would disappear.

see depth because you merely *know* that converging lines mean increasing distance, you should be able to perceive the depth from the converging lines at equiluminance. Thus at a relatively low level in the visual system some simple interactions must initiate the automatic interpretation of a two-dimensional image into three-dimensional information; moreover, these operations seem to be performed only in the achromatic magno system, not in the parvo system.

Figure-ground discrimination and linking features

Why should the depth and movement functions described above all be carried by the magno system and not by the parvo system? We at first assumed that it was because they might all be performed best by a system with the special characteristics of the magno system. But later we wondered if these various functions might be more related than they seemed at first—whether they could be parts of a more global function. We were struck by the similarity between the list of functions we had ascribed to the magno system and the Gestalt psychologists' list of features used to discriminate objects from each other and from the background—namely, figure-ground discrimination (Rubin, 1916). Most scenes contain a huge amount of visual information—information about light intensity and color at every point on the retina and the presence and orientation of discontinuities in the light pattern. The Gestalt psychologists recognized that one important step in making sense of an image must be to correlate related pieces of visual information; that is, to decide whether a series of light-dark discontinuities forms a single edge, whether adjacent edges belong to the same object, whether two parts of an occluded edge are related, and so on. They determined that several kinds of cues are used in this way and to organize the visual elements in a scene into discrete objects. Horace Barlow (1981) has called these "linking features" because they are used to link or join related elements. These linking features include: *common movement* (contours moving in the same direction and velocity are likely to belong to the same object, even if they are different in orientation or not contiguous); *common depth* (contours at the same distance from the observer are likely to belong to the same object); *collinearity* (if a straight or continuously curved contour is interrupted or occluded by another object, it is still seen as a single contour); and *common color or lightness.* The results described below suggest, however, that only luminance contrast and not color differences can be used to link parts together.

Linking by collinearity, as shown in Figure 12.7, also breaks down when the lines are equiluminant with the background; the figure then just looks like a jumble of lines instead of a pile of blocks.

Linking by collinearity is seen in the phenomenon of illusory contours, figures that produce a vivid perception of an edge in the absence of any real discontinuity (Figure 12.8). When these figures are drawn in equiluminant colors, the illusory borders disappear, even though the elements defining them (the pacmen, the spokes, the lines, or the circles) remain perfectly visible. Because the perception of illusory borders also manifests fast temporal resolution, high contrast sensitivity, and low spatial resolution, we suspect that it too may represent a magno function. Illusory borders have been called "cognitive contours" because of the suggestion that the perception of the border is due to a high-level deduction that there must be an object occluding a partially visible figure (Gregory, 1972). We suspect that this is not the case because the illusory borders disappear at

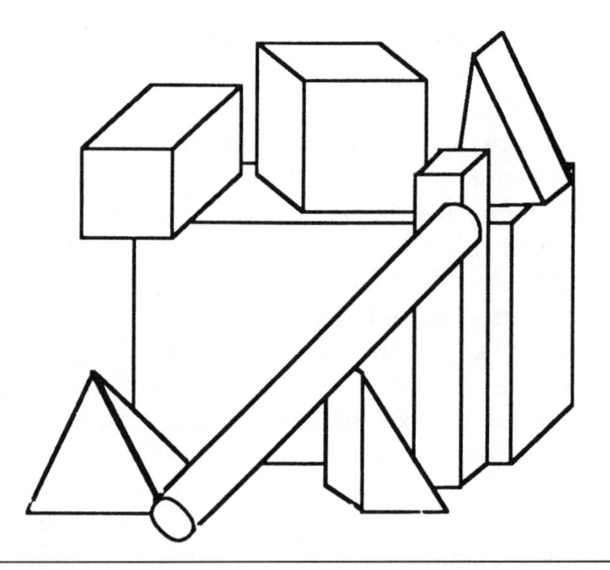

Figure 12.7 Linking by collinearity. It is clear which edges are part of the same object, even when they are occluded by another object. At equiluminance this linking disappears, and it looks like a jumble of lines instead of a pile of blocks. (After Barlow, 1981.)

equiluminance, even though the real parts of the figure are still perfectly visible.

Fifty years ago the Gestalt psychologists observed that figure-ground discrimination and the ability to organize the elements in a scene decrease at equiluminance. Equiluminant figures have been described as "jazzy," "unstable," "jelly-like," or "disorganized" (Liebmann, 1926; Gregory, 1977). Koffka (1935) pointed out that luminance differences are strikingly more important than color differences for figure-ground segregation: "Thus two greys which look very similar will give a perfectly stable organization if one is used for the figure and the other for the ground, whereas a deeply saturated blue and a grey of the same luminosity which look very different indeed will produce practically no such organization."

Edgar Rubin's popular demonstration of the problem of figure-ground discrimination is the vase/faces (Figure 12.9). At non-equiluminance the percept is bistable, so that one sees either the faces or the vase, but usually not both at the same time. At equiluminance the two percepts reverse rapidly, and one can occasionally see both

Figure 12.8 Illusory borders, which disappear at equiluminance. (After Kanizsa, 1979.)

the vase and the faces simultaneously. The distinction between figure and ground thus gets weaker or even disappears entirely.

Why Should the Visual System Be Subdivided?

Electrophysiological studies suggest that the magno system is responsible for carrying information about movement and depth. We extended our ideas about the possible functions of the magno system

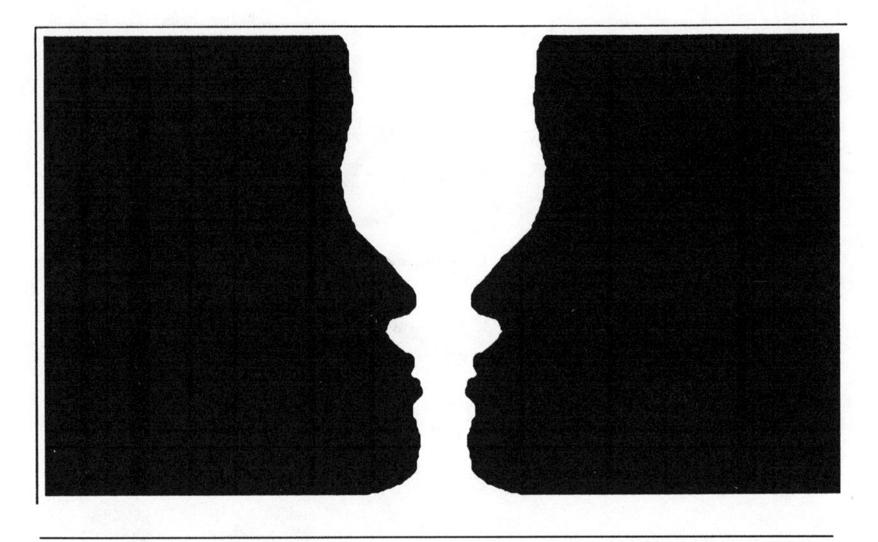

Figure 12.9 Rubin's figure-ground discrimination problem.

with perceptual studies and concluded that the magno system may have a more global function of interpreting spatial organization. Magno functions may include deciding which visual elements, such as edges and discontinuities, belong to and define individual objects in the scene, and they may determine the overall three-dimensional organization of the scene and the positions of objects in space and movements of objects.

If the magno system covers such a broad range of functions, then what is the function of the tenfold more massive parvo system? The color selectivity of the parvo system should enable us to see borders using color information alone, borders that might be camouflaged to the colorblind magno system. But defeating camouflage may be only a small part of what the parvo system is specialized for. Experiments with fading low-contrast images (for references see Livingstone and Hubel, 1987b) indicate that the magno system is not capable of sustained scrutiny, since images that can be seen only by the magno system disappear after a few seconds of voluntary fixation. Thus, while the magno system is sensitive primarily to moving objects and carries information about the overall organization of the visual world, the parvo system seems to be important for analyzing the scene in much greater and more leisurely detail. These postulated functions would be consistent with the evolutionary relationship of the two systems: the magno system seems to be more primitive than the parvo system (Guillery, 1979; Sherman, 1985) and is possibly homologous to the entire visual system of nonprimate mammals. If so, it should not be surprising that the magno system is capable of what seem to be the essential functions of vision for an animal that uses vision to navigate in its environment, catch prey, and avoid predators. The parvo system, which is well developed only in primates, seems to have the added ability to scrutinize in much more detail the shape, color, and surface properties of objects. This ability creates the possibility of assigning multiple visual attributes to a single object and correlating its parts. Indeed, if the magno system needs to use the various visual attributes of an object in order to link its parts together, this need could preclude its being able to analyze the attributes independently. It thus seems reasonable to us that the parvo–temporal lobe system is especially suited for visual identification and association.

Is the existence of separate pathways an accident of evolution or a useful design principle? Segregating the processing of different types of information into separate pathways might facilitate the interactions between cells carrying the same type of information. It might also allow each system to develop functions particularly suited to its specialization. If the parvo system did evolve after the magno system, by duplication of previously existing structures, it should nevertheless not be surprising to find some redundancy in the properties of the two systems. Indeed, both seem to carry information about orientation, and perceptual experiments indicate that both systems can be used to determine shape.

Hints That the Magno System Is Defective in Dyslexics

It has been suggested that dyslexic subjects are defective in fast, or phasic, visual information processing (Lovegrove et al., 1986; Martin and Lovegrove, 1987). In the primate, fast, transient visual information is carried by the magno subdivision of the visual pathway. The observations that dyslexics often have poor stereoacuity (Stein et al., 1987), visual instability and problems in visual localization (Stein et al., 1989), and poor double-flash order discrimination (Williams and LeCluyse, 1989) are all consistent with a selective deficit in the magnocellular pathway. I hypothesize that at least a subset of dyslexics, in particular those with visual disturbances, will show defects in visual functions carried by the magnocellular pathway.

Other sensory and motor systems are also functionally subdivided (Mountcastle, 1957; Abeles and Goldstein, 1970; Woolsey and van der Loos, 1970; Goldman and Nauta, 1977), and it is likely that the columnar architecture in these areas, like the visual system, is segregated into fast and slow subdivisions. This is particularly likely in light of the observations of McGuire et al. (1989), who found that an antibody, CAT 301, selectively stains the magnocellular subdivision of the visual pathway, from the geniculate through primary and secondary visual cortices up to higher parietal visual areas. They also found that this same antibody also stains many other cortical areas, including some somatosensory areas (but not others), a subset of the motor areas, and many other less well defined areas. Most of these areas differ from areas that do not stain with CAT 301 in that they are heavily myelinated. This suggests that these areas all have in common that they process information rapidly. Therefore, the neuronal subdivisions involved in the fastest information processing in each modality may share some particular molecular entities.

Dyslexia has been thought to be a very high-level, even cognitive, defect, since dyslexics have been shown to do poorly in some auditory, somatosensory, visual, and motor tasks. But Tallal et al. (1985) found that dyslexics failed in each of these modalities only in tasks that required very rapid processing of information. I suggest therefore that dyslexics have a specific defect in the rapid subdivisions, the magnocellular homologues, of all neuronal systems.

Acknowledgments

Visual stimuli were programmed by David Freeman. Histology was done by Janet Robbins. The work was supported by NIH grant EY00605, Office of Naval Research Grant N00014 88 K 0200, and a Presidential Young Investigator Award from the NSF.

13

The Neurobiology of Learning Disabilities: Potential Contributions from Magnetic Resonance Imaging

Verne S. Caviness, Jr.
Pauline A. Filipek
David N. Kennedy

Learning, the capacity to modify behavior as a result of experience, is the essential adaptive function of the nervous system. Whereas the term *learning* might be applied to any acquired behavior, a better definition for our context is the acquisition of linguistic, cognitive, and socially directed behaviors. These functions more than all others define the individual character and opportunities of the child.

Our quest, seen most generally, is an exploration of the role of the central nervous system in the acquisition and mediation of these behaviors. We begin with the following general hypotheses. Linguistic, cognitive, and socially adaptive behaviors represent the operation of specific information-processing algorithms. These algorithms are served by discrete and interconnected processing subcomponents composed of local neuronal networks (Kosslyn, 1988). These local networks, in turn, are distributed to finite and specific locations within the brain (Mountcastle, 1978).

Cognitive and clinical neuroscience have provided the conceptual framework and the testing instruments for analysis in the domain of behavior. We consider here certain potential contributions of magnetic resonance imaging (MRI), complementary to those of cognitive and clinical neuroscience, to an examination of the structure and function of the brain.

The Forebrain and Disordered Learning

Focal lesions

The forebrain, in particular the cerebral hemispheres, are indispensable to the acquisition and execution of linguistic, cognitive, and socially directed behaviors in the course of development. Thus, in the child as in the adult, acquired focal lesions of the left hemisphere impair language acquisition and operation (Aram, 1988; Riva and Cazzaniga, 1987; Vargha-Khadem et al., 1985). Precentral lesions as well as lesions of the central white matter and basal ganglia have been associated with impairments of language articulation and disabilities of linguistic semantic and syntactic operations (Aram, 1988). Lesions in left postcentral regions—specifically, focal lesions in the posterior temporal and adjacent occipital and parietal opercular regions—have been associated with fluent language disorders in which there is a disturbance of perceptual, syntactic, and semantic functions (Visch-Brink and Van de Sandt-Koenderman, 1984). Focal lesions of the right cerebral convexity of the developing child, as in the brain of the adult, on the other hand, are followed by degradation of visual-spatial integration and of mathematical abilities (Nass and Peterson, 1989; Stiles-Davis et al., 1985, 1988; Voeller, 1986; Woods, 1980).

The general impression emerges from analyses of behavior and focal lesions that the circuits that support the information-processing algorithms of the cerebrum are established in the earliest phases of development (Vargha-Khadem et al., 1985; O'Gorman and Watters, 1985). They may be "hard-wired" early into the cerebrum as its primary connections are established. Thus, left cerebral lesions acquired in intra-uterine life may be followed by enduring deficiencies in semantic and syntactic processing (Aram, 1988; Vargha-Khadem et al., 1985; Nass and Peterson, 1989). However, there is probably substantial adaptive plasticity in circuit development as well as compensation for damaged circuits by those which are intact, particularly when an insult occurs before birth or even as early as the first year of life (Aram, 1988; Nass and Peterson, 1989; Woods and Carey, 1979; Woods and Teuber, 1978).

Nonfocal cerebral disorders

Disabilities, certainly those in the language domain, do not necessarily reflect the consequences of focal lesions or otherwise destructive processes affecting the developing cerebrum. A range of variations in the degree of cerebral symmetry have been correlated with dyslexia and other developmental language disorders (Geschwind and Behan, 1982; Haslam et al., 1981; Hier et al., 1978; Jernigan et al., 1989; Rosenberger and Hier, 1980). Even more problematic for our understanding of the structural basis of cerebral information processing are a spectrum of behavioral disorders that have only inconstantly been associated with morphological correlates in the cerebrum. In particular, attentional deficits and autism have variably been associated with diffuse or symmetrical reduction in volume of one or more cerebral regions, anomalies of cerebral symmetry, or even no recognizable morphological abnormality of the cerebrum (Creasey et al., 1986; Galaburda et al., 1985; Geschwind and Behan, 1982).

A role for MRI

Magnetic resonance imaging promises to open a new chapter in the study of the structural basis of cerebral organization in developmental disorders of the human brain. It brings expectations for advances to be made in three general areas of structure-function analysis: morphometry, architectonic localization, and mapping of physiologic phenomena, when undertaken in conjunction with suitable methods (Caviness et al., 1989). These physiologic methods include positron emission tomography (PET), electroencephalography (EEG) or magnetoencephalography, magnetic resonance spectroscopy (MRS), and echoplanar MRI. Plans for the use of MRI have sound theoretical supports and, it should be underscored, are feasible from the point of view of practical considerations.

Theoretical considerations

Magnetic resonance imaging provides a view of the entire brain with all regions of the brain represented with more or less uniform fidelity.

This feature of MRI is in contrast to the image obtained with computerized tomography (CT), which is much degraded by X-ray scatter at bone-brain interfaces.

Two issues of central theoretic importance merit special emphasis. First of all, the MR image is a rational subdivision of the human brain in that it generates differential signal contrast values for gray- and white-matter structures (Caviness et al., 1989). It thus distinguishes those regions of the nervous system where analytic operations are performed by local neuronal circuits from the masses of fiber tracts providing connections between such circuits.

The second major theoretical consideration is that the magnetic resonance imaging data set, when acquired with well-tuned, state-of-the-art instruments, is approximately a linear transform of the brain. Morphometric studies of the brain are, therefore, in principle feasible with MRI (Caviness et al., 1989).

Magnetic resonance imaging systems operate in multiple modes. The most commonly used are the so-called T1- and T2-weighted modes. The T1-weighted images are particularly sensitive to resonant energy levels and provide optimum contrast between gray- and white-matter structures. Spatial resolution approaching 1–2 mm is routinely achieved under optimum imaging conditions. The T2-weighted mode, sensitive to the coherency of the resonant signal, is complementary to the T1. The T2 signal is augmented in any pathologic process that increases the amount of water in tissue. Gliosis following injury is an example (Filipek et al., 1990a,b).

Practical considerations

Several practical issues favor application of magnetic resonance imaging for the study of developmental disorders of the human brain (Filipek et al., 1990a,b). Foremost among these is the advantage that magnetic resonance imaging is noninvasive. At magnetic field strengths currently employed, the procedure poses no risk to biological tissues (Budinger, 1981). Second, imaging sequences are generally of the order of 10–15 minutes, and a full imaging session is 30–40 minutes. In general, a child tolerates testing for these lengths of time. Finally to be mentioned here, instruments of high operating capacity are widely available. The current cost of a scan may vary from $700 to $1,200, depending upon the sequences employed. The data may be conveniently entered upon magnetic tape for transfer for further analysis off line.

Analysis

We review here a strategy for correlative anatomic-behavioral analysis based upon magnetic resonance images. These operations must begin with a suitable imaging data set. Optimally, the imaging data set is three-dimensional and provides a representation of the entire brain. Resolution is inversely related to the imaging section thickness, and state-of-the-art instruments allow acquisition of sections that are 1 mm or even less in thickness (Caviness et al., 1989; Filipek et al., 1989). However, there is also an inverse relation between section thickness and the time required for image acquisition when the entire brain is to be included in the data set. We have adopted as a satisfactory compromise a T1-weighted gradient echo pulse sequence achieved in approximately 10 minutes, which provides 60 coronal sections that are 3 mm in thickness (Caviness et al., 1989; Filipek et al., 1989). When this sequence is used, the aggregate error of volume determinations, based upon estimates with complex phantom forms, is between 5 and 10 percent of the actual volume of the imaged object.

The initial two analytic operation are, in order, (1) gray matter–white matter segmentation and (2) rotation of the image data set to a standard alignment in stereotactic space.

Segmentation

Segmentation is the process of constructing contours that define boundaries at gray matter–white matter interfaces. This operation formalizes the separation of the structures, which are visually distinguishable in terms of differential signal intensity values. Depending upon the objectives of analysis, the segmentation operation in the forebrain may include the separate cortical regions, basal ganglia, and thalamic masses. Because the volumetric units of image acquisitions (volume picture elements, or *voxels*) will have a dimension corresponding to the image section thickness, and because this is substantially greater than the in-plane voxel dimensions, the precision of the segmentation operation is greatest when performed on the plane of imaging (Filipek et al., 1990b). It is for this reason that segmentation is antecedent to image rotation. The segmentation operation must be undertaken one image plane at a time. The process is tedious, requiring close and sustained attention by an investigator or technician with the appropriate knowledge of neuroanatomy. As practiced in our laboratories, computer-executed, semi-automated algorithms, requiring substantial editing by the investigator, allow the 60 image planes of a complete brain to be segmented within a day's time (Caviness et al., 1989; Filipek et al., 1989).

Rotation

The plane of address of imaging systems is variable under general application, as when the subject's head is not fixed to standard alignment. Once the brain data set has been segmented, the imaged brain is aligned with reference to internal anatomic landmarks according to established practices now standard in stereotactic surgery and radiation therapy (Filipek et al., 1989). The anterior-posterior commissure line is designated the Y axis, a plane elevated from the Y axis within the interhemispheric fissure the Z plane, and the orthogonal or coronal plane the X plane (Talairach and Tournoux, 1988).

Once segmented and rotated to a standard alignment within stereotactic space, the data set is suitable for a series of computational options. These include not only computation of the volumes of specified anatomic structures but also several options for the measurement of shape (Kennedy et al., 1990; Sacks et al., 1990). These morphometric options place emphasis upon the study of structures that are directly visible within the differential signal contrast range of the primary imaging data set.

Morphometric Application

Measures of volume

The morphometric values for the brain derived from MRI images obtained in living subjects appear reasonably in accord with those obtained by direct morphometry of specimens obtained after death. Estimates of volumes of the brain or brain subdivisions from image data sets from seven normal adults (Filipek et al., 1989; Caviness et al., 1989) deviated less than a few percent from values taken previously by direct planimetric measurements from unfixed anatomic specimens (Wessely, 1970; Paul, 1971; Kretschmann et al., 1979). In addition to the volume of the whole brain, the measures included the separate volumes of cerebral and cerebellar hemispheres and a set of cerebral and cerebellar cortical and subcortical gray and white matter structures. Although the numbers of MRI and postmortem brains upon which these determinations were based were relatively small, the exercise establishes that, comparatively, the MRI-based analysis provides a valid range of absolute volumetric measures.

For this analysis, the imaged forebrains of two normal siblings, an eighteen-year-old male and his seventeen-year-old sister, were subdivided into precallosal, paracallosal, and retrocallosal compartments, and the paracallosal region was further subdivided into suprasylvian, infrasylvian, and "central nuclear" compartments (Filipek et al., 1988; Caviness et al., 1989). The volumetric determinations for these compartments in the two brains were variant by less than several percentage points throughout the full set of regions. The brains of the two siblings were similarly asymmetric, with the left frontal region larger than the right but the right posterior region larger than the left. This pattern of asymmetry, essentially identical in the two siblings, is the reverse the right frontal but left posterior predominance characteristic of 80 percent of the general population (Bear et al., 1986).

Developmental verbal auditory agnosia

A scheme of compartmental analysis identical to that cited above, used for comparison of the brains of the normal siblings, has been applied in the study of four adult male patients with developmental verbal auditory agnosia (Filipek et al., 1987; Caviness et al., 1989). In the four patients there was selectively a decrease bilaterally in the volumes of the posterior temporal regions. The reduction was greater in the right than in the left hemisphere. This finding is, in principle, consistent with the observations of Jernigan and Tallal (1990), who documented volumetric reduction in approximately the same bilateral posterior hemispheric regions in children with developmental impairment of language and learning.

Measures of shape

Each human brain is unique in its configuration. For example, the patterns of tertiary gyral folding in a normal brain are substantially different from normal brain to normal brain. Even the primary and secondary convolutions, though represented in each normal brain, are variable in the specific details of their size, alignment, and lengths (Bailey and Von Bonin, 1951). Furthermore, the shape of subcortical structures and their angular relations to each other may vary as much as gyral configuration from individual to individual. Thus, when normal brains are "warped" upon a standard brain, a uniform linear adjustment in size achieves no more than 65 percent of the correction. There is an additional nonlinear shape correction for each brain that amounts to 35 percent of the shape variance (Evans et al., personal communication).

By extrapolation from physiological and hodological studies in nonhuman primates, one assumes that shape variations in the normal human brain are not indicative of variations in the topology of the functional organization of the brain. Thus, with probably only rare exceptions, each normal brain has the same analytic elements. These probably have the same neighborhood relations to each other in every brain and are approximately identically aligned with respect to the principal geometric axes of the brain. The utility of stereotactically guided applications is a validation of this generalization.

Where the course of brain development has not been deflected from normal by a focally acting destructive process, one assumes that variations in brain shape reflect variations in the fine details of "wiring diagrams." Shape variations might be a manifestation of regional differences from brain to brain in neuronal or glial cell numbers, in the abundance and relative patterns of distribution of certain connections, or in the richness of myelination within and external to gray-matter structures.

The view that variations in shape reflect circuit variations is not implausible, given the behavioral variations associated with variations in cerebral symmetry in man. Evidence from experimental animals suggests that hemispheric symmetry, as opposed to asymmetry, represents the failure of the regressive developmental processes that normally result in differential elimination of neurons and/or certain patterns of connectivity (see Chapter 5). Optimally, it appears, these regressive processes occur to differential degrees in the two hemispheres.

Preliminarily, we and others have begun to explore two general strategies for metric characterization of the shape of the human brain. The first of these, based upon Fourier mathematics, is suitable for characterization of brain-surface contour. The overall gyral pattern within a given MRI plane or within a larger multiplane volume may be decomposed into its two-dimensional or three-dimensional spatial coefficients, respectively. In an exploratory application of this method in two dimensions, the concordance of the coefficient profile of two siblings was found to be substantially greater than that of a small series of age-matched normal controls (Filipek et al., 1988; Caviness et al., 1989; Kennedy et al., 1988, 1990). This exercise makes the point that the Fourier method is usefully sensitive to differences in cerebral-surface conformation. It also illustrates the importance of genetics as a determinant of cerebral shape (Caviness et al., 1989).

The second approach, also thus far pursued in only a preliminary fashion, is the application of finite element analysis as a measure of the degrees of warp that differentiate specified regions of different brains. Initial applications of this method to corresponding single-image planes in different brains has provided a measure of the differential linear and nonlinear warp across a group of normal brains. As noted above, these components have been estimated at 65 and 35 percent, respectively, in normal adult brains (Evans et al., personal communication).

Architectonic specification and physiologic localization

The gray-matter masses visualized in MRI data sets are actually composed of architectonic subdivisions *not* visible with MRI. These architectonic subdivisions may be defined confidently only when the brain is studied after death, with the aid of general cell and axonal or myelin stains. These "MRI-invisible" subdivisions correspond to the different physiologic representations of the brain. The white-matter masses of the brain, similarly, are composed of discrete, MRI-invisible axonal fascicles, each representing the specific interconnections between cortical regions and between cortical and subcortical structures. Although not explicitly definable in magnetic resonance images, the location of the architectonic and fiber fascicle structures of the normal brain may be predicted by the imaging method (Jouandet et al., 1989; Steinmetz et al., 1989). This is because these individual structural components are substantially alike in their sizes and relative positions from normal brain to normal brain. A method for making localization estimates, currently under exploration in our laboratory, is essentially a matter of adapting the MRI data set to the stereotactic spatial coordinate system of Talairach and Tournoux (1988). The constant and predictable relationship of major architectonic subdivisions and fiber tracts as well as blood vessels has been abundantly validated through years of application of the Talairach atlas in neurosurgery and radiation therapy.

The system formalized by Talairach and Tournoux (1988) is based upon an X-Y-Z coordinate system, described in an earlier section, in which the anterior commissure–posterior commissure line (AC-PC line) defines the Y axis. In the MRI data set, these and the other requisite anatomic landmarks can be directly identified in the midsagittal image plane. The Talairach-Tournoux system fits the brain to a standard three-dimensional grid in which the scaling factors make stepwise jumps in crossing the equatorial plane at the AC-PC line and in transition to the coronal sectors rostral to, including, and posterior to the AC-PC line. These stepwise jumps in scaling accommodate the nonlinearity of brain-to-brain warp, estimated to be of the order of 35 percent among normal brains (Evans et al., personal communication).

Any given coordinate reading in the Talairach-Tournoux system predicts a specific architectonic position within the brain. For surface landmarks readily identified in all normal brains, such as the principal primary and secondary fissures, the error of the prediction has been approximately plus or minus 7 percent for about 80 percent of a large number of brains studied anatomically and at surgery (Talairach and Tournoux, 1988). A comparable level of precision appears to operate when the method is applied to magnetic resonance images.

By extrapolations from studies of connectivity between homologous structures in rhesus (Pandya and Yeterian, 1985) and from the known course and relations of the principal forebrain axonal fascicles in the human brain (Krieg, 1973), we have developed a provisional atlas of connectivity for the human brain. This connectivity atlas will be accommodated by the (Talairach and Tournoux, 1988) coordinate system.

Where functional imaging studies, such as those derived from PET, EEG, or MRS, are registered with the MRI-based coordinates, the MRI data set provides the anatomic scaffold for physiologic-architectonic and connectivity correlations (Petersen et al., 1988; Posner et al., 1988). In an initial application of this MRI-based stereotactic atlas to a brain with a moderately sized left centro-sylvian infarct associated with language and visual processing deficits, we have found that the predicted zone of ipsilateral cortical disconnections corresponded closely to a zone of cortex determined by PET to have a significant reduction in blood flow.

Prospectus

The full realization of the potential of magnetic resonance imaging in developmental brain science will reflect the convergent contributions of state-of-the-art behavioral and clinical studies on the one hand and the coordinate application of the technology of anatomic and physiologic analyses on the other. The potential of the coordinate approach is massively greater than the contributions of its component methods and approaches. Only pedestrian achievement is likely to issue from analyses that are not multidisciplinary across behavior and technology and that are not convergent in application.

Current and future efforts will be continually concerned with the development of tools of analysis of a suitable degree of selectivity and sensitivity. This holds from the perspective of behavioral measures as probes for the individual components of information-processing algorithms of the brain. It also holds for applications of technology to the analysis of anatomic and physiologic parameters and for brain structure localization.

Initially, energies will be predominantly concentrated on the exploration of normal brain biology. Pathologic states may provide a range of anatomic and physiologic variations that will be interpretable only after the normal ranges have been substantially defined. As methodologic development allows, one may expect to have in hand the methods to search for two classes of anomalies of brain processing elements in association with development disorders of learning. These two hypothetical classes of abnormality are:

(1) Topologic abnormalities of the brain computational map. Examples would include elimination, conjunction, or other patterns of relative rearrangement of components. Such anomalies, to be expected as sequelae to any focally destructive process, might underlie the phenomena collectively referred to as "plasticity."

(2) Reductions in the size and analytic power of individual processing components. This class of anomaly, to be expected as a correlate of any pathologic process unfavorable to forebrain development, might in addition underlie the phenomenon of "crowding" seen after massively destructive lesions.

14

Studies of Handedness and Anomalous Dominance: Problems and Progress

Steven C. Schachter

One of the most challenging puzzles in neuroscience is to account for the development of human cerebral dominance. Defining the anatomic bases and biological associations of cerebral dominance in groups of normals and in subjects with anomalous dominance and then deducing possible mechanisms governing the development of lateralized dominance has led to considerable advances (Schachter and Galaburda, 1986). Anatomical methods characterize patterns of brain asymmetries through gross dissection, microscopic examination, or radiological imaging. Patterns of asymmetries are then correlated with functional dominance, such as handedness, and independently with anomalous functional dominance, such as learning disabilities (LD).

Many studies of the biological associations of cerebral dominance correlate handedness with quantifiable attributes, including LD. While these two current methods share certain target populations, anatomical and handedness correlation studies have largely developed in parallel; therefore, the full significance of handedness studies in clarifying the developmental origins of cerebral dominance has yet to be realized. Further compounding this problem is a lack of standardization in handedness measurement and analysis.

The objective of this chapter is to address these issues in greater detail. First, the current hypothesis of Albert Galaburda and the late

Norman Geschwind and their colleagues regarding the development of lateralized dominance is outlined. Next, the measurement and analysis of handedness are discussed. (More exhaustive treatments and genetic considerations may be found in Annett, 1978, 1985; Annett and Kilshaw, 1984; McManus, 1984; Porac and Coren, 1981.) Finally, a critical review of pertinent handedness studies is presented.

The Geschwind-Galaburda Hypothesis and Associated Predictions

The precise controls for determining asymmetries in the developing human brain are not known. Interested readers are encouraged to pursue other sources for a more complete discussion (Galaburda et al., 1978; Geschwind and Galaburda, 1984, 1985a,b; Schachter and Galaburda, 1986). One comprehensive approach, here called the Geschwind-Galaburda hypothesis, is based on two general sets of observations. First, anatomic asymmetries in brain structure, designated as standard structural dominance, are found in approximately 70 percent of normal subjects (Geschwind and Behan, 1984; Schachter and Galaburda, 1986). Any deviation from standard structural dominance is called anomalous structural dominance. Second, there are quantifiable biological attributes that correlate with anomalous structural dominance, including non-right-handedness (NRH), right- or mixed-hemisphere language dominance, and specific LD. These attributes and others are examples of anomalous functional dominance.

Besides these two observations, there are three assumptions central to the Geschwind-Galaburda hypothesis. First, intrauterine factors operating at critical times in fetal cortical development influence genetically programmed neuronal migration and maturation. Because of lateralized differences in the rate of hemisphereal formation, intrauterine factors affect brain development asymmetrically, especially late in fetal life, when left-hemisphere growth accelerates. Proposed intrauterine influences, such as testosterone and immune system factors, may therefore be associated with anomalous structural and functional dominance. Second, intrauterine factors, especially testosterone, affect the developing fetal immune system, perhaps via the thymus (Geschwind and Behan, 1984; Lahita, 1988) or hypothalamus (Wofsy, 1984), resulting in elevated frequencies of immune-mediated diseases in subjects with anomalous structural and functional dominance. Third, intrauterine factors that slow or disrupt focal development may foster additional growth or neural connectivity in other cortical or subcortical foci (Goldman and Galkin, 1978; Schneider, 1981). This may lead to unusual cognitive abilities in association with anomalous structural and functional dominance (Geschwind and Galaburda, 1985a; Marx, 1982). Corollaries to these assumptions would anticipate interrelationships between these effects, such as elevated frequencies of allergies in dyslexics (Pennington et al., 1987) and stutterers (Diehl, 1958) or increased immune disorders in dyslexics (Chapter 7, this volume; Urion, 1988) and subjects with unusual cognitive abilities (Benbow, 1986).

Hand Preference Measures

A readily measured aspect of functional dominance is handedness. Two methods of measuring handedness are generally used: (1) determining hand preference for everyday tasks and (2) measuring manual performance in a defined task, such as moving pegs. The cross-correlations of these two forms of measurement, preference and performance, are reviewed elsewhere (Bishop, 1989; Chapman and Chapman, 1987; Johnstone et al., 1979; Raczkowski and Kalat, 1974). Most studies in the adult literature use preference measures and therefore measures of manual skills and performance will not be discussed.

Handedness batteries measure hand usage for commonly performed tasks through subject report or interview or by direct observation of the subject. Several questionnaires have been developed and adapted to a variety of cultural groups and age ranges. The Edinburgh Handedness Inventory (EHI; Oldfield, 1971), shown in Figure 14.1, is widely used because of its ease of administration and brevity. Initially, Oldfield submitted a questionnaire with 20 handedness preference items based on a thesis of M. Humphrey of Oxford University (Humphrey, 1951) to 1,100 undergraduate psychology students in England and Scotland. After statistical analysis, 10 items were selected for the final questionnaire. Some items left off were culturally specific, such as using a cricket bat. The subject places a

EDINBURGH HANDEDNESS INVENTORY

Surname__________ Given Name__________
Date of Birth__________ Sex__________

Please indicate your preference in the use of hands in the following activities by putting + in the appropriate column. When the preference is so strong that you would never try to use the other hand unless absolutely forced to, put ++. If in any case you are really indifferent, put + in both columns.
Some of the activities require both hands. In these cases the part of the task, or object for which hand preference is wanted is indicated in brackets.
Please try to answer all the questions, and only leave a blank if you have no experience at all of the object or task.

		LEFT	RIGHT
1	Writing		
2	Drawing		
3	Throwing		
4	Scissors		
5	Toothbrush		
6	Knife (without fork)		
7	Spoon		
8	Broom (upper hand)		
9	Striking Match (match)		
10	Opening box (lid)		
i	Which foot do you prefer to kick with?		
ii	Which eye do you use when using only one?		

Figure 14.1 The Edinburgh Handedness Inventory (Oldfield, 1971).

plus sign in the left or right column for each item according to which hand is usually preferred. When one hand is always used for an item, or "where the preference is so strong that you [the subject] would never try to use the other hand unless absolutely forced to," then the subject puts two pluses in the appropriate column. If there is no preference, a plus is put in both columns. The extensive use of this questionnaire in the literature has substantiated Oldfield's call for "a measure of hand laterality . . . simply applied and widely used." Another popular questionnaire, published by Annett (1970), contains 12 tasks. The subject indicates hand usage for each item without regard to degree of preference.

Many investigators rely on self-described handedness—the subject states which hand is used for writing, or whether the subject considers him- or herself to be right-handed (RH), left-handed (LH), or ambidextrous. Some studies leave out ambidextrous as a choice.

Scoring Methods

Laterality quotient

The scoring method recommended by Oldfield (1971) for the EHI was based on an accepted index of handedness, (R − L)/(R + L) − 100, and results in the laterality quotient (LQ, range −100 to +100). In Oldfield's words (1971) "to calculate the L.Q., all that has to be done is to add all the +'s for each hand, subtract the sum for the left from that for the right, divide by the sum of both and multiply by 100." This scoring method has also been used with Annett's questionnaire (Lindesay, 1987).

Laterality score (Geschwind score)

An alternate method for scoring the EHI, first described by White and Ashton (1976), has been used in several other studies (Bryden, 1977; Messinger et al., 1988; Schachter et al., 1987). In one method (Schachter et al., 1987), each of the five possible responses to an item was renamed and given associated scores: "always left" (−10), "usually left" (−5), "no preference" (0), "usually right" (+5), and "always right" (+10). Scores for the 10 items are summed and the total is the laterality score (LS); the range varies from −100 to +100 by integral units of 5. The result is quickly and easily tabulated. Both the LQ and LS methods agree in the separation of subjects above and below 0. Schachter et al. (1987) emphasized the advantages of this scoring method, as illustrated by the following examples.

	Left	*Right*
1. Writing		++
2. Drawing		++
3. Throwing		++
4. Scissors		++
5. Toothbrush		++
6. Knife		++
7. Spoon		++

8. Broom (upper hand)		++
9. Striking match		++
10. Opening box		++

Laterality quotient = (20 − 0)/20 × 100 = +100
Laterality score = (10 × 10) = +100

In the example given above, the subject *always* uses the right hand for all ten items. Both the LQ and the LS are the same, +100. In the next example, however, the subject *usually* uses the right hand for all items. The LQ is still +100, but the LS is +50, thereby reflecting the difference in degree of preference.

	Left	*Right*
1. Writing		+
2. Drawing		+
3. Throwing		+
4. Scissors		+
5. Toothbrush		+
6. Knife		+
7. Spoon		+
8. Broom (upper hand)		+
9. Striking match		+
10. Opening box		+

Laterality quotient = (10 − 0)/10 × 100 = +100
Laterality score = (10 × 5) = +50

These two examples show that the LQ score of +100, often called "complete right-handedness," combines subjects with an LS of +50 together with those with an LS of +100. In effect, as pointed out by McMeekan and Lishman (1975), if the subject uses the same hand for all 10 items (whether + or ++) the LQ is always −100 or +100 (for left or right hand usage); for those subjects, LQ scores do not reflect *degrees* of preference, whether one hand is *usually* or *always* used for an item. Use of LQ assumes that individuals with equivalent scores have identical patterns of structural and functional dominance. However, evidence to the contrary will be presented later. To confuse the issue further, one study used the LS method except for those subjects who checked "usually right" for all 10 items—they were scored +100 (Messinger et al., 1988).

The example below points out the counter-intuitiveness of the Oldfield scoring method. A subject who *always* (++) uses the right hand for nine items and usually the left hand for the tenth item has a lower LQ (+89) than the subject who *usually* (+) uses the right hand for every item (LQ = +100). The respective LS scores are +85 and +50, which again more accurately reflects the degree of preference. This example is particularly noteworthy because the item "top hand on a broom" weakly correlates with the other items (Oldfield, 1971; Williams, 1986).

	Left	*Right*
1. Writing		++
2. Drawing		++
3. Throwing		++
4. Scissors		++
5. Toothbrush		++
6. Knife		++
7. Spoon		++
8. Broom (upper hand)	+	
9. Striking match		++
10. Opening box		++

Laterality quotient = $(18 - 1)/19 \times 100 = +89$
Laterality score = $(9 \times 10) + (-5) = +85$

Given the advantages of the LS, which became apparent during a project in the early 1980s (Schachter et al., 1987), Geschwind recommended this scoring method to others (Messinger et al., 1988; Meyers and Janowitz, 1985; Tan, 1988). Tan (1988) referred to the LS as the "Geschwind score." A similar scoring modification for Annett's questionnaire (1970) was described by Briggs and Nebes (1975).

Self-described handedness

Compared with quantitative measures, self-described handedness (SDH) may be less informative, perhaps because of an overlap of self-described RH in the strong and weak RH ranges of quantified handedness distributions. For example, the distribution of subjects in the study of Schachter et al. (1987) was 86 percent self-described

RH (N = 958), 4 percent ambidextrous (N = 46), and 10 percent LH (N = 113). Among all subjects with an LS of 0 to +70 (N = 152), most were self-described RH (85 percent) and yet the frequencies of blond hair and learning disabilities were significantly more common in the subjects with laterality scores in that range than in the group with LS greater than +70 (which was 99 percent self-described RH). These results suggest that the self-description of RH may not reflect biological associations as sensitively as preference measurements. This is analogous to the problem with the LQ scoring method discussed above. Chapman and Chapman (1987) found that SDH correlated better with handedness scores when subjects said whether they were "strongly" LH or RH. Similarly, Oldfield (1971) asked his subjects if they had a tendency toward LH and found that the answer underestimated the degree of departure from strong RH. In addition, SDH may not correlate with actual performance on manual dexterity tasks (Benton et al., 1962; Satz et al., 1967).

Utilizing the criterion "hand used for writing" as a handedness measure may also obscure potential associations of handedness, especially in populations with widely varying ages (see below), though perhaps not in narrow age ranges (Hugdahl et al., 1989; van Strien et al., 1987). Schachter et al. (1987) found a significant correlation between NRH, whether defined as self-described LH and ambidextrous or as LS $\leqslant$ +70, and blond hair color. Yet the association between left hand used for writing and blond hair did not reach statistical significance—14 percent of blonds wrote LH compared with 11 percent of nonblonds.

There are implications beyond those studies in which the handedness of the target population is measured by self-report or parental report. Often the family history of handedness is obtained by inquiring about hand used for writing (Bishop, 1980). The usefulness of quantifying family handedness has been demonstrated (Benbow, 1986; Urion, 1988).

Test/retest reliability

Few studies have evaluated the test/retest reliability of handedness preference measurements. Raczkowski and Kalat (1974) asked approximately 650 undergraduates to fill out a 23-item questionnaire and selected 47 for further study in order to include a large percentage of apparent LH. Subjects were given three possible responses for each item—right, left, or both. Degree of preference was not studied. One month later 27 of the 47 subjects filled out the same questionnaire a second time and the authors evaluated how often the responses to a given item were the same, though data were not included for subjects who answered "both" for a particular item on one questionnaire and "right" or "left" on the other questionnaire for that same item. For 17 of 23 items, there was agreement in the responses for at least 90 percent of subjects. The other 6 items showed less than 90 percent agreement, and one of these—"Which hand is on top of the handle when you sweep the floor with a straight broom?"—showed agreement in only 74 percent of subjects. That is, 1 in 4 subjects who answered this item "right" or "left" the first time gave the opposite response the second time. A very similar item appears on the EHI. Chapman and Chapman (1987) performed a similar study using 14 items found by Raczkowski and Kalat (1974) to have the highest test/retest agreement and showed 97 percent correlation in responses to items for males (N = 79) and 96 percent agreement for females (N = 187) when these subjects answered two identical questionnaires 6 weeks apart. Six of these 14 items appear in similar form on the EHI.

The EHI was subjected to reliability analysis by McMeekan and Lishman (1975) in 73 LH and ambidextrous volunteers. Sixty-two subjects changed the strength of hand preference without changing the laterality of hand preference on at least one item at retest, and 47 changed the laterality of hand preference ("right"/"left"/"either") on at least one item.

Methods of Analysis

In addition to selecting appropriate methods to measure and score handedness, handedness researchers must avoid an information bias (Chavance et al., 1990), control for many variables, including age, and use suitable statistical methods for describing handedness distributions and differences between groups.

Age

Advancing age correlates with increasing RH (Fleminger et al., 1977; Lansky et al., 1988; Plato et al., 1984; Schachter et al., 1987). Plato et al. (1984) noted that 11.8 percent of adult white men under 40 taken from an aging study wrote with the left hand, compared with 3.5 percent of men over 60. Similarly, Schachter et al. (1987) found that 40 percent of subjects under 30 had a NRH LS ($N = 58$), whereas 25 percent of subjects over 30 ($N = 1059$) scored as NRH. Further, the younger group used the left hand for writing more than twice as often as the older group; both age-related differences were statistically significant. One possible explanation is a decrease in cultural pressures to use the right hand for writing, which means that more naturally LH subjects in the younger age groups may be allowed to write with the left hand. There has not been a longitudinal study of handedness to exclude other factors beyond the early elementary years, however, and at least one study found an opposite relationship between age and handedness (Lahita, 1988). Age is therefore an important variable to control, particularly if handedness is assessed by the hand used to write.

Distribution

The distribution of handedness is strikingly asymmetric when assessment is based on preference measures. Annett (1970) and others (Cernacek, 1964; Raczkowski and Kalat, 1974) have pointed out the continuum of discrete measurement intervals in the distribution of handedness between complete LH and RH. Figure 14.2 shows a frequency polygon of a handedness distribution taken from a study of 1,100 professionals (Schachter et al., 1987). Handedness was measured with the EHI and LS were tabulated, as described above. The figure shows the deviation from a bell-shaped curve that is called "J-shaped" in the literature. Other investigators have found similarly shaped bimodal distributions (Briggs and Nebes, 1975; Dellatolas et al., 1990; Johnstone et al., 1979; Oldfield, 1971). Figure 14.3 shows the handedness distribution of 90 subjects on a 12-item questionnaire (Johnston et al., 1979), and Table 14.1 shows frequencies of normal subjects by LQ and LS taken from several studies.

Statistical implications

Because handedness distributions are bimodal, parametric statistical methods may not be applicable for their analysis. That is, measures of a central tendency, such as a mean, do not accurately describe

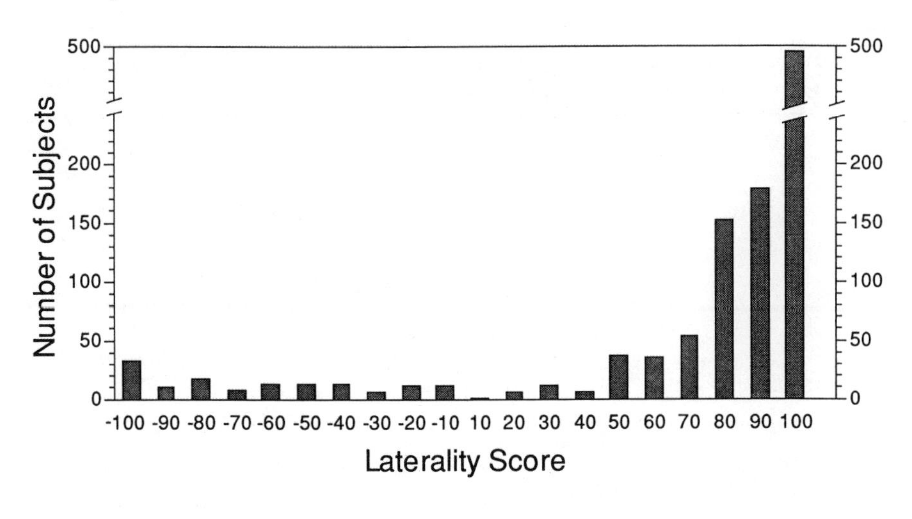

Figure 14.2 A frequency histogram of a handedness distribution taken from a study of 1,100 professionals (Schachter et al., 1987).

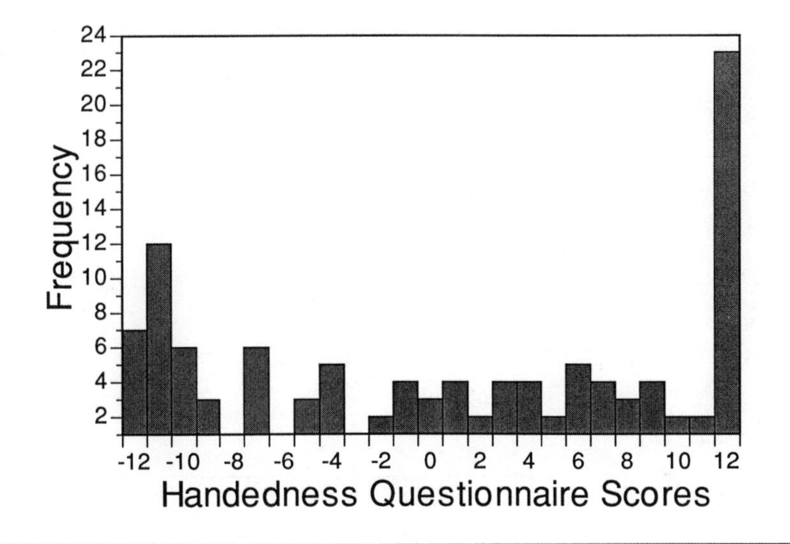

Figure 14.3 The handedness distribution derived from the score of 90 subjects on a 12-item questionnaire (Johnstone et al., 1979).

Table 14.1. Frequencies of handedness scores in normal subjects

LQ	Frequency (%)	Reference
−100	7	Schur (1986)
	8.3	Meyers and Janowitz (1985)
−100–0	6.2	Cosi et al. (1988)
	7	Weinstein and Pieper (1988)
	9	Smith (1987)
	10	Bear et al. (1986)
	11	Lahita (1988)
	13	Meyers and Janowitz (1985)
	16	Oldfield (1971)
	17	Schur (1986)
0–50	56	Lahita (1988)
50–100	34	Lahita (1988)
70 or less	24	Bear et al. (1986)
Over 70	76	Bear et al. (1986)
100	47	Bear et al. (1986)
	~50	Cosi et al. (1988)
	73	Schur (1986)

LS	Frequency (%)	Reference (Schachter et al., 1987, except where noted)
−100	2.6	
−100–0	12	
	9.8	Betancur et al. (1990)
0–50	6	
50–100	82	
Less than 70	27.3	Betancur et al. (1990)
70 or less	26	
0–70	14	
Over 70	74	
100	36	

typical handedness distributions. Despite this, many studies analyze preference score means (Dellatolas et al., 1990; Nass et al., 1987; Neils and Arams, 1987; Pennington et al., 1987; Schur, 1986; Searleman and Fugagli, 1987; Weinstein and Pieper, 1988; White and Ashton, 1976), and most do not show the handedness distributions for the study and control populations.

Other studies often rely on chi-square statistics. Yet many have insufficient numbers of subjects or controls to show significant differences between the study and control groups (Galaburda, 1987). For example, if the frequency of LH (irrespective of definition) in a control group is 10 percent and a study group has a threefold increase (30 percent) in LH, then there must be a minimum of 71 subjects and 71 controls to demonstrate statistical significance at the 0.05 level with a power of 0.80 (20 percent chance of missing the effect) (Fleiss, 1981). If the actual elevation in the study group is less than three times the control group, or the frequency of LH in controls is greater than 10 percent, then even larger numbers are necessary. Yet some studies conclude that LH is not increased when as few as 14 study subjects and equal numbers of controls are used (Bender et al., 1983), in which case LH would have to be six times more frequent in the study population (with a power of 0.80) to show statistical significance at the 0.05 level (assuming 10 percent frequency of LH in controls). Therefore, studies should include sufficient numbers of subjects and controls, based on the anticipated difference in handedness frequencies, and statistics such as chi-square should be used when the handedness distributions are not normal.

Definition of right- and left-handedness

Additional methodological problems in laterality research are created when authors adopt different definitions for RH and LH. One approach used the EHI but ignored the actual scores and counted as a left-hander anyone who expressed preference for the left hand on any item (McGee and Cozad, 1980). Geschwind and Behan (1982, 1984) used LQ = +100 and −100 as the criteria for RH and LH, respectively. Others (Behan and Geschwind, 1985; Benbow, 1986; Cosi et al., 1988; Dellatolas et al., 1990; Geschwind and Behan, 1982; Koff et al., 1986; Messinger et al., 1988; Meyers and Janowitz, 1985; Schur, 1986; Smith, 1987; Weinstein and Pieper, 1988) define

LH as LQ less than 0, after Oldfield (1971). As noted above, separating RH from LH by using a score of 0 has the advantage of identifying the same groups irrespective of the scoring method (LS or LQ) and, to some extent, regardless of the questionnaire used. As a result, this approach simplifies cross-study comparison (Dellatolas et al., 1990). In addition, it readily distinguishes laterality of hand preference (LH vs. RH) and agrees well with SDH. Schachter et al. (1987) found that 97 percent of self-described LH score 0 or less (LS) and 99 percent of self-described RH scored over 0. When LH is defined as a score below 0, there is no advantage to using the LS method over the LQ, although this point is not generally stated (Meyers and Janowitz, 1988).

The term *non-right-handed* (NRH) may designate all individuals other than strong RH and has been defined as LS or LQ less than or equal to +60 or +70, after Schachter et al. (1987) (Bear et al., 1986; Betancur et al., 1990; Dellatolas et al., 1990; Urion, 1988). Schachter et al. (1987) selected +70 as a cutoff point on the basis of the cross-correlation of LS and SDH in 1,117 professionals. Among self-described RH ($N = 958$), the distribution of LS was 86 percent over +70, 13 percent between 0 and +70, and 1 percent less than 0. Further, among those with LS over +70 ($N = 826$), 99 percent were self-described RH. Therefore, the region of LS over +70 was designated as RH, and LS less than or equal to +70 as NRH. This method of defining handedness emphasizes *degrees* of hand preference rather than direction.

In some studies, the definitions used for RH and LH exclude some of the subjects from entering into the analysis because only a limited portion of the entire handedness range is evaluated. Such methods typically emphasize direction, but not degree, of hand preference. For instance, if an investigator defines RH and LH as LQ = +100 and LQ = −100, respectively, and then compares the frequencies of these two scores in subjects and controls, those with LQs between these extremes will not be included. Similarly, if the same battery and scoring method are used and subjects with an LQ less than or equal to 0 are defined as LH and compared with controls, then those with LQ greater than 0 will be excluded. Potentially useful information may be lost in this manner if a biological trait is overrepresented in the segment of the handedness range excluded from analysis or if there are insufficient numbers of individuals (e.g., LQ = −100). In support of this objection, several studies suggest that *degree* of hand preference may be more important than *direction* of hand preference with respect to anomalous functional dominance. In a study of handedness and hair color using the EHI and LS (Schachter et al., 1987), there was no significant correlation of blond hair with LS less than 0. However, the percentage of blonds among those with LS between 0 and +70 was almost 2.5 times greater than the percentage of nonblonds. As a result, blond hair color correlated with LS less than +70 because the upper-middle range strongly contributed to the statistical association. Further, the frequency of learning disability was higher in the those with LS between 0 and +70 (11 percent) than in the those with LS less than 0 (8 percent).

Two other studies further illustrate the usefulness of considering the full handedness spectrum—both show elevated frequencies of biological traits in the upper-middle RH ranges of study groups compared with controls. Dellatolas et al. (1990) found stuttering was more than doubled in those with handedness scores equivalent to LS of +70 or less, compared with the frequency in those with LS over +70, and Lahita (1988) established that the percentage of patients with systemic lupus erythematosus who had LQ between 0 and +50 was significantly greater than the percentage in controls.

Other scoring methods may present additional problems in defining handedness. For instance, Pennington et al. (1987) measured LQ on the EHI and defined handedness in a way that would classify a subject with LS of +80 as mixed-handed while a subject with LS of +45 would be RH. As mentioned earlier, the use of the LQ scoring method combines LS = +50 (usually right-handed for all ten items) with LS = +100 (both have an LQ of +100). Yet the above examples suggest that the lower LS (LS = +50) is associated with different frequencies of certain biological traits than are found among those with the higher LS. Therefore, use of the LQ scoring method or analysis of restricted segments of the handedness range (such as examining only those with scores less than 0) could potentially obscure biological correlations through exclusion of data from pertinent domains of the laterality range.

Recommendations

The methodological issues raised in the above sections may be reformulated into the following recommendations. Handed studies should include an appropriate number of study and control subjects to demonstrate statistical significance and must control for age and gender as well as other factors not discussed, including educational background and familial handedness (McGee and Cozad, 1980). A sensitive, validated measuring instrument (e.g., the EHI, rather than SDH) should be given in the same manner to subjects and controls and scored quantitatively with a method, such as LS, that assigns a score parallel with the degree and direction of handedness. The distribution of the entire range of scores should be tested for normality, and segments of the handedness range should be selected and assessed for differences between study and control groups using the appropriate statistics based on the distribution of scores. Finally, the selection of segments along the handedness spectrum that include as many subjects as possible will enable investigators to further explore which segments are most informative for particular biological variables. To date, some studies support a selection in the upper-middle RH range as an important cutoff point. Perhaps further work will show the value of +50 (rather than +70) as a cutoff; this score, like the score of 0, may be more consistently defined across questionnaires and across scoring methods.

Handedness Studies and the Geschwind-Galaburda Hypothesis

There are two general categories of handedness studies. *Population studies* assess handedness in a group (usually large) of individuals who collectively have no known diseases or conditions and compare the frequencies of certain measurable traits within specified domains of the handedness range. *Condition-specific studies* compare the handedness distributions of a defined group of experimental subjects (for example, patients with systemic lupus erythematosus) and a control group. The reports by Geschwind and Behan (1982, 1984) illustrate both types of studies.

The remainder of this chapter will consist of examples from the handedness literature pertinent to the Geschwind-Galaburda hypothesis. These examples are separated by type of study and approached from the standpoint of study design as discussed above, though variations in study design may hamper the presentation of data in a consistent format.

Handedness and Cerebral Asymmetry

The first task of handedness studies is to affirm an association between structural brain asymmetries and laterality or degree of handedness. Although the existence of cerebral asymmetries has been known for over 100 years (Eberstaller, 1884), gross anatomical cerebral asymmetries have not been clearly correlated with handedness. Geschwind and Levitsky (1968), who noted a larger left planum temporale in 65 of 100 brains studied, were unable to obtain handedness information. The study of Rubens et al. (1976) did not include any left-handers. Other studies that have documented cerebral (usually temporal lobe) asymmetries in adults and infants (Beck, 1955; Campain and Minckler, 1976; Chi et al., 1977; Falzi et al., 1982; Teszner et al., 1972; Wada et al., 1975; Weinberger et al., 1982; Witelson and Pallie, 1973) have not evaluated handedness. McRae et al. (1968), who evaluated the dominant hand used for a variety of tasks in 100 consecutive patients undergoing pneumoencephalograms and ventriculograms, had too few LH and ambidextrous subjects to test whether these groups showed patterns of ventricular asymmetry different from those of right-handers.

Cerebral arteriography

LeMay and Culebras (1972) retrospectively compared the cerebral arteriograms of 18 patients stated in their hospital records to be LH with 44 consecutive arteriograms from other patients. The authors assumed that 90 percent of these controls were right-handed. They found a more highly developed parietal operculum on the left in 38 of 44 controls, whereas 15 of 18 LH had bilaterally equal parietal opercularization. Hochberg and LeMay (1975) defined handedness according to self-report or the hand used for writing and analyzed bilateral cerebral angiograms in 123 RH and 38 LH. In 67 percent of RH, sylvian point angles were over 10 degrees more on the right than the left; the same finding occurred in only 21 percent of LH.

Computed tomography

Several studies have evaluated handedness and cerebral asymmetries seen on computed tomography (CT). LeMay and Kido (1978) correlated frontal and occipital lobe widths with SDH in 165 patients. Among the 80 RH, 75 percent had longer left occipital lobes and only 9 percent had larger right occipital lobes. The frontal lobe was wider on the right than on the left in 58 percent of RH and equal in width on both sides in 30 percent whereas, only 12 percent of RH had wider left frontal lobes. The LH group ($N = 85$) showed an almost equal distribution of subjects with wider right occipital lobes (32 percent), wider left occipital lobes (34 percent), and occipital lobes of equal size (34 percent). Similarly, there were nearly equal numbers of LH with wider right frontal lobes (35 percent), wider left frontal lobes (31 percent), or symmetric frontal lobes (34 percent). Although the pattern of cerebral asymmetries seemed to differ between the two handedness groups, no statistical analysis was given. In another CT study, LeMay (1977) found similar trends in lobar asymmetries related to handedness, as well as skull asymmetries in RH analogous to the CT findings, but again, no statistical analysis was made.

Koff et al. (1986) measured handedness with a questionnaire in 146 dextral and 26 sinistral patients aged 16–80. The method used to define handedness was quantitative, but the scores lumped together strong RH with subjects in the upper-middle range. No correlation was found between handedness and lobar asymmetries (frontal and occipital), nor between degree of handedness and degree of asymmetry. Similarly, Chui and Damasio (1980) found no relation between handedness and frontal and occipital asymmetries, except for left occipital width predominance in NRH compared to RH. Deuel and Moran (1980) studied 94 children and could not correlate parent-reported hand preference with CT asymmetries.

Bear et al. (1986) used the EHI and the Oldfield scoring method in 66 patients. Non-right-handers ($N = 16$; LQ of +70 or less) had a reduction or reversal of left occipital predominance, especially for width, compared with RH ($N = 50$; LQ over +70), but no difference in frontal asymmetries.

Magnetic resonance imaging

Kertesz et al. (1986) used preference and performance measures to classify 20 subjects as either RH or LH, though the scoring method, statistical findings, and classification criteria were not detailed. The opercular parietal demarcation of sulci on MRI scan was sharper on the right side in 60 percent of RH, as compared with 10 percent of LH. Occipital width was larger on the left in 90 percent of RH and 30 percent of LH. The anterior frontal width, parietal width, and sulcal demarcations taken together predicted handedness in 19 of 20 subjects.

Some of the above evidence lends support to an association between handedness and cerebral asymmetries. In general, LH have less interhemispheric structural differences than RH, and among RH, the right frontal and left occipital lobes tend to be larger than the corresponding lobes of the other hemisphere. However, some of the studies find no statistical association, and therefore the goal of correlating RH with standard structural dominance and LH or NRH with anomalous structural dominance has not been completely achieved. Few studies have used a quantified handedness measure and more are needed, especially as new *in vivo* methods for quantifying cerebral anatomy are developed (Jouandet et al., 1990) and as methodological issues in quantifying cerebral morphology are resolved (Weis et al., 1989). With improvements in these areas, additional handedness studies may identify those measures that correlate best with specific lobar or even gyral asymmetries.

Handedness and Gender

A corollary of the Geschwind-Galaburda hypothesis is that intrauterine exposure to testosterone has a growth-retarding effect on fetal brain development and that, because of the asymmetric rate of development of the two hemispheres, it particular affects left-hemisphere maturation. This predicts that NRH would correlate with male gender and other masculinizing conditions of fetal life.

Population studies

The report of Oldfield (1971) found that more males than females are left-handed. Two other large population studies confirmed Oldfield's findings. In a sample of over 11,000 children born in one week of March 1958, Calnan and Richardson (1976a) found a greater rate of left- and mixed-handedness in boys than in girls. Lansky et al. (1988) chose 2,083 adults of age 18–80 at random and measured handedness using 5 preference items. Males were less RH than females, even when the subjects were controlled for age and race. Further studies have been reviewed by Annett (1985).

Condition-specific studies

Boys with an extra X chromosome (a condition referred to as 47,XXY or Klinefelter's syndrome) have deficient verbal skills (Walzer, 1985). Netley and Rovet (1984) studied hemispheric lateralization and handedness in 32 boys with 47,XXY and showed significant shifts toward right-hemisphere dominance for both verbal and nonverbal stimuli. Boys with 47,XXY were twice as likely (21 percent) to be NRH than controls (10 percent); this statistic is not significant, however, because of low numbers of subjects in the sample. NRH was defined as 7 or fewer right-hand preferences on 10 simple manual activities. The same authors (Netley and Rovet, 1982) had earlier reported that 8 of 33 boys (24 percent) with 47,XXY were NRH, whereas only 10 percent of controls were NRH, a statistically significant difference. These were apparently the same boys who comprised the 1984 study; one NRH boy was presumably left out of the later report, leaving 7 of 32 of the sample with NRH (approximately 21 percent).

Another study tested an equally low number of subjects and used nonquantitative laterality measures (Ratcliffe and Tierney, 1982). No difference in handedness was found. Of interest is one NRH boy with 47,XXY who had high neonatal testosterone levels and a severe language delay. Similarly, Bender et al. (1983) did not find elevated NRH in fourteen 47,XXY boys.

In congenital adrenal hyperplasia, low glucocorticoid levels disinhibit the hypothalamo-pituitary-adrenal axis, resulting in increased testosterone production. Nass et al. (1987) compared mean LQ on the EHI in 37 patients with congenital adrenal hyperplasia with scores from same-sex siblings. The females were significantly more LH than their sisters; the males were no different from their brothers.

The biological origins of homosexuality remain obscure. Suggested factors include abnormal intrauterine testosterone levels and maternal stress. James (1989) presented evidence to account for both in accordance with the Geschwind-Galaburda hypothesis (Geschwind and Galaburda, 1985a). Rosenstein and Bigler (1987) used the EHI and compared the handedness of 38 nonheterosexual psychology students with that of 51 heterosexual students. Although no differences were found, only 5 members of the study group, all male, were exclusively homosexual. Lindesay (1987) applied the LQ scoring method to Annett's questionnaire (1970) and found elevations of NRH in 94 homosexual men, compared with 100 heterosexual men. McCormick et al. (1990) gave the same questionnaire to 38 male and 32 female homosexual members of a homophile organization. A subject who used the left hand for any item was classified as having a non-consistent right-hand preference (non-CRH). Among the women (age 19–45), 69 percent were non-CRH and 19 percent wrote left-handed. Both criteria were significantly more common in this group than in control groups from other studies. The corresponding figures for the men were 45 and 11 percent, respectively.

Thus, the evidence from population studies reviewed above attests to a correlation between NRH and the male gender. Furthermore, the condition-specific studies of fetal masculinizing conditions, while limited in number, are also suggestive.

Handedness and Learning Disabilities

Intrauterine factors that adversely affect the growth and organization of the fetal brain, especially the left hemisphere, may disrupt focal cortical development of language centers and be reflected behaviorally as learning disabilities. Anomalous structural dominance in dyslexics was first reported by Drake (1968) in an ambidextrous 12-year-old and has been substantiated through additional postmortem gross and microscopic studies by Galaburda and colleagues (Galaburda, 1983; Galaburda and Kemper, 1979; Galaburda et al., 1984; Humphreys et al., 1990; Kaufmann and Galaburda, 1989). Among 24 selected dyslexics, Hier et al. (1978) found that reversal of usual asymmetry (42 percent), discovered by head CT scans, correlated with low mean verbal IQ scores. Using MRI, Hynd et al. (1990) noted right anterior widths and left plana temporale smaller in 10 dyslexics than in age- and sex-matched controls.

Another possible outcome of intrauterine abnormalities is exuberant neural growth or neural connectivity in other cortical or subcortical foci (Goldman and Galkin, 1978; Schneider, 1981), potentially resulting in unusual cognitive abilities. Thus, both learning disabilities and intellectual giftedness may be predicted to be associated with anomalous structural dominance and NRH.

Population studies

In a large study of children identified by doctors, teachers, or a speech test to have a speech disorder, Calnan and Richardson (1976b) did not find an excess of LH (as related by parents). However, LH children did significantly worse than RH children on tests of reading comprehension (Calnan and Richardson, 1976a). Similarly, van Strien et al. (1987) found significantly more reading and arithmetic difficulties in 241 LH university students than in 178 of their RH schoolmates. Handedness was determined by the hand used for writing.

In three separate studies, Geschwind and Behan (1982, 1984) found increased frequencies of dyslexia and stuttering in subjects with LQ = −100 compared with those with +100. Furthermore, dyslexia and stuttering were more common in the close relatives of the LH than in relatives of the RH (Geschwind and Behan, 1982). Dellatolas et al. (1990) evaluated handedness in 698 volunteers with a method similar to LS. Stuttering was more than doubled in those with scores equivalent to LS of +70 or less than in those with LS over +70.

Schachter et al. (1987) received the EHI from 1,117 randomly selected college-educated professionals and calculated LS. Overall, 4 percent of the respondents acknowledged a history of LD (dyslexia, hyperactivity, delayed speech, difficulty learning math, autism, Tourette's syndrome). The frequency of LD was three times higher in subjects with an LS of +70 or less (9 percent) than in those with LS above +70 (3 percent). The correlations were observed in both women and men.

A potential confounding issue in studies of reading disabilities and handedness is the preponderance of males. In Orton's 1925 report, 14 of 15 children with reading disability were boys. Regardless of whether males are overrepresented because of an identification bias by school personnel (Shaywitz et al., 1990), gender should be controlled in comparisons of different groups.

Condition-specific studies

Over 50 years ago, Orton (1937) drew attention to the association of LH and speech and learning disorders. Not all agree with this conclusion: after a review of the evidence, Satz and Soper (1986) termed the association of dyslexia and LH a "myth."

Annett and Kilshaw (1984) found elevated LH writing in dyslexic boys (19 percent) and girls (15 percent) attending a remedial clinic; only 8.2 percent of controls wrote LH. Neils and Aram (1986) correlated handedness on a 5-item preference inventory with type or severity of developmental language problem in 75 children of age 4–6 with an IQ over 85. Mean handedness scores were calculated for the patients and for 36 normal controls and no differences were found. However, severely impaired children were more likely to be LH than mildly impaired children.

Thomson (1975) gave 60 children with reading ages at least 18 months below their chronological age a 5-item questionnaire and compared the results with the scores of 60 controls. The gender distributions were not given. There were no significant differences in the frequencies of LH writing or completely LH children. Still, mixed-handedness was increased in the reading-impaired group. Connolly (1983) defined handedness of 91 selected children with LD according to an infrequently used battery and compared the results with those of subjects from other studies. No difference in mixed-handedness was seen, and frequencies of RH and LH were not given.

Several studies have shown elevated NRH in autistic children (for a review, see Bryson 1990), and two provide evidence for anomalous structural dominance (Hauser et al., 1975; Tsai et al., 1983). In an uncontrolled series of 17 autistic children, Hauser et al. (1975) noted that 8 were LH (including one who was "remarkably skilled in puzzles" and another who had "faultless rhythm on drums"). In another uncontrolled study, Tsai (1983) assessed handedness in 70 patients admitted to the Iowa Autism Program by observing performance on 5 tasks. Right- and left-handedness were defined by lack of use of the other hand for any task. Among autistic children over age 8, 21 percent were LH and 43 percent were mixed. In 40 consecutive autistic patients, Tsai and Stewart (1982) found NRH in 57 percent of autistic children. Colby and Parkison (1977) compared 20 autistic children with 25 normal children on a series of simple manual tasks. The percentage of children using the left hand for at least one activity was significantly higher (65 percent) among autistic children than among controls (12 percent).

Behan and Geschwind (1985) gave the EHI to 83 children with severe hyperactivity. In boys, the frequency of LQ less than 0 was 19 percent. The corresponding frequency in girls was 21 percent. While these figures suggest increased sinistrality (see Table 14.1), the study was not controlled.

Benbow (1986) studied handedness and intellectual giftedness. He gave the EHI to 416 students who were estimated to be the top 1 in 10,000 of their age group in mathematical or verbal reasoning ability. The precocious students were significantly more LH (LQ less than 0; 15.1 percent) than controls (10.2 percent). Other studies have shown increased LH in student architects and musicians (Gotestam, 1990) and in professional groups such as mathematicians, architects, artists, and musicians (Annett and Kilshaw, 1982; Peterson, 1979; Peterson and Lansky, 1974). Annett and Manning (1990) have recently correlated left-handed writing with arithmetic ability in normal 9- to 11-year-old children.

The savant syndrome refers to persons who are mentally handicapped but who have some extraordinary abilities, including calendar calculating and musical skills. Limited evidence suggests a pattern of anomalous structural dominance, with left-hemisphere damage predisposing the person to excessive right-hemisphere function; furthermore, the 6-to-1 ratio of males to females implies a role for testosterone (Treffert, 1988). However, clear patterns of handedness have not been described.

In summary, the available evidence from population studies presented above, which is derived from quantified handedness measurement, confirms an association between NRH and learning disorders. Condition-specific studies have produced conflicting findings but are limited by problems in study design.

Handedness and Immune-Mediated Disease

As noted earlier, the Geschwind-Galaburda hypothesis states that intrauterine factors, perhaps including testosterone, may affect the developing fetal brain and immune system in parallel and in turn lead to both anomalous structural dominance and susceptibility to immune-mediated disease. This would predict elevated frequencies of immune illnesses in NRH, particularly in males.

Population studies

In a landmark paper that stimulated much of the work reviewed in this chapter, Norman Geschwind and Peter Behan (1982) reported an association between LH and immune disease. The authors compared 253 shoppers at a London store for left-handers with LQ = −100 with 253 age- and sex-matched controls from the general population of Glasgow with LQ = +100. The frequency of immune disorders reported by the study subjects was 2.7 times higher in the sinistrals, who showed particular increases in thyroid and bowel diseases. Furthermore, the frequency of immune conditions was higher in the first- and second-degree relatives of LH than in relatives of RH subjects. These results were nearly duplicated in 247 LH and 647 RH Glasgow citizens with hospital-diagnosed immune disorders and in another study of 304 patients attending Scottish clinics (Geschwind and Behan, 1984). The methods employed in these studies have been criticized (Bishop, 1986; Satz and Soper, 1986).

Defining handedness according to the preferred hand for writing, van Strien et al. (1987) found no excess of autoimmune diseases in 138 LH university students or in their families compared with 89 RH classmates. Dellatolas et al. (1990) scored handedness questionnaires received from 698 volunteers with a method similar to the LS. When LH was defined as a score of 30 or more (range 10–50; equivalent to an LS less than or equal to 0), there was no difference in frequencies of allergic disorders among LH and RH groups. However, individuals assessed as extreme RH (equivalent to an LS of +95 to +100) had significantly lower frequencies of asthma and hay fever than were reported by other subjects. Finally, Betancur et al. (1990) found no differences in LS of less than 0 or less than +70 in 325 patients with allergies attending a clinic compared with 205 age- and sex-matched controls, nor in patients with IgE-mediated symptoms contrasted to patients without.

Condition-specific studies

Weinstein and Pieper (1988) compared the mean LQ and frequency of LQ less than 0 on the EHI of 390 atopic patients visiting a private allergy practice with those of 429 controls attending a health screening program. Patients with asthma and controls with self-defined allergies (SDA) had lower mean LQ than controls without SDA. Overall, 12 percent of patients had LQ less than 0, whereas 6.5 percent of controls had negative LQ. Age and gender were not controlled. Smith (1987) found increased LH (LQ less than 0) in 313 patients at an allergy clinic compared with 350 age- and sex-matched controls selected at random from a railway station, particularly patients with atopy and IgE-mediated symptoms. Among 10 patients with IgE-mediated urticaria, half had LQ less than 0. There was no difference in LH between atopic and nonatopic patients, though the middle handedness ranges were not analyzed.

Systemic lupus erythematosus (SLE) is an autoimmune disease of connective tissue. Salcedo et al. (1985) looked at which hand was used for writing in 54 pediatric patients with SLE. Approximately 15 percent were LH, which was not significantly different from two control groups taken from other studies, one of which was unpublished. Schur (1986) compared LQ of 88 adults with SLE (9 percent males) to 120 normal controls (18 percent males) and 130 relatives of the SLE patients (43 percent males). Distributions for the LQ were not presented. LQ means were calculated for each group and no significant differences were found. Furthermore, the percentage of patients with LQ less than 0 in each group was not significantly different. On the other hand, Lahita (1988) measured LQ of 114 patients with SLE and found that the percentages of individuals with LQ less than 0 and those with LQ between 0 and +50 were significantly greater among patients than among a control population of artistic individuals and employees of an advertising agency.

Type-I diabetes (insulin-dependent) may be immune-mediated. Searleman and Fugagli (1987) solicited responses to a newspaper ad from individuals with diabetes and measured handedness on a 7-item battery using a method similar to the LS. Strong LH was nearly three times more frequent among males with type-I diabetes than among subjects with type-II diabetes (not dependent on insulin). Additionally, the mean handedness score of male type-I diabetics was significantly shifted toward LH.

There are several studies of handedness and autoimmune bowel diseases. Meyers and Janowitz (1985) compared the frequency of LS less than 0 in patients with ulcerative colitis and Crohn's disease with that in controls (all employees on one floor of a medical school) and found no differences. The percentage of subjects in the middle range was not given. Geschwind and Behan (1984) used the same definition of LH and showed significantly increased sinistrality in patients with Crohn's disease ($N = 30$), celiac disease ($N = 36$), and ulcerative colitis ($N = 32$). Searleman and Fugagli (1987) studied the handedness of 152 subjects with Crohn's disease, 46 patients with ulcerative colitis, and 279 respondents to a college campus ad. They found higher rates of LH in females with either condition and in men with both conditions than in the control group. Persson and Ahlbom (1988) reanalyzed these data and found that LH had relative risks of 2.9 for ulcerative colitis and 1.5 (males) or 3.3 (females) for Crohn's disease.

Myasthenia gravis is an autoimmune disease of muscle tissue. In two studies, Geschwind and Behan (1982, 1984) found that 13–19 percent of patients with myasthenia gravis visiting neurological clinics in Glasgow had LQ less than 0 ($N = 134$), compared with 7–11 percent in two groups of general population controls; both studies were statistically significant. Cosi et al. (1988) compared 102 patients with acquired generalized myasthenia gravis, of age 13–78, with 178 clinical staff members, attendants, and management in an Institute of Neurology for LQ less than 0 and found no differences. Yet, the frequency of LQ scores of +100 was approximately 30 percent lower in myasthenics than in controls.

Thus, the current evidence for an association of NRH with immune-mediated disease is mixed. The strongest association appears to exist in subjects with asthma, atopy, and type-I diabetes. However, these data, as well as the results for SLE, autoimmune bowel diseases, and myasthenia gravis, must be interpreted in light of the issue of study design discussed earlier. Furthermore, there are many immune-mediated diseases that have not yet been tested for handedness associations.

Conclusion

The essential assumption of the Geschwind-Galaburda hypothesis asserts that intrauterine factors influence fetal brain development. This provoking and exciting conceptual breakthrough to the nature-nurture argument has stimulated a great deal of work. Yet to this point, the focus has largely been limited to two putative influences —testosterone and the immune system and whether these two are completely independent. Furthermore, the available evidence for effects on brain development have been limited to indirect measures of cortical organization, whether structural or functional. The studies reviewed in this chapter neither refute nor fully confirm the influence of testosterone and the immune system and have not determined whether the action of one may be independent of the other. Further gains will depend on reducing inconsistencies in study design and establishing accepted and widely applied methods for defining, measuring, and analyzing handedness.

The Geschwind-Galaburda hypothesis gives special importance to handedness correlation studies by seeking to increase the understanding of the interrelationships between structural and functional cerebral dominance. Additional population studies of handedness and gender, learning disabilities, and immune-mediated disease are needed. More condition-specific studies, particularly those which specify clear inclusion and exclusion criteria to ensure study group homogeneity, are needed to complement recent work and to expand the list of conditions tested for handedness correlations.

Even if well-designed studies fail to affirm a role for either testosterone or immune factors in fetal brain development, this failure would not deny the core assumption of the Geschwind-Galaburda hypothesis. New associations of handedness or anomalous structural dominance may yet be identified, and hence other potential intrauterine influences on brain development will emerge, continuing the search for the solution to the puzzle of cerebral dominance.

Acknowledgments

The author wishes to dedicate this chapter to the memory of Norman Geschwind and to express deep appreciation to Albert M. Galaburda for his enthusiastic support and encouragement.

References

Contributors

Index

References

1. Regressive Events in Early Cortical Maturation

Angevine, J. B., Jr., and R. L. Sidman. 1961. Autoradiographic study of cell migration during histogenesis of cerebral cortex in the mouse. *Nature* 192:766–768.

——— 1962. Autoradiographic study of histogenesis in the cerebral cortex of the mouse. *Anatomical Record* 142:210–222.

Bates, C. A., and H. P. Killackey. 1984. The emergence of a discretely distributed pattern of corticospinal projection neurons. *Developmental Brain Research* 13:265–273.

Bear, M. F., L. N. Cooper, and F. F. Ebner. 1987. A physiological basis for a theory of synapse modification. *Science* 237:42–48.

Bruce, L. L., and B. E. Stein. 1988. Transient projections from the lateral geniculate to the posteromedial lateral suprasylvian visual cortex in kittens. *Journal of Comparative Neurology* 278:287–302.

Caminiti, R., and G. M. Innocenti. 1981. The postnatal development of somatosensory-callosal connections after partial lesions of somatosensory areas. *Experimental Brain Research* 42:53–62.

Changeux, J. P., and A. Danchin. 1976. Selective stabilisation of developing synapses as a mechanism for the specification of neuronal networks. *Nature* 264:705–712.

Crandall, J. E., and V. S. Caviness. 1984. Thalamocortical connections in newborn mice. *Journal of Comparative Neurology* 228:542–556.

Cusick, C. G., and R. D. Lund. 1982. Modification of visual callosal projections in rats. *Journal of Comparative Neurology* 212:385–398.

Dawson, D. R., and H. P. Killackey. 1985. Distinguishing topography and somatotopy in the thalamocortical projections of the developing rat. *Developmental Brain Research* 17:309–313.

Dwyer, B. E., R. N. Nishimura, C. L. Powell, and S. L. Mailheau. 1987. Focal protein synthesis inhibition in a model of neonatal hypoxic-ischemic brain injury. *Experimental Neurology* 95:277–289.

Finlay, B. L., B. Miller, S. Nioka, D. Nagy, A. Zaman, and B. Chance. 1990. Neonatal hypoxia increases the number of visual callosal projections in the cat. *Neuroscience Abstracts* 16:850.

Finlay, B. L., and S. L. Pallas. 1989. Control of cell number in the developing visual system. *Progress in Neurobiology* 32:207–234.

Finlay, B. L., and M. Slattery. 1983. Local differences in amount of early cell death in neocortex predict adult local specializations. *Science* 219:1349–1351.

Frost, D. O., and Y. P. Moy. 1989. Effects of dark rearing on the development of visual callosal connections. *Experimental Brain Research* 78:203–213.

Gu, Q., M. F. Bear, and W. Singer. 1989. Blockade of NMDA-receptors prevents ocularity changes in kitten visual cortex after reversed monocular deprivation. *Developmental Brain Research* 47:281–288.

Hagberg, H., P. Andersson, I. Kjellmer, K. Thiringer, and M. Thordstein. 1987. Extracellular overflow of glutamate, aspartate, GABA and taurine in the cortex and basal ganglia of fetal lambs during hypoxia-ischemia. *Neuroscience Letters* 78:311–317.

Hamburger, V., and R. Levi-Montalcini. 1949. Proliferation, differentiation and degeneration in the spinal ganglia of the chick embryo under normal and experimental conditions. *Journal of Experimental Zoology* 111:457–502.

Herkenham, M. 1986. New perspectives on the organization and evolution of nonspecific thalamocortical projections. In *Cerebral Cortex,* vol. 5, ed. E. G. Jones and A. Peters. New York: Plenum Press.

Heumann, D., and G. Leuba. 1983. Neuronal death in the development and aging of the cerebral cortex of the mouse. *Neuropathology and Applied Neurobiology* 9:297–311.

Hollyday, M., and V. Hamburger. 1976. Reduction of naturally occurring motor neuron loss by enlargement of the periphery. *Journal of Comparative Neurology* 170:311–320.

Howard, B., B. Miller, and B. L. Finlay. 1989. Reorganization of visual callosal projections after early thalamic lesions in the golden hamster. *Society of Neuroscience Abstracts* 15:1339.

Innocenti, G. M. 1981. Growth and reshaping of axons in the establishment of visual callosal connections. *Science* 312:824–827.

——— 1986. General organization of callosal connections in the cerebral cortex. In *Cerebral Cortex,* vol. 5, ed. E. G. Jones and A. Peters. New York: Plenum Press.

Innocenti, G. M., P. Berbel, and S. Clarke. 1988. Development of projections from auditory to visual areas in the cat. *Journal of Comparative Neurology* 272:242–259.

Innocenti, G. M., and R. Caminiti. 1980. Postnatal shaping of callosal connections from sensory areas. *Experimental Brain Research* 38:381–294.

Innocenti, G. M., and S. Clarke. 1984a. Bilateral transitory projections to visual areas from auditory cortex. *Developmental Brain Research* 14:143–148.

——— 1984b. The organization of immature callosal connections. *Journal of Comparative Neurology* 230:287–309.

Innocenti, G. M., S. Clarke, and R. Kraftsik. 1986. Interchange of callosal and association projections in the developing visual cortex. *Journal of Neuroscience* 6:1384–1409.

Innocenti, G. M., and D. O. Frost. 1979. Effect of visual experience on the maturation of the efferent system to the corpus callosum. *Nature* 280:231–234.

——— 1980. The postnatal development of visual callosal connections in the absence of visual experience or of the eyes. *Experimental Brain Research* 39:365–375.

Ivy, G. O., R. M. Akers, and H. P. Killackey. 1979. Differential distribution of callosal projection neurons in the neonatal and adult rate. *Brain Research* 173:532–537.

Ivy, G. O., and H. P. Killackey. 1981. The ontogeny of the distribution of callosal projection neurons in the rat parietal cortex. *Journal of Comparative Neurology* 195:367–389.

——— 1982. Ontogenetic changes in the projections of neocortical neurons. *Journal of Neuroscience* 2:735–743.

Johnston, M. V., and F. S. Silverstein. 1987. Perinatal hypoxia. In *Animal Models of Dementia,* ed. J. T. Coyle. New York: A. R. Liss.

Killackey, H. P., and L. M. Chalupa. 1986. Ontogenic change in the distribution of callosal projection neurons in the postcentral gyrus of the fetal rhesus monkey. *Journal of Comparative Neurology* 244:331–348.

Kostovic, I., and P. Rakic. 1980. Cytology and time of origin of interstitial neurons in the white matter in infant and adult human and monkey telencephalon. *Journal of Neurocytology* 9:219–242.

Laemle, L. K., and S. C. Sharma. 1986. Bilateral projections of neurons in the lateral geniculate nucleus and nucleus lateralis posterior to the visual cortex in the neonatal rat. *Neuroscience Letters* 63:207–214.

Lamb, A. H. 1984. Motoneuron death in the embryo. *Critical Reviews in Clincal Neurobiology* 1:141–179.

Luskin, M. B., and C. J. Shatz. 1985. Studies of the earliest generated cells of the cat's visual cortex: Cogeneration of subplate and marginal zones. *Journal of Neuroscience* 5:1062–1075.

Marín-Padilla, M. 1978. Dual origin of the mammalian neocortex and evolution of the cortical plate. *Anatomical Embryology* 152:109–126.

McConnell, S. K. 1988. Development and decision-making in the mammalian cerebral cortex. *Brain Research Review* 13:1–23.

Miller, B., M. S. Windrem, L. Anllo-Vento, and B. L. Finlay. 1987. Minor reorganization of thalamocortical projections following large neonatal thalamic lesions in the golden hamster. *Society of Neuroscience Abstracts* 13:1419.

Miller, B., M. S. Windrem, and B. L. Finlay. 1991. Thalamic ablations and neocortical development: Alterations in thalamic and callosal connectivity. *Cerebral Cortex* 1:25–49.

Naegele, J. R., S. Jhaveri, and G. E. Schneider. 1988. Sharpening of topographical projections and maturation of geniculocortical axon arbors in the hamster. *Journal of Comparative Neurology* 281:1–12.

Olavarria, J., and R. C. Van Sluyters. 1985. Organization of callosal connections in the visual cortex of the rat. *Journal of Comparative Neurology* 239:1–26.

O'Leary, D. D. M. 1989. Do cortical areas emerge from a protocortex? *Trends in Neuroscience* 12:400–406.

O'Leary, D. D. M., and W. M. Cowan. 1984. Survival of isthmo-optic neurons after early removal of one eye. *Developmental Brain Research* 12:293–310.

O'Leary, D. D. M., and B. B. Stanfield. 1985. Occipital cortical neurons with transient pyramidal tract axons extend and maintain collaterals to subcortical but not intracortical targets. *Brain Research* 336:326–333.

——— 1986. A transient pyramidal tract projection from the visual cortex in the hamster and its removal by selective collateral elimination. *Developmental Brain Research* 27:97–99.

O'Leary, D. D. M., B. B. Stanfield, and W. M. Cowan. 1981. Evidence that the early postnatal restriction of the cells of origin of the callosal projection is due to the elimination of axonal collaterals rather than to the death of neurons. *Developmental Brain Research* 1:607–617.

Pallas, S. L., S. Gilmour, and B. L. Finlay. 1988. Control of cell number in the developing neocortex: I. Effects of early tectal ablation. *Developmental Brain Research* 43:1–11.

Parnavelas, J. G., and A. Chatzissavidou. 1981. The development of the thalamic projections to layer I of the visual cortex of the rat. *Anatomical Embryology* 163:71–75.

Payne, B., H. Pearson, and P. Cornwell. 1988. Development of visual and auditory cortical connections in the cat. In *Cerebral Cortex*, vol. 7, ed. A. Peters and E. G. Jones. New York: Plenum Press.

Pilar, G., L. Landmesser, and L. C. Burstein. 1980. Competition for survival among developing ciliary ganglion cells. *Journal of Neurophysiology* 43:233–254.

Rakic, P. 1971. Guidance of neurons migrating to the fetal monkey neocortex. *Brain Research* 33:471–476.

——— 1972. Mode of cell migration to the superficial layers of fetal monkey neocortex. *Journal of Comparative Neurology* 145:61–84.

——— 1974. Neurons in the rhesus monkey visual cortex: Systematic relation between time of origin and eventual disposition. *Science* 183:425–427.

——— 1988. Specification of cerebral cortical areas. *Science* 241:170–176.

——— 1990. Critical cellular events during cortical evolution: Radial unit hypothesis. In *The Neocortex: Ontogeny and Phylogeny,* ed. B. L. Finlay, G. M. Innocenti, and H. Scheich. New York: Plenum Press.

Ramirez, L. F., and K. Kalil. 1985. Critical stages for growth in the development of cortical neurons. *Journal of Comparative Neurology* 237:506–518.

Ross, M. H., A. M. Galaburda, and T. L. Kemper. 1984. Down's syndrome: Is there a decreased population of neurons? *Neurology* 34:909–916.

Schlaggar, B. L., and D. D. M. O'Leary. 1989. Embryonic rat neocortex transplanted homotopically into newborn neocortex develops area appropriate features. *Society of Neuroscience Abstracts* 15:1050.

Shimada, M., and J. Langman. 1970. Cell proliferation, migration and differentiation in the cerebral cortex of the golden hamster. *Journal of Comparative Neurology* 139:227–244.

Sidman, R. L., I. L. Miale, and N. Feder. 1959. Cell proliferation and migration in the primitive ependymal zone: An autoradiographic study of histogenesis in the nervous system. *Experimental Neurology* 1:322–333.

Stanfield, B. B., and D. D. M. O'Leary. 1985. The transient corticospinal projection from the occipital cortex during the postnatal development of the rat. *Journal of Comparative Neurology* 238:236–248.

Stanfield, B. B., D. D. M. O'Leary, and C. Fricks. 1982. Selective collateral elimination in early postnatal development restricts cortical distribution of rat pyramidal tract neurones. *Nature* 29:371–373.

Tolbert, D. L. 1987. Intrinsically directed pruning as a mechanism regulating the elimination of transient collateral pathways. *Developmental Brain Research* 33:11–21.

Tolbert, D. L., and T. Der. 1987. Redirected growth of pyramidal tract axons following neonatal pyramidotomy in cats. *Journal of Comparative Neurology* 260:299–311.

Valverde, F., and M. V. Facal-Valverde. 1987. Transitory population of cells in the temporal cortex of kittens. *Developmental Brain Research* 32:283–288.

Windrem, M. S., and B. L. Finlay. 1991. Thalamic ablations and neocortical development: Alterations in neocortical cell number and cytoarchitecture. *Cerebral Cortex* 1:1–24.

Windrem, M. S., S. Jan de Beur, and B. L. Finlay. 1988. Control of cell number in the developing neocortex: II. Effects of corpus callosum transection. *Developmental Brain Research* 43:13–22.

Wise, S. P., and E. G. Jones. 1978. Developmental studies of thalamocortical and commisural connections in the rat somatic sensory cortex. *Journal of Comparative Neurology* 178:87–208.

Woo, T. U., J. M. Beale, and B. L. Finlay. 1991. Dual fate of subplate neurons in the rodent. *Cerebral Cortex* 1:173–200.

2. Androgens and Brain Development

Arnold, A. 1984. Androgen regulation of motor neuron size and number. *Trends in Neuroscience* 7:239–242.

Arnold, A., F. Nottebohm, and D. Pfaff. 1976. Hormone concentrating cells in vocal control and other areas of the brain of the zebra finch *(Poephilia guttata). Journal of Comparative Neurology* 165:487–512.

Bleisch, W., and A. Harrelson. 1989. Androgens modulate endplate size and receptor density at synapses in rat levator ani muscle. *Journal of Neurobiology* 20:189–202.

Breedlove, S. M. 1986. Cellular analysis of hormone influence on motoneuronal development and function. *Journal of Neurobiology* 17:157–176.

Breedlove, S. M., and A. Arnold. 1980. Hormone accumulation in a sexually dimorphic motor nucleus in the rat spinal cord. *Science* 210:564–566.

——— 1981. Sexually dimorphic motor nucleus in the rat lumbar spinal cord: Response to adult hormone manipulation, absence in androgen-insensitive rats. *Brain Research* 225:297–307.

Breedlove, S. M., C. Jordan, and A. Arnold. 1983. Neurogenesis of motoneurons in the sexually dimorphic spinal nucleus of the bulbocavernosus in rats. *Developmental Brain Research* 9:39–43.

Clark, A., N. MacLusky, and P. Goldman-Rakic. 1988. Androgen binding and metabolism in the cerebral cortex of the developing rhesus monkey. *Endocrinology* 123:932–940.

DeVoogd, T., and F. Nottebohm. 1981. Sex difference in dendritic morphology of a song control nucleus in the canary: A quantitative Golgi study. *Journal of Comparative Neurology* 196:309–316.

Fahrbach, S., and J. Truman. 1989. Autoradiographic identification of ecdysteroid-binding cells in the nervous system of the moth. *Manduca sexta. Journal of Neurobiology* 20:681–702.

Fishman, R., L. Chism, G. Firestone, and S. M. Breedlove. 1990. Evidence for androgen receptors in sexually dimorphic perineal muscles of neonatal male rats: Absence of androgen accumulation by the motoneurons which innervate them. *Journal of Neurobiology* 21:694–704.

Gahr, M. 1990. Delineation of a brain nucleus: Comparisons of cytochemical, hodological and cytoarchitectural views of the song control nucleus HVc of the adult canary. *Journal of Comparative Neurology* 294:30–36.

Galaburda, A. M., F. Aboitiz, G. D. Rosen, and G. F. Sherman. 1986. Histological asymmetry in the primary visual cortex of the rat: Implications for mechanisms of cerebral asymmetry. *Cortex* 22:151–160.

Galaburda, A. M., J. Corsiglia, G. D. Rosen, and G. F. Sherman. 1987. Planum temporale asymmetry: Reappraisal since Geschwind and Levitsky. *Neuropsychologia* 25:853–868.

Galaburda, A. M., F. Sanides, and N. Geschwind. 1978. Human brain: Cytoarchitectonic left-right asymmetries in the temporal speech region. *Archives of Neurology* 35:812–817.

Geschwind, N., and P. Behan. 1982. Left handedness: Association with immune disease, migraine and developmental learning disorder. *Proceedings of the National Academy of Sciences* (USA) 79:5097–5100.

Geschwind, N., and A. M. Galaburda. 1985a. Cerebral lateralization: I. Biological mechanisms, association and pathology. *Archives of Neurology* 42:428–462.

——— 1985b. Cerebral lateralization: II. Biological mechanisms, association and pathology. *Archives of Neurology* 42:521–556.

——— 1985c. Cerebral lateralization: III. Biological mechanisms, association and pathology. *Archives of Neurology* 42:634–654.

Geschwind, N., and W. Levitsky. 1968. Human brain: Left-right asymmetries in temporal speech region. *Science* 161:186–187.

Godemet, P., J. Vanselow, S. Thanos, and F. Bonhoeffer. 1987. A study in developing visual system with a new method of staining neurons and their processes in fixed tissue. *Development* 101:697–713.

Goldman, S., and F. Nottebohm. 1983. Neuronal production, migration, and differentiation in a vocal control nucleus of the adult female canary brain. *Proceedings of the National Academy of Sciences* (USA) 80:2390–2394.

Gorlick, D., and D. Kelley. 1986. The ontogeny of androgen receptors in the CNS of *Xenopus laevis*. *Developmental Brain Research* 26:193–201.

——— 1987. Neurogenesis in the vocalization pathway of *Xenopus laevis*. *Journal of Comparative Neurology* 257:614–627.

Greenough, W., C. Carter, C. Steerman, and T. DeVoogd. 1977. Sex differences in dendritic patterns in hamster preoptic area. *Brain Research* 126:63–72.

Gurney, M. 1981. Hormonal control of cell form and number in the zebra finch song system. *Journal of Neuroscience* 1:658–673.

Joubert, Y., and C. Tobin. 1989. Satellite cell proliferation and increase in the number of myonuclei induced by testosterone in the levator ani muscle of the adult female rat. *Developmental Biology* 131:550–557.

Jung, I., and E. Baulieu. 1972. Testosterone cytosol receptor in the rat levator ani muscle. *Nature: New Biology* 237:24–26.

Kelley, D. 1978. Neuroanatomical correlates of hormone sensitive behaviors in birds and frogs. *American Zoologist* 18:477–488.

——— 1980. Auditory and vocal nuclei of frog brain concentrate sex hormones. *Science* 207:553–555.

——— 1981. Locations of androgen-concentrating cells in the brain of *Xenopus laevis:* Autoradiography with 3H-dihydrotestosterone. *Journal of Comparative Neurology* 199:221–231.

——— 1986. Genesis of male and female brains. *Trends in Neuroscience* 9:499–502.

——— 1988. Sexually dimorphic behaviors. *Anual Review of Neuroscience* 11:225–252.

Kelley, D., and J. Dennison. 1990. The vocal motor neurons of *Xenopus laevis:* Development of sex differences in axon number. *Journal of Neurobiology* 21:869–882.

Kelley, D., and D. Gorlick. 1990. Sexual selection and the nervous system. *Bioscience* 40:275–283.

Kelley, D., J. Morrell, and D. Pfaff. 1975. Autoradiographic localization of hormone-concentrating cells in the brain of an amphibian, *Xenopus laevis*. I. Testosterone. *Journal of Comparative Neurology* 164:63–78.

Kelley, D., and D. Pfaff. 1978. Generalizations from comparative studies on neuroanatomical and endocrine mechanisms of sexual behavior. In *Biological Determinants of Sexual Behavior,* ed. J. Hutchison, 225–254. Chichester: Wiley.

Kelley, D., D. Sassoon, N. Segil, and M. Scudder. 1989. Development and hormone regulation of androgen receptor levels in the sexually dimorphic larynx of *Xenopus laevis. Developmental Biology* 131:111–118.

Kelley, D., and M. Tobias. 1989. The genesis of courtship song: Cellular and molecular control of a sexually differentiated behavior. In *Perspectives in Neural Systems and Behavior,* ed. T. J. Carew and D. B. Kelley, 175–194. New York: A. Liss.

Konishi, M., and E. Akutagawa. 1985. Neuronal growth, atrophy and death in a sexually dimorphic song nucleus in the zebra finch brain. *Nature* 315:145–147.

Lieberburg, I., and F. Nottebohm. 1979. High affinity androgen binding proteins in syringeal tissues of song birds. *General and Comparative Endocrinology* 27:286–293.

Marin, M., M. Tobias, and D. Kelley. 1990. Hormone sensitive stages in the sexual differentiation of laryngeal muscle fiber number in *Xenopus laevis. Development* 110:703–711.

McLusky, N., I. Lieberburg, and B. McEwen. 1979. The development of estrogen receptor systems in the rat brain: Perinatal development. *Brain Research* 76:129–142.

Michael, R., R. Bonsall, and H. Rees. 1986. The nuclear accumulation of [3H] testosterone and [3H] estradiol in the brain of the female primate: Evidence for the aromatization hypothesis. *Endocrinology* 118:1935–1944.

Money, J., and A. Erhardt. 1972. *Man and Woman: Boy and Girl.* Baltimore: Johns Hopkins Press.

Morrell, J., D. Kelley, and D. Pfaff. 1975a. Sex steroid binding in the brain of vertebrates: Studies with light microscopic autoradiography. In *The Ventricular System in Neuroendocrine Mechanisms: Proceedings of the Second Brain-Endocrine Interaction Symposium,* ed. K. M. Knigge, D. E. Scott, and M. Kobayashi, 230–256. Basel: Karger.

——— 1975b. Autoradiographic localization of hormone-concentrating cells in the brain of an amphibian, *Xenopus laevis.* II. Estradiol. *Journal of Comparative Neurology* 164:63–78.

Nordeen, K., and E. Nordeen. 1988. Projection neurons within a vocal motor pathway are born during song learning in zebra finches. *Nature* 334:149–151.

Nordeen, E., K. Nordeen, and A. Arnold. 1987. Sexual differentiation of androgen accumulation within the zebra finch through selective cell loss and addition. *Journal of Comparative Neurology* 259:393–399.

Nordeen, E., K. Nordeen, D. Sengelaub, and A. Arnold. 1985. Androgens prevent normally occurring cell death in a sexually dimorphic spinal nucleus. *Science* 299:671–673.

Nottebohm, F. 1981. A brain for all seasons: Cyclical anatomical changes in song control nuclei of the canary brain. *Science* 214:1368–1370.

Nottebohm, F., and A. Arnold. 1976. Sexual dimorphism in vocal control areas of the song bird brain. *Science* 194:211–213.

Nottebohm, F., T. Stokes, and C. Leonard. 1976. Central control of song in the canary, *Serinus canarius. Journal of Comparative Neurology* 207:344–357.

Oppenheim, R. 1985. Naturally occurring cell death during neural development. *Trends in Neuroscience* 8:487–493.

Pfaff, D., and M. Keiner. 1973. Atlas of estradiol-concentrating cells in the central nervous system of the female rat. *Journal of Comparative Neurology* 151:121–158.

Rosen, G. D., G. F. Sherman, and A. M. Galaburda. 1989. Interhemispheric connections differ between symmetrical and asymmetrical brain regions. *Neuroscience* 33:525–533.

——— 1991. Ontogenesis of cortical asymmetry: A [3H]thymidine study. *Neuroscience* 41:779–790.

Sar, M., and W. Stumpf. 1977. Androgen concentration in motor neurons of cranial nerves and spinal cord. *Science* 197:77–80.

Sassoon, D., G. Gray, and D. Kelley. 1987. Androgen regulation of muscle fiber type in the sexually dimorphic larynx of *Xenopus laevis. Journal of Neuroscience* 7:3198–3206.

Sassoon, D., and D. Kelley. 1986. The sexually dimorphic larynx of *Xenopus laevis:* Development and androgen regulation. *American Journal of Anatomy* 177:455–472.

Sassoon, D., N. Segil, and D. Kelley. 1986. Androgen-induced myogenesis and chondrogenesis in the larynx of *Xenopus laevis. Developmental Biology* 113:135–145.

Segil, N., L. Silverman, and D. Kelley. 1987. Androgen binding levels in a sexually dimorphic muscle of *Xenopus laevis. General and Comparative Endocrinology* 66:95–101.

Simerly, R., C. Chang, M. Muramatsu, and L. Swanson. 1990. Distribution of androgen and estrogen receptor mRNA-containing cells in the rat brain: An in situ hybridization study. *Journal of Comparative Neurology* 294:76–95.

Tobias, M., and D. Kelley. 1987. Vocalizations of a sexually dimorphic isolated larynx: Peripheral constraints on vocal expression. *Journal of Neuroscience* 7:3191–3197.

——— 1988. Electrophysiology and dye coupling are sexually dimorphic characteristics of individual laryngeal muscle fibers in *Xenopus laevis. Journal of Neuroscience* 8:2422–2429.

Tobin, C., and Y. Joubert. 1988. The levator ani of the female rat: A suitable model for studying the effects of testosterone on the development of mammalian muscles. *Biological and Structural Morphogenesis* 1:28–33.

Venable, J. 1966. Morphology of the cells of normal, testosterone deprived and testosterone stimulated levator ani muscles. *American Journal of Anatomy* 112:271–302.

Watson, J., and E. Adkins-Regan. 1989. Neuroanatomical localization of sex-steroid concentrating cells in the Japanese quail *(Coturnix japonica):* Autoradiography with 3H testosterone, 3H estradiol and 3H dihydrotestosterone. *Neuroendocrinology* 49:51–64.

Weisz, J., and I. Ward. 1980. Plasma testosterone and progesterone titers of pregnant rats, their male and female fetuses and neonatal offspring. *Endocrinology* 106:306–316.

Wetzel, D., U. Haerter, and D. Kelley. 1985. A proposed efferent pathway for mate calling in South African clawed frogs, *Xenopus laevis:* Tracing afferents to laryngeal motor neurons with HRP-WGA. *Journal of Comparative Physiology* 157:749–761.

Wetzel, D., and D. Kelley. 1983. Androgen and gonadotropin control of the mate calls of male South African clawed frogs, *Xenopus laevis. Hormones and Behavior* 17:388–404.

Yamamoto, K. 1985. Steroid hormone regulated transcription of specific genes and gene networks. *Annual Review of Genetics* 19:209–252.

3. Peptidergic Neurons in the Hypothalamus

Akaishi, T., and Y. Sakuma. 1985. Estrogen excites oxytocinergic but not vasopressinergic cells in the paraventricular nucleus of female rat hypothalamus. *Brain Research* 335:302–307.

Alonso, G., and I. Assenmacher. 1979. Three-dimensional organization of the endoplasmic reticulum in supraoptic neurons of the rat. A structural functional correlation. *Brain Research* 170:247–258.

——— 1981. Radioautographic studies on the neurohypophysial projections of the supraoptic and paraventricular nuclei in the rat. *Cell Tissue Research* 219:525–534.

Ambach, G., and M. Palkovits. 1979. The blood supply of the hypothalamus in the rat. In *Anatomy of the Hypothalamus,* vol. 1, ed. P. J. Morgane and J. Panksepp, 263–377. New York: Marcell Dekker.

Andrew, R. D., B. A. MacVicar, F. E. Dudek, and G. I. Hatton. 1981. Dye transfer through gap junctions between neuroendocrine cells of rat hypothalamus. *Science* 211:1187–1189.

Aravich, P. F., and J. R. Sladek. 1987. Aging of rodent vasopressin systems: Morphometric and functional considerations. In *Vasopressin: Principles and Properties,* ed. D. M. Gash and G. J. Boer, 579–610. New York: Plenum.

Armstrong, W. E., and G. I. Hatton. 1978. Morphological changes in the rat supraoptic and paraventricular nuclei during the diurnal cycle. *Brain Research* 157:407–413.

Bargmann, W., and B. Scharrer. 1951. The site of origin of hormones of the posterior pituitary. *American Scientist* 39:255.

Beroukas, D., J. O. Willoughby, and W. W. Blessing. 1989. Neuropeptide Y–like immunoreactivity is present in boutons synapsing on vasopressin-containing neurons in rabbit supraoptic nucleus. *Neuroendocrinology* 50:222–228.

Brimble, M. J., R. E. J. Dyball, and M. L. Forsling. 1978. Oxytocin release following osmotic activation of oxytocin neurons in the paraventricular and supraoptic nuclei. *Journal of Physiology* (London) 278:69–78.

Broadwell, R. D., and M. W. Brightman. 1979. Enzyme cytochemical and immunocytochemical investigations of the secretory process in peptide-secreting cells. In *Current Methods in Cellular Neurobiology,* vol. 1, ed. J. L. Barker and J. F. McKelvy, 175–202. New York: Wiley-Interscience.

Brownstein, M. J., J. A. Russell, and H. Gainer. 1980. Synthesis, transport, and release of posterior pituitary hormones. *Science* 207:373–368.

Buijs, R. M., M. Geffard, C. W. Pool, and E. M. D. Hoorneman. 1984. The dopaminergic innervation of the supraoptic and paraventricular nucleus. A light and electron microscopical study. *Brain Research* 323:65–75.

Buijs, R. M., E. H. S. VanVulpen, and M. Geffard. 1987. Ultrastructural localization of GABA in the supraoptic nucleus and neural lobe. *Neuroscience* 20:247–355.

Castel, M., H. Gainer, and H. D. Dellmann. 1984. Neuronal secretory systems. *International Review of Cytology* 28:303–459.

Chapman, D. B., D. T. Theodosis, C. Montagnese, D. A. Poulain, and J. F. Morris. 1986. Osmotic stimulation causes structural plasticity of the neuron-glia relationships of the oxytocin but not vasopressin secreting neurons in the hypothalamic supraoptic nucleus. *Neuroscience* 17:670–686.

Cowan, W. M., J. W. Fawcett, D. D. M. O'Leary, and B. B. Stanfield. 1984. Regressive events in neurogenesis. *Science* 225:1258–1265.

Crespo, D., and C. Fernández-Viadero. 1989. The microvascular system of the optic nerve in control and enucleated rats. *Microvascular Research* 38:237–242.

Crespo, D., C. Fernández-Viadero, and S. Cos. 1989. Responses of neurosecretory neurons of the rat supraoptic nucleus after ovariectomy. *European Journal of Neuroscience* 15:20.

Crespo, D., J. Ramos, C. González, and C. Fernández-Viadero. 1990. The supraoptic nucleus: A morphological and quantitative study in control and hypophysectomized rats. *Journal of Anatomy* 169:115–123.

Crespo, D., C. Viadero, J. Villegas, and M. Lafarga. 1988. Nucleoli numbers and neuronal growth in supraoptic nucleus neurons during postnatal development in the rat. *Brain Research* 44:151–155.

Defendini, R., and E. A. Zimmerman. 1978. The magnocellular neurosecretory system of the mammalian hypothalmus. In *Hypothalamus,* ed. S. Reichlin, R. J. Baldessarin, and J. B. Martin, 137–154. New York: Raven.

Dierickx, K., F. Vandensande, and N. Goessens. 1978. The one neuron–one neurohypophysial hormone hypothesis and the hypothalamic magnocellular neurosecretory system of the vertebrate. In *Biologie Cellulaire des Processus Neurosécrétoires Hypothalamiques,* vol. 280, ed. J. D. Vincent and C. Kordon, 391–398. Paris: CNRS.

Ezzarani, E. A., J. P. Laulin, and R. Brudieux. 1985. Effects de la privation d'eau sur la production de corticostérone chez le rat male Brattleboro génétiquement privé de vasopressine. *Journal de Physiologie* (Paris) 80:157–163.

Gainer, H., M. Alstein, and Y. Hara. 1988. Oxytocin and vasopressin: After the genes, what next? In *Neurosecretion: Cellular Aspects of the Production and Release of Neuropeptides,* ed. B. T. Pickerin, J. B. Wakerley, and A. J. S. Summerlee, 1–11. New York: Plenum Press.

Gainer, H., Y. Sarne, and M. J. Bronstein. 1977. Biosynthesis and axonal transport of rat neurohypophysial proteins and peptides. *Journal of Cell Biology* 73:366–381.

Goessens, G. 1984. Nucleolar structure. *International Review of Cytology* 87:107–157.

Gross, P. M., N. M. Sposito, S. E. Pettersen, and J. D. Fenstermacher. 1986. Differences in function and structure of the capillary endothelium in the supraoptic nucleus and pituitary neural lobe of rats. *Neuroendocrinology* 44:401–407.

Hatton, G. I., J. I. Johnson, and C. Z. Malatesta. 1972. Supraoptic nuclei of rodents adapted from mesic and xeric enviroments: Numbers of cells, multiple nucleoli, and their distributions. *Journal of Comparative Neurology* 145:43–60.

Ikeda, K., M. Kinoshita, Y. Iwasaki, T. Shiojima, and K. Takamiya. 1989. Neurotrophic effect of angiotensin. II. Vasopressin and oxytocin on the ventral spinal cord of rat embryo. *International Journal of Neuroscience* 48:19–23.

Ivell, R., and D. Richter. 1984. Structure and comparison of the oxytocin and vasopressin genes from rat. *Proceedings of the National Academy of Science* (USA) 81:2006.

Kawata, M., J. T. McCabe, and D. Harrington. 1988. In situ hybridization analysis of osmotic stimulus-induced changes in ribosomal RNA in rat supraoptic nucleus. *Journal of Comparative Neurology* 270:528–536.

Lafarga, M., J. P. Hervás, D. Crespo, and J. C. Villegas. 1980. Ciliated neurons of rat hypothalamus during neonatal period. *Anatomy and Embryology* 160:29–38.

Larramendi, L. M. H. 1969. Analysis of synaptogenesis in the cerebellum of the mouse. In *Neurobiology of Cerebellar Evolution and Development,* ed. R. Llinas, 803–843. Chicago: American Medical Association.

Léránth, Cs., L. Záborszky, J. Matron, and M. Pakovits. 1975. Quantitative studies on the supraoptic nucleus in the rat. I. Synaptic organization. *Experimental Brain Research* 22:509–523.

Loewy, A. D., J. H. Wallach, and S. McKellar. 1981. Efferent connections of the ventral medulla oblongata in the rat. *Brain Research Review* 3:63–80.

Majzoub, J. A., A. Rich, J. Van Boom, and J. F. Habener. 1983. Vasopressin and oxytocin mRNA regulation in the rat assessed by hybridization with synthetic oligonucleolides. *Journal of Biological Chemistry* 258:14016–14064.

McEwen, B. S. 1981. Neuronal gonadal steroid action. *Science* 211:1303–1311.

McNeill, T. H., C. J. Clayton, and J. R. Sladek. 1980. Immunocytochemical characteristics of the hypothalamic magnocellular neurons in the aged rat. *Journal of Histochemistry and Cytochemistry* 28:611–612.

Modney, B. K., and G. I. Hatton. 1989. Multiple synapse formation: A possible compensatory mechanism for increased cell size in rat supraoptic nucleus. *Journal of Neuroendocrinology* 1:21–27.

Montagnese, C. M., D. A. Poulain, J. D. Vincent, and D. T. Theodosis. 1987. Structural plasticity in the supraoptic nucleus during gestation, post-partum, lactation and suckling-induced pseudogestation and lactation. *Journal of Endocrinology* 115:97–105.

Moos, F., D. A. Poulain, F. Rodriguez, Y. Guerné, J. D. Vincent, and P. Richard. 1989. Release of oxytocin within the supraoptic nucleus during the milk ejection reflex in rats. *Experimental Brain Research* 76:593–602.

Mugnaini, E. 1982. Membrane specializations in neuroglial cells and at neuron-glia contacts. In *Neuronal-Glial Cell Interrelationships,* ed. T. A. Sears, 39–56. Berlin: Springer-Verlag.

North, W. G. 1987. Biosynthesis of vasopressin and neurophysins. In *Vasopressin,* ed. D. M. Gash and G. J. Boer, 175–209. New York: Plenum.

North, W. G., H. Valtin, S. Cheng, and G. R. Hardy. 1983. The neurophysins: Production and turnover. *Progress in Brain Research* 60:217–225.

Perlmutter, L. S., C. D. Tweedle, and G. I. Hatton. 1985. Neuronal-glial plasticity in the supraoptic dendritic zone in response to acute and chronic dehydration. *Brain Research* 361:225–232.

Poulain, D. A., and J. B. Wakerley. 1982. Electrophysiology of hypothalamic neurons secreting oxytocin and vasopressin. *Neuroscience* 7:773–808.

Pow, D. V., and J. F. Morris. 1988. Exocytotic release of oxytocin and vasopressin in the central nervous system. *Annals of Endocrinology* 49:42N.

Ray, P. K., and S. R. Choudhury. 1990. Vasopressinergic axon collaterals and axon terminals in the magnocellular neurosecretory nuclei of the rat hypothalamus. *Acta Anatomica* 137:37–44.

Russell, J. A. 1980. Water deprivation in lactating rats: Changes in nucleolar dry mass of paraventricular and supraoptic neurons. *Cell Tissue Research* 212:315–331.

Scharrer, E. 1928. Die lichtempfindlichkeit blinder ellritzen *(Phoxinus laevis)* untersuchungen über das zwischenhirn der fische. *Zeitschrift Vergleichende Physiologie* 7:1–38.

Scharrer, E., and B. Scharrer. 1940. Secretory cells within the hypothalamus. In *Hypothalamus,* vol. 20, 170–174. New York: Hafner Publishing Co.

Sherman, T. G., O. Civelli, J. Douglass, E. Herbert, and S. J. Watson. 1986. Coordinate expression of hypothalamic pro-dynorphin and pro-vasopressin mRNAs with osmotic stimulation. *Neuroendocrinology* 44:211–228.

Sikora, K. C., and H. D. Dellmann. 1980. Pre- and postnatal synaptogenesis in the rat supraoptic nucleus. *Peptides* 1:229–238.

Silverman, A. J., and Zimmerman, E. A. 1983. Magnocellular neurosecretory system. *Annual Review of Neuroscience* 6:357–380.

Siveraag, A. M., J. E. Black, D. Shafron, and W. T. Greenough. 1988. Direct evidence that complex experience increases capillary branching and surface area in visual cortex of young rats. *Development in Brain Research* 43:199–304.

Sofroniew, M. V., and A. Weindl. 1981. Central nervous system distribution of vasopressin, oxytocin, and neurophysin. In *Endogenous Peptides and Learning and Memory Processes,* ed. J. L. Martinez, 327–367. New York: Academic Press.

Sommerville, J. 1986. Nucleolar structure and ribosome biogenesis. *TIBS* 11:438–442.

Sposito, N. M., and P. M. Gross. 1987. Morphometry of individual capillary beds in the hypothalamo-neurohypophysial system of rats. *Brain Research* 403:375–379.

Summerlee, A. J. S. 1981. Extracellular recordings from oxytocin neurons during the expulsive phase of birth in unanaesthetized rats. *Journal of Physiology* (London) 321:1–9.

Swanson, L. W., and P. E. Sawchenko. 1983. Hypothalamic integration: Organization of the paraventricular and supraoptic nuclei. *Annual Review of Neuroscience* 6:269–324.

Theodosis. D. T., and D. A. Poulain. 1984a. Evidence for structural plasticity in the supraoptic nucleus of the rat hypothalamus in relation to gestation and lactation. *Neuroscience* 11:183–193.

——— 1984b. Oxytocin-secreting neurons: A physiological model for structural plasticity in the adult mammalian brain. *Trends in Neurosciences* 10:426–430.

Theodosis, D. T., D. B. Chapman, C. Montagnese, D. A. Poulain, and J. F. Morris. 1986. Structural plasticity in the hypothalamic supraoptic nucleus at lactation affects oxytocin-, but not vasopressin-secreting neurons. *Neuroscience* 17:661–678.

Theodosis, D. T., L. Paut, and M. L. Tappaz. 1986. Immunocytochemical analysis of the gabaergic innervation of oxytocin- and vasopressin secreting neurons in the rat supraoptic nucleus. *Neuroscience* 19:207–222.

Tweedle, C. D., and G. I. Hatton. 1984. Synapse formation and disappearance in adult rat supraoptic nucleus during different hydration states. *Brain Research* 309:373–376.

Wachtler, F., M. Hartung, M. Devictor, J. Wiegant, A. Stahl, and H. G. Schwarzacher. 1989. Ribosomal DNA is located and transcribed in the dense fibrillar component of human Serotili cell nucleoli. *Experimental Cell Research* 1:61–70.

Willoughby, J. O., and W. W. Blessing. 1987. Neuropeptide Y injected into the supraoptic nucleus causes secretion of vasopressin in the unanaesthetized rat. *Neuroscience Letters* 75:17–22.

4. Pathogenesis of Late-Acquired Leptomeningeal Heterotopias and Secondary Cortical Alterations

Andres, K. H. 1967. Uber die Feinstruktur der Arachnoidea und Dura mater von Mammalia. *Zeitschrift für Zellforschung* 79:272–295.

Barth, P. G. 1987. Disorders of neuronal migration. *Canadian Journal of Neurological Sciences* 14:1–16.

Bayer, S. A., and J. Altman. 1990. Development of layer I and the subplate in the rat neocortex. *Experimental Neurology* 107:48–62.

Bignami, A., G. Palladini, and M. Zappella. 1968. Unilateral megalencephaly with nerve cell hypertrophy. An anatomical and quantitative histochemical study. *Brain Research* 9:103–114.

Brun, A., 1965a. Marginal glioneuronal heterotopias of the central nervous system. *Acta Pathologica et Microbiologica Scandinavica* 65:221–233.

——— 1965b. The subpial granular layer of the foetal cerebral cortex. *Acta Pathologica et Microbiologica Scandinavica* (Supplement 179) 79:1–89.

Cajal, S. Ramón y. 1968. *Degeneration and Regeneration of the Nervous System,* vol. 2. Trans. R. M. May. New York: Hafner Publishing Company.

Caviness, V. S., P. Evrad, and G. Lyon. 1978. Radial neuronal assemblies, ectopia and necrosis of developing cortex: A case analysis. *Acta Neuropathologica* 41:67–72.

Caviness, V. S., J.-P. Misson, and J.-F. Gadisseux. 1989. Abnormal neuronal patterns and disorders of neocortical development. In *From Reading to Neurons,* ed. A. M. Galaburda, 405–442. Cambridge, Mass.: MIT Press.

Choi, B. H., and S. C. Matthias. 1987. Cortical dysplasia associated with massive ectopia of neurons and glial cells within the subarachnoid space. *Acta Neuropathologica* 73:105–109.

Clarren, S. K., E. C. Alvord, S. M. Sumi, A. P. Streissguth, and D. W. Smith. 1978. Brain malformations related to prenatal exposure to ethanol. *Journal of Pediatrics* 92:64–67.

Clarren, S. K., and D. W. Smith. 1978. The fetal alcohol syndrome. *New England Journal of Medicine* 298:1063–1068.

Cowan, W. M., J. W. Fawcett, D. D. O'Leary, and B. B. Stanfield. 1984. Regressive events in neurogenesis. *Science* 225:1258–1265.

Dodd, J., and T. M. Jessel. 1988. Axon guidance and the patterning of neuronal projections in vertebrates. *Science* 242:692–699.

Easter, S. S., P. Purves, P. Rakic, and N. C. Spitzer. 1985. The changing view of neural plasticity. *Science* 230:507–511.

Evrard, P., P. Saint-George, H. J. Kadhim, and J.-F. Gadisseux. 1989. Pathology of prenatal encephalopathies. In *Child Neurology and Developmental Disabilities: Selected Proceedings of the Fourth International Child Neurology Congress,* ed. J. H. French et al., 153–176. Baltimore: Paul H. Brookes.

Farbman, A. I., and L. M. Squinto. 1985. Early development of olfactory receptor cell axons. *Developmental Brain Research* 19:205–213.

Folkman, J. 1982. Angiogenesis: Initiation and control. *Annals of the New York Academy of Sciences* 401:212–227.

Freeman, J. M. 1985. *Prenatal and Perinatal Factors Associated with Brain Disorders.* National Institute of Child Health and Human Development NIH Publication #85-1149.

Friede, R. L. 1989. *Developmental Neuropathology,* 2d ed. Berlin: Springer-Verlag.

Friede, R. L., and J. Mikolasek. 1978. Postencephalitic porencephaly, hydranencephaly or polymicrogyria. A review. *Acta Neuropathologica* 43:161–168.

Galaburda, A. M., and T. L. Kemper. 1979. Cytoarchitectonic abnormalities in development dyslexia: A case study. *Annals of Neurology* 6:94–100.

Galaburda, A. M., G. F. Sherman, G. D. Rosen, F. Aboitiz, and N. Geschwind. 1985. Developmental dyslexia: Four consecutive patients with cortical anomalies. *Annals of Neurology* 18:222–233.

Huttenlocher, P. R., C. DeCourten, L. J. Garey, and H. Van der Loos. 1982. Synaptogenesis in human visual cortex: Evidence for synaptic elimination during normal development. *Neurosciences Letters* 33:247–252.

Huttenlocher, P. R., and R. L. Wollmann. 1980. The fine structure of cerebral cortex in tuberous sclerosis. A Golgi study. *Annals of Neurology* 8:223–243.

Kaplan, H. A., and D. H. Ford, eds. 1966. *The Brain Vascular System.* Amsterdam: Elsevier Publishing Company.

Krisch, B., H. Leonhardt, and A. Oksche. 1982. The meningeal compartment of the median eminence and the cortex. A comparative analysis in the rat. *Cells and Tissue Research* 228:597–640.

——— 1983. Compartments and vascular arrangement of the meninges covering the cerebral cortex of the rat. *Cells and Tissue Research* 238:459–474.

Larroche, J.-C. 1981. The marginal zone in the neocortex of a 7-week-old human embryo. *Anatomy and Embryology* 162:301–312.

Larroche, J.-C., and F. Razavi-Encha. 1987. Cytoarchitectonic abnormalities. In *Malformations,* vol. 50 (revised), ed. N. C. Myrianthopoulos, *Handbook of Clinical Neurology,* 245–266. Amsterdam: Elsevier Science Publisher.

Luskin, M. B., and C. J. Shatz. 1985. Studies of the earliest generated cells in the cat's visual cortex: Cogeneration of subplate and marginal zones. *Journal of Neuroscience* 5:1062–1075.

Marchal, G., F. Andermann, D. Tampieri, Y. Robitaille, D. Melanson, B. Sinclair, A. Olivier, K. Silver, and P. Langevin. 1989. Generalized cortical dysplasia manifested by diffusely thick cerebral cortex. *Archives of Neurology* 46:430–434.

Marín-Padilla, M. 1970a. Prenatal and early postnatal ontogenesis of the human motor cortex. A Golgi study. I. The sequential development of the cortical layers. *Brain Research* 23:167–183.

——— 1970b. Prenatal and early postnatal development of the human motor cortex. A Golgi study. II. The basket-pyramidal system. *Brain Research* 23:185–191.

——— 1971. Early prenatal ontogenesis of the cerebral cortex (neocortex) of the cat *(Felis domestica)*. A Golgi study. I. The primordial neocortical organization. *Zeitschrift für Anatomie und Entwicklungsgeschichte* 134:117–145.

——— 1972. Prenatal ontogenetic history of the principal neurons of the neocortex of the cat *(Felis domestica)*. A Golgi study. II. Developmental differences and their significance. *Zeitschrift für Anatomie und Entwicklungsgeschichte* 136:125–142.

——— 1978. Dual origin of the mammlian neocortex and evolution of the cortical plate. *Anatomy and Embryology* 152:109–126.

——— 1979. Cortical organization in prenatal encephaloclastic porencephaly. *Society for Neuroscience Abstracts* 5:631.

——— 1983. Structural organization of the human cerebral cortex prior to the appearance of the cortical plate. *Anatomy and Embryology* 168:21–40.

——— 1985. Early vascularization of the embryonic cerebral cortex: Golgi and electron microscopic studies. *Journal of Comparative Neurology* 241:237–249.

——— 1987. Embryogenesis of the early vascularization of the central nervous system. In *Microneurosurgery: Clinical Considerations and Microsurgery of Racemous Angiomas,* vol. 3, ed. M. G. Yasargil, 23–47. Stuttgart: Thieme-Verlag.

——— 1988a. Early ontogenesis of the human cerebral cortex. In *Cerebral Cortex,* vol. 7, ed. A. Peters and E. G. Jones, 1–34. New York: Plenum.

——— 1988b. Embryonic vascularization of the mammalian cerebral cortex. In *Cerebral Cortex,* vol. 7, ed. A. Peters and E. G. Jones, 479–509. New York: Plenum.

——— 1988c. Structure and function of meganeurons in cortical (marginal) heterotopias. A Golgi study. *Society of Neuroscience Abstracts* 14:1174.

——— 1990a. Origin, formation, and prenatal maturation of the human cerebral cortex: An overview. *Journal of Craniofacial Genetics and Developmental Biology* 10:137–146.

——— 1990b. Three-dimensional structural organization of layer I of the human cerebral cortex: A Golgi study. *Journal of Comparative Neurology* 299:89–105.

Marín-Padilla, M., and M. R. Amieva. 1989. Early neurogenesis of the mouse olfactory nerve: Golgi and electron microscopic studies. *Journal of Comparative Neurology* 288:339–352.

Marín-Padilla, M., and T. M. Marín-Padilla. 1982. Origin, prenatal development and structural organization of layer I of the human cerebral (motor) cortex. A Golgi study. *Anatomy and Embryology* 164:161–206.

Mervis, R. F., and A. J. Yates. 1980. Ectopic dendritic growth and meganeurite formation in porencephaly with polymicrogyria: A Golgi study. *Society of Neuroscience Abstracts* 6:738.

Müller, F., and R. O'Rahilly. 1988a. The first appearance of the future cerebral hemispheres in the human embryo at stage 14. *Anatomy and Embryology* 177:495–511.

——— 1988b. The development of the human brain, including the longitudinal zoning of the diencephalon at stage 15. *Anatomy and Embryology* 179:55–71.

O'Rahilly, R., and F. Müller. 1986. The meninges in human development. *Journal of Neuropathology and Experimental Neurology* 45:588–608.

——— 1987. *Developmental Stages in Human Embryos.* Carnegie Institute of Washington, Publication #637.

Pape, E. K., and J. S. Wigglesworth. 1979. Haemorrhage, ischaemia, and the perinatal brain. *Spastics International Medical Publications,* ed. E. K. Pape and J. S. Wiggiesworth, 11–38. London: Spastics International.

Rakic, P. 1989. Competitive interactions during neuronal and synaptic development. In *From Reading to Neurons,* ed. A. M. Galaburda, 243–462. Cambridge, Mass.: MIT Press.

Rakic, P., J.-P. Bourgeois, M. F. Eckenhoff, N. Zezecic, and P. S. Goldman-Rakic. 1986. Concurrent overproduction of synapses in diverse regions of primate cerebral cortex. *Science* 232:232–235.

Sarnat, H. B. 1987. Disturbances of late neuronal migration interest perinatal period. *American Journal of Diseases of Children* 141:969–980.

Taylor, D. C., M. A. Falconer, C. J. Bruton, and J. A. N. Corsellis. 1971. Focal dysplasia of the cerebral cortex in epilepsy. *Journal of Neurology, Neurosurgery and Psychiatry* 34:369–387.

Townsend, J. J., S. L. Nielsen, and N. Malamud. 1975. Unilateral megalencephaly: Hamartoma or neoplasm? *Neurology* 25:448–453.

Vigevano, F., E. Bertini, R. Boldrini, C. Bosman, D. Claps, M. di Capua, C. di Rocco, and G. F. Rossi. 1989. Hemimegalencephaly and intractable epilepsy: Benefits of hemispherectomy. *Epilepsia* 30:833–843.

Volpe, J. J. 1987. *Neurology of the Newborn.* Philadelphia: W. B. Saunders Company.

Williams, R. S., and V. S. Caviness. 1984. Normal and abnormal development of the brain. In *Advances in Clinical Neuropsychology,* ed. R. E. Tarter and G. Goldstein, 1–62. New York: Plenum.

Williams, R. W., and K. Herrup. 1988. Control of neuron number. *Annual Reviews of Neurosciences* 11:423–453.

Wisniewski, K., M. Dambska, J. H. Sher, and Q. Qazi. 1983. A clinical neuropathological study of fetal alcohol syndrome. *Neuropediatrics* 4:197–201.

Zezevic, N., J. P. Bourgeois, and P. Rakic. 1989. Changes in synpatic density in motor cortex of rhesus monkey during fetal and postnatal life. *Developmental Brain Research* 50:11–32.

5. Dyslexia and Brain Pathology

Borit, A., and R. M. Herndon. 1970. The fine structure of plaques fibromyélinques in ulegyria and in status marmoratus. *Acta Neuropathologica* (Berlin) 14:304–311.

Caviness, V. S., Jr. 1976. Patterns of cell and fiber distribution in the neocortex of the reeler mutant mouse. *Journal of Comparative Neurology* 170:435–448.

——— 1982. Neocortical histogenesis and reeler mice: A developmental study based upon [3H]thymidine autoradiography. *Developmental Brain Research* 4:293–302.

Caviness, V. S., Jr., and D. O. Frost. 1983. Thalamocortical projections in the reeler mutant mouse. *Journal of Comparative Neurology* 219:182–202.

Caviness, V. S., Jr., and C. H. Yorke, Jr. 1976. Interhemispheric neocortical connections of the corpus callosum in the reeler mutant mouse: A study based on anterograde and retrograde methods. *Journal of Comparative Neurology* 170:449–460.

Chi, J. G., E. C. Dooling, and F. H. Gilles. 1977. Gyral development of the human brain. *Annals of Neurology* 1:86–93.

Cowan, W. M., J. W. Fawcett, D. D. M. O'Leary, and B. B. Stanfield. 1984. Regressive events in neurogenesis. *Science* 225:1258–1265.

DeFries, J. C., D. W. Julker, and M. C. LaBuda. 1987. Evidence for a genetic aetiology in reading disability of twins. *Nature* 329:537–539.

Denenberg, V. H. 1981. Hemispheric laterality in animals and the effects of early experience. *Behavioral and Brain Sciences* 4:1–49.

Denenberg, V. H., J. Garbanati, G. Sherman, D. A. Yutzey, and R. Kaplan. 1978. Infantile stimulation induces brain lateralization in rats. *Science* 201:1150–1152.

Denenberg, V. H., G. F. Sherman, G. D. Rosen, and A. M. Galaburda. 1988. Learning and laterality differences in BXSB mice as a function of neocortical anomaly. *Society of Neuroscience Abstracts* 14:1260.

Denenberg, V. H., G. F. Sherman, L. M. Schrott, G. D. Rosen, and A. M. Galaburda. 1991. Spatial learning, discrimination learning, paw preference, and neocortical ectopias in two autoimmune strains of mice. *Brain Research* 562:98–104.

Diamond, M. C., R. E. Johnson, and C. A. Ingham. 1975. Morphological changes in the young, adult and aging rat cerebral cortex, hippocampus, and diencephalon. *Behavioral Biology* 14:163–174.

Diamond, M. C., D. Young, S. S. Singh, and R. E. Johnson. 1981. Age-related morphological differences in the rat cerebral cortex and hippocampus: Male-female; right-left. *Experimental Neurology* 81:1–13.

Dlugosz, L. J., T. Byers, M. E. Msall, J. Marchall, A. Lesswing, and R. E.

Cooke. 1988. Relationships between laterality of congenital upper limb reduction defects and school performance. *Clinical Pediatrics* 27:319–324.

Drake, W. E. 1968. Clinical and pathological findings in a child with a developmental learning disability. *Journal of Learning Disabilities* 1:9–25.

Dvorák, K., and J. Feit. 1977. Migration of neuroblasts through partial necrosis of the cerebral cortex in newborn rats—Contribution to the problems of morphological development and developmental period of cerebral microgyria. *Acta Neuropathologica* (Berlin) 38:203–212.

Dvorák, K., J. Feit, and Z. Juránková. 1978. Experimentally induced focal microgyria and status verrucosus deformis in rats—Pathogenesis and interrelation histological and autoradiographical study. *Acta Neuropathologica* (Berlin) 44:121–129.

Eidelberg, D., and A. M. Galaburda. 1984. Inferior parietal lobule. Divergent architectonic asymmetries in the human brain. *Archives of Neurology* 41:843–852.

Entus, A. K. 1977. Hemispheric asymmetry in processing of dichotically presented speech and nonspeech stimuli by infants. In *Language Development and Neurological Theory,* ed. S. J. Segalowitz and F. A. Gruber. New York: Academic Press.

Falk, D., J. Cheverud, M. W. Vannier, and G. C. Conroy. 1986. Advanced computer graphics technology reveals cortical asymmetry in endocasts of rhesus monkeys. *Folia Primatology* (Basel) 46:98–103.

Fink, R. P., and L. Heimer. 1967. Two methods for selective silver impregnation of degenerating axons and their synaptic endings in the central nervous system. *Brain Research* 4:369–374.

Finlay, B. L., K. G. Wilson, and G. E. Schneider. 1979. Anomalous ipsilateral retinotectal projections in Syrian hamsters with early lesions: Topography and functional capacity. *Journal of Comparative Neurology* 183:721–740.

Finucci, J. M., S. D. Isaacs, C. C. Whitehouse, and B. Childs. 1983. Classification of spelling errors and their relationship to reading ability, sex, grade placement, and intelligence. *Brain and Language* 20:340–345.

Galaburda, A. M. 1980. La région de Broca: Observations anatomiques faites un siècle après la mort de son découvreur. *Revue de Neurologie* (Paris) 136:609–616.

——— 1989. Ordinary and extraordinary brain development: Anatomical variation in developmental dyslexia. *Annals of Dyslexia* 39:67–80.

Galaburda, A. M., F. Aboitiz, G. D. Rosen, and G. F. Sherman. 1986. Histological asymmetry in the primary visual cortex of the rat: Implications for mechanisms of cerebral asymmetry. *Cortex* 22:151–160.

Galaburda, A. M., J. Corsiglia, G. D. Rosen, and G. F. Sherman. 1987. Planum temporale asymmetry: Reappraisal since Geschwind and Levitsky. *Neuropsychologia* 25:853–868.

Galaburda, A. M., and T. L. Kemper. 1979. Cytoarchitectonic abnormalities in developmental dyslexia: A case Study. *Annals of Neurology* 6:94–100.

Galaburda, A. M., M. LeMay, T. L. Kemper, and N. Geschwind. 1978. Right-left asymmetries in the brain. *Science* 199:852–856.

Galaburda, A. M., F. Sanides, and N. Geschwind. 1978. Human brain: Cytoarchitectonic left-right asymmetries in the temporal speech region. *Archives of Neurology* 35:812–817.

Galaburda, A. M., G. F. Sherman, G. D. Rosen, F. Aboitiz, and N. Geschwind. 1985. Developmental dyslexia: Four consecutive cases with cortical anomalies. *Annals of Neurology* 18:222–233.

Geschwind, N., and P. O. Behan. 1982. Left-handedness: Association with immune disease, migraine, and developmental disorder. *Proceedings of the National Academy of Sciences* (USA) 79:5097–5100.

——— 1984. Laterality, hormones, and immunity. In *Cerebral Dominance: The Biological Foundations,* ed. N. Geschwind and A. M. Galaburda. Cambridge, Mass.: Harvard University Press.

Geschwind, N., and W. Levitsky. 1968. Human brain: Left-right asymmetries in temporal speech region. *Science* 161:186–187.

Goldman, P. S., and T. W. Galkin. 1978. Prenatal removal of frontal association cortex in the fetal rhesus monkey: Anatomical and functional consequences in postnatal life. *Brain Research* 152:451–485.

Goldman-Rakic, P. S., and P. Rakic. 1984. Experimentally modified convolutional patterns in nonhuman primates: Possible relevance of connections to cerebral dominance in humans. In *Biological Foundations of Cerebral Dominance,* ed. N. Geschwind and A. M. Galaburda. Cambridge, Mass.: Harvard University Press.

Gundara, N., and S. Zivanovic. 1968. Asymmetry in East African skulls. *American Journal of Physical Anthropology* 28:331–338.

Hamburger, V., and J. W. Yip. 1984. Reduction of experimentally induced neuronal death in spinal ganglia of the chick embryo by nerve growth factor. *Journal of Neuroscience* 4:764–774.

Hamilton, C. R., and B. A. Vermeire. 1988. Complementary hemispheric specialization in monkeys. *Science* 242:1691–1694.

Humphreys, P., W. E. Kaufmann, and A. M. Galaburda. 1990. Developmental dyslexia in women: Neuropathological findings in three cases. *Annals of Neurology* 28:727–738.

Humphreys, P., G. D. Rosen, D. M. Press, G. F. Sherman, and A. M. Galaburda. 1991. Freezing lesions of the newborn rat brain: A model for cerebrocortical microgyria. *Journal of Neuropathology and Experimental Neurology* 50:145–160.

Hynd, G. W., and M. Semrud-Clikeman. 1989. Dyslexia and brain morphology. *Psychological Bulletin* 106:447–82.

Innocenti, G. M., and P. Berbel. 1991a. Analysis of an experimental cortical network: i) Architectonics of visual areas 17 and 18 after neonatal injections of ibotenic acid; similarities with human microgyria. *Journal of Neural Transplantation and Plasticity* 2:1–28.

——— 1991b. Analysis of an experimental cortical network: ii) Connections of visual areas 17 and 18 after neonatal injections of ibotenic acid. *Journal of Neural Transplantation and Plasticity* 2:29–54.

Innocenti, G. M., and S. Clarke. 1983. Multiple sets of visual cortical neurons projecting transitorily through the corpus callosum. *Neuroscience Letters* 41:27–32.

Ivy, G. O., R. M. Akers, and H. P. Killackey. 1979. Differential distribution of callosal projections in the neonatal and adult rat. *Brain Research* 173:532–537.

Ivy, G. O., and H. P. Killackey. 1981. The ontogeny of the distribution of callosal projection neurons in the rat parietal cortex. *Journal of Comparative Neurology* 195:367–389.

——— 1982. Ontogenetic changes in the projections of neocortical neurons. *Journal of Neuroscience* 2:735–743.

Jacobson, S., and J. Q. Trojanowski. 1974. The cells of origin of the corpus callosum in the rat, cat and rhesus monkey. *Brain Research* 74:149–155.

Kaufmann, W. E., and A. M. Galaburda. 1989. Cerebrocortical microdysgenesis in neurologically normal subjects: A histopathologic study. *Neurology* 39:238–244.

Kean, M. L. 1984. The question of linguistic anomaly in developmental dyslexia. *Annals of Dyslexia* 34:137–151.

Kertesz, A., and N. Geschwind. 1971. Patterns of pyramidal decussation and their relationship to handedness. *Archives of Neurology* 24:326–332.

Kolb, B., R. J. Sutherland, A. J. Nonneman, and I. Q. Whishaw. 1982. Asymmetry in the cerebral hemispheres of the rat, mouse, rabbit, and cat: The right hemisphere is larger. *Experimental Neurology* 78:348–359.

Kromer, L. F. 1987. Nerve growth factor treatment after brain injury prevents neuronal death. *Science* 235:214–216.

Lahita, R. G. 1988. Systemic lupus erythematosus: Learning disability in the male offspring of female patients and relationship to laterality. *Psychoneuroendocrinology* 13:385–396.

Larroche, J. C. 1986. Fetal encephalopathies of circulatory origin. *Biology of the Neonate.* 50:61–74.

Larsen, J., T. Håien, I. Lundberg, and H. Odegaard. 1990. MRI evaluation of the size and symmetry of the planum temporale in adolescents with developmental dyslexia. *Brain and Language* 39:289–301.

LeMay, M., and A. Culebras. 1972. Human brain: Morphologic differences in the hemispheres demonstratable by carotid arteriography. *New England Journal of Medicine* 287:168–170.

LeMay, M., and N. Geschwind. 1975. Hemispheric differences in the brains of great apes. *Brian, Behavior and Evolution* 11:48–52.

Malamud, N. 1950. Status marmoratus: A form of cerebral palsy following either birth injury or inflammation of the central nervous system. *Journal of Pediatrics* 37:610–619.

McBride, M. C., and T. L. Kemper. 1982. Pathogenesis of four-layered microgyric cortex in man. *Acta Neuropathologica* (Berlin) 57:93–98.

Michel, G. F. 1981. Right-handedness: A consequence of infant supine head-orientation preference. *Science* 212:685–687.

Molfese, D. L., R. B. Freeman, and D. S. Palermo. 1975. The ontogeny of brain lateralization for speech and nonspeech stimuli. *Brain and Language* 2:356–368.

Myers, R. E. 1969. Atrophic cortical sclerosis associated with status marmoratus in a perinatally damaged monkey. *Neurology* 19:1177–1188.

Nandy, K., H. Lal, M. Bennett, and D. Bennett. 1983. Correlation between a learning disorder and elevated brain-reactive antibodies in aged C57BL/6 and young NZB mice. *Life Sciences* 33:1499–1503.

Needels, D. L., M. Nieto-Sampedro, and C. W. Cotman. 1986. Induction of a neurite-promoting factor in rat brain following injury or deafferentiation. *Neuroscience* 18:517–526.

Nieto-Sampedro, M., E. R. Lewis, C. W. Cotman, M. Manthorpe, S. D. Staper, G. Barbin, F. M. Longo, and S. Varon. 1982. Brain injury causes a time-dependent increase in neuronotrophic activity at the lesion site. *Science* 217:860–861.

Norman, M. G. 1981. On the morphogenesis of ulegyria. *Acta Neuropathologica* (Berlin) 53:331–332.

Nottebohm, F. 1977. Neural lateralization of vocal control in a passerine bird. *Journal of Experimental Neurology* 177:229–262.

Nottebohm, F., and M. E. Nottebohm. 1976. Left hypoglossal dominance in the control of canary and white-crowned sparrow song. *Journal of Comparative Physiology* 108:171–192.

Olavarria, J., and R. C. van Sluyters. 1985. Organization and postnatal development of callosal connections in the visual cortex of the rat. *Journal of Comparative Neurology* 239:1–26.

O'Leary, D. D. M., B. B. Stanfield, and W. M. Cowan. 1981. Evidence that the early postnatal restriction of the cells of origin of the callosal projection is due to the elimination of axonal collaterals rather than to the death of neurons. *Brain Research* 227:607–617.

Pennington, B. F., S. D. Smith, W. J. Kimberling, P. A. Green, and M. M. Haith. 1987. Left-handedness and immune disorders in familial dyslexics. *Archives of Neurology* 44:634–639.

Petersen, M. R., M. D. Beecher, S. Zoloth, D. B. Moody, and W. C. Stebbins. 1978. Neural lateralization of species-specific vocalizations by Japanese macaques *(Macaca fuscata)*. *Science* 202:324–327.

Pieniadz, J. M., and M. A. Naeser. 1984. Computed tomographic scan cerebral asymmetries and morphological brain asymmetries: Correlation in the same cases post mortem. *Archives of Neurology* 41:403–409.

Rakic, P. 1988. Specification of cerebral cortical areas. *Science* 241:170–176.

Rakic, P., and R. W. Williams. 1986. Thalamic regulation of cortical parcellation: An experimental perturbation of the striate cortex in rhesus monkeys. *Society for Neuroscience Abstracts* 12:1499.

Richman, D. P., R. M. Stewart, and V. S. Caviness. 1974. Cerebral microgyria in a 27-week fetus: An architectonic and topographic analysis. *Journal of Neuropathology and Experimental Neurology* 33:374–384.

Rosen, G. D., A. S. Berrebi, D. A. Yutzey, and V. H. Denenberg. 1983.

Prenatal testosterone causes shift of asymmetry in neonatal tail posture of the rat. *Developmental Brain Research* 9:99–101.

Rosen, G. D., S. Finklestein, A. L. Stoll, D. A. Yutzey, and V. H. Denenberg. 1984. Neonatal tail posture and its relationship to striatal dopamine asymmetry in the rat. *Brain Research* 297:305–308.

Rosen, G. D., A. M. Galaburda, and G. F. Sherman. 1989. Cerebrocortical microdysgenesis with anomalous callosal connections: A case study in the rat. *International Journal of Neuroscience* 47:237–247.

Rosen, G. D., G. F. Sherman, and A. M. Galaburda. 1989. Interhemispheric connections differ between symmetrical and asymmetrical brain regions. *Neuroscience* 33:525–533.

——— 1991. Ontogenesis of neocortical asymmetry: A [3H]thymidine study. *Neuroscience* 41:779–790.

Rosen, G. D., G. F. Sherman, C. Mehler, K. Emsbo, and A. M. Galaburda. 1989. The effect of developmental neuropathology on neocortical asymmetry in New Zealand Black mice. *International Journal of Neuroscience* 45:247–254.

Ross, D. A., S. D. Glick, and R. C. Meibach. 1981. Sexually dimorphic brain and behavioral asymmetries in the neonatal rat. *Proceedings of the National Academy of Sciences* (USA) 78:1958–1961.

——— 1982. Sexually dimorphic cerebral asymmetries in 2-deoxy-D-glucose uptake during postnatal development of the rat: Correlations with age and relative activity. *Developmental Brain Research* 3:341–347.

Rubens, A. B., M. W. Mahowald, and J. T. Hutton. 1976. Asymmetry of lateral (sylvian) fissures in man. *Neurology* 26:620–624.

Salcedo, J. R., B. J. Spiegler, E. Gibson, and D. B. Magilavy. 1985. The autoimmune disease systemic lupus erythematosus is not associated with left-handedness. *Cortex* 21:645–647.

Satz, P., and H. V. Soper. 1986. Left-handedness, dyslexia and autoimmune disorder: A critique. *Journal of Clinical and Experimental Neuropsychology* 8:453–458.

Schachter, S. C., B. J. Ransil, and N. Geschwind. 1987. Associations of handedness with hair color and learning disabilities. *Neuropsychologia* 25:269–276.

Schneider, G. E. 1981. Early lesions and abnormal neuronal connections. *Trends in Neuroscience* 4:187–192.

Schneider, G. F. 1979. Is it really better to have your brain lesion early? A revision of the "Kennard principle." *Neuropsychologia* 17:557–583.

Schur, P. H. 1986. Handedness in systemic lupus erythematosus. *Arthritis and Rheumatism* 29:419–420.

Searleman, A., and A. K. Fugagli. 1988. Left-handedness and suspected autoimmune disorders: A reply to Persson and Ahlbom. *Neuropsychologia* 26:739–740.

Sherman, G. F., and A. M. Galaburda. 1984. Neocortical asymmetry and open-field behavior in the rat. *Experimental Neurology* 86:473–482.

Sherman, G. F., A. M. Galaburda, P. O. Behan, and G. D. Rosen. 1987.

Neuroanatomical anomalies in autoimmune mice. *Acta Neuropathologica* (Berlin) 74:239–242.

Sherman, G. F., L. Morrison, G. D. Rosen, P. O. Behan, and A. M. Galaburda. 1990a. Brain abnormalities in immune defective mice. *Brain Research* 532:25–33.

Sherman, G. F., J. S. Stone, D. M. Press, G. D. Rosen, and A. M. Galaburda. 1990b. Abnormal architecture and connections disclosed by neurofilament staining in the cerebral cortex of autoimmune mice. *Brain Research* 529:202–207.

Sherman, G. F., J. S. Stone, G. D. Rosen, and A. M. Galaburda. 1990c. Neocortical VIP neurons are increased in the hemisphere containing focal cerebrocortical microdysgenesis in New Zealand Black mice. *Brain Research* 532:232–236.

Smith, S. D., W. J. Kimberling, B. F. Pennington, and H. A. Lubs. 1983. Specific reading disability: Identification of an inherited form through linkage analysis. *Science* 219:1345–1347.

Spencer, D. G., K. Humphries, D. Mathis, and H. Lal. 1986. Behavioral impairments related to cognitive dysfunction in the autoimmune New Zealand Black mouse. *Behavioral Neuroscience* 100:353–358.

Spencer, D. G., and H. Lal. 1983. Specific behavioral impairments in association tasks with an autoimmune mouse. *Society of Neuroscience Abstracts* 9:96.

Sperry, R. W. 1974. Lateral specialization in the surgically separated hemispheres. In *Hemispheric Specialization and Interaction,* ed. B. Milner. Cambridge, Mass.: MIT Press.

Springer, S. P., and G. Deutsch. 1981. *Left Brain, Right Brain.* San Francisco: W. H. Freeman and Company.

Steinberg, A. D., D. P. Huston, J. D. Taurog, J. S. Cowdery, and E. S. Racheche. 1981. The cellular and genetic basis of murine lupus. *Immunological Review* 55:121–154.

Teszner, D., A. Tzavaras, J. Gruner, and H. Hécaen. 1972. L'asymmétrie droite-gauche du planum temporale: A propos de l'étude anatomique de 100 cervaeux. *Revue de Neurologie* (Paris) 126:444–449.

Theofilopoulos, A. N., and F. J. Dixon. 1981. Etiopathogenesis of murine SLE. *Immunological Review* 55:179–216.

Urion, D. K. 1988. Nondextrality and autoimmune disorders among relatives of language-disabled boys. *Annals of Neurology* 24:267–269.

van Eden, C. G., H. B. M. Uylings, and J. van Pelt. 1984. Sex-difference and left-right asymmetries in the prefrontal cortex during postnatal development in the rat. *Developmental Brain Research* 12:146–153.

Wada, J. A., R. Clarke, and A. Hamm. 1975. Cerebral hemispheric asymmetry in humans. *Archives of Neurology* 32:239–246.

Weiskrantz, L. 1977. On the role of cerebral commissures in animals. In *Structure and Function of Cerebral Commissures,* ed. I. S. Russell, M. W. van Hof, and G. Berlucchi. Baltimore: University Park Press.

Wise, S. P., and E. G. Jones. 1976. The organization and postnatal development of the commissural projection of the rat somatic sensory cortex. *Journal of Comparative Neurology* 168:313–344.

Witelson, S. F., and W. Pallie. 1973. Left hemisphere specialization for language in the newborn: Neuroanatomical evidence of asymmetry. *Brain* 96:641–646.

Yakovlev, P. I., and P. Rakic. 1966. Patterns of decussation of bulbar pyramids and distribution of pyramidal tracts on two sides of the spinal cord. *Transactions of the American Neurological Association* 91:366–367.

Yeni-Komshian, G. H., and D. A. Benson. 1976. Anatomical study of cerebral asymmetry in the temporal lobe of humans, chimpanzees, and rhesus monkeys. *Science* 192:387–389.

Zaborszky, L., and J. R. Wolff. 1982. Distributional patterns and individual variations of callosal connections in the albino rat. *Anatomy and Embryology* 165:213–232.

6. Anatomical and Functional Aspects of an Experimental Visual Microcortex That Resembles Human Microgyria

Assal, F., Y. Melzer, and G. M. Innocenti. 1989. Functional analysis of a visual cortical circuit resembling human microgyria. *European Journal of Neuroscience Supplement* 2:256.

Benevento, L. A., O. D. Creutzfeldt, and U. Kuhnt. 1972. Significance of intracortical inhibition in the visual cortex. *Nature: New Biology* 238:124–125.

Bolz, J., and C. D. Gilbert. 1989. The role of horizontal connections in generating long receptive fields in the cat visual cortex. *European Journal of Neuroscience Supplement* 1:263–268.

Clarke, S., and G. M. Innocenti. 1990. Auditory neurons with transitory axons to visual areas form short permanent projections. *European Journal of Neuroscience Supplement* 2:227–242.

Eugster, C. H. 1968. Wirkstoffe aus dem Fliegenpilz. *Naturwissenschaften* 7:305–313.

Evrard, P., P. de Saint-Georges, H. J. Kadhim, and J.-F. Gadisseux. 1989. Pathology of prenatal encephalopathies. In *Child Neurology and Developmental Disabilities: Selected Proceedings of the Fourth International Child Neurology Congress,* 153–176. Baltimore: Paul H. Brookes.

Frost, D. O. 1986. Development of anomalous retinal projections to nonvisual thalamic nuclei in Syrian hamsters: A quantitative study. *Journal of Comparative Neurology* 252:95–105.

Gilbert, C. D., and T. N. Wiesel. 1979. Morphology and intracortical projections of functionally characterised neurones in the cat visual cortex. *Nature* 280:120–125.

——— 1989. Columnar specificity of intrinsic horizontal and corticocortical connections in cat visual cortex. *Journal of Neuroscience* 9:2432–2442.

Hornung, J. P., F. Assal, and G. M. Innocenti. 1989. Distribution of diffuse afferents and interneurons in experimentally induced microcortex in cat visual cortex. *European Journal of Neuroscience Supplement* 2:105.

Hubel, D. H., and T. N. Wiesel. 1962. Receptive fields, binocular interaction and functional architecture in the cat's visual cortex. *Journal of Physiology* (London) 160:106–154.

Humphreys, P., G. D. Rosen, D. M. Press, G. F. Sherman, and A. M. Galaburda. 1991. Freezing lesions of the developing rat brain. I. A model for cerebrocortical microgyria. *Journal of Neuropathology and Experimental Neurology* 50:145–160.

Innocenti, G. M. 1991. The development of projections from cerebral cortex. *Progress in Sensory Physiology* 12:65–114.

Innocenti, G. M., and P. Berbel. 1991a. Analysis of an experimental cortical network: i) Architectonics of visual areas 17 and 18 after neonatal injections of ibotenic acid: Similarities with human microgyria. *Journal of Neural Transplantation* 2:1–28.

——— 1991b. Analysis of an experimental cortical network: ii) Connections of areas 17 and 18 after neonatal injections of ibotenic acid. *Journal of Neural Transplantation* 2:29–52.

Innocenti, G. M., P. Berbel, and S. Clarke. 1988. Development of projections from auditory to visual areas in the cat. *Journal of Comparative Neurology* 272:242–259.

Luhmann, H. J., W. Singer, and L. Martínez-Millán. 1990a. Horizontal interactions in cat striate cortex: I. Anatomical substrate and postnatal development. *European Journal of Neuroscience* 2:344–357.

Luhmann, H. J., J. M. Greuel, and W. Singer. 1990b. Horizontal interactions in cat striate cortex: II. A current source-density analysis. *European Journal of Neuroscience* 2:358–368.

Malpeli, J. G. 1983. Activity of cells in area 17 of the cat in absence of input from layer A of lateral geniculate nucleus. *Journal of Neurophysiology* 49:595–610.

Matsubara, J., M. Cynader, N. V. Swindale, and M. P. Stryker. 1985. Intrinsic projections within visual cortex: Evidence for orientation-specific local connections. *Proceedings of the National Academy of Sciences* (USA) 82:935–939.

Métin, C., and D. O. Frost. 1989. Visual responses of neurons in somatosensory cortex of hamsters with experimentally induced retinal projections to somatosensory thalamus. *Proceedings of the National Academy of Sciences* (USA) 86:357–361.

Pearce, B., J. Albrecht, C. Morrow, and S. Murphy. 1986. Astrocyte glutamate receptor activation promotes inositol phospholipid turnover and calcium flux. *Neuroscience Letters* 72:335–340.

Schwarcz, R., T. Hökfelt, S. Fuxe, G. Jonsson, M. Goldstein, and L. Terenius. 1979. Ibotenic acid–induced neuronal degeneration: A morphological and neurochemical study. *Experimental Brain Research* 37:199–216.

Shatz, C. J., and M. B. Luskin. 1986. The relationship between the geniculocortical afferents and their cortical target cells during development of the cat's primary visual cortex. *Journal of Neuroscience* 6:3655–3668.

Shou, T., and A. G. Levanthal. 1989. Organized arrangement of orientation-sensitive relay cells in the cat's dorsal lateral geniculate nucleus. *Journal of Neuroscience* 9:4287–4302.

Simon, R. P., J. H. Swan, T. Griffiths, and B. S. Meldrum. 1984. Blockade of N-methyl-D-aspartate receptors may protect against ischemic damage in the brain. *Science* 226:850–852.

Symonds, L. L., and A. C. Rosenquist. 1984. Corticocortical connections among visual areas in the cat. *Journal of Comparative Neurology* 229:1–38.

Tusa, R. J., L. A. Palmer, and A. C. Rosenquist. 1978. The retinotopic organization of area 17 (striate cortex) in the cat. *Journal of Comparative Neurology* 177:213–236.

7. Functional Brain Asymmetry, Dyslexia, and Immune Disorders

Annett, M., ed. 1985. *Left, Right, Hand and Brain: The Right Shift Theory.* London: Lawrence Erlbaum.

Annett, M., and A. Turner. 1974. Laterality and the growth of intellectual abilities. *British Journal of Educational Psychology* 44:37–46.

Beaumont, J. G., M. Thomson, and M. Rugg. 1981. An interhemispheric integration deficit in dyslexia. *Current Psychological Research* 1:185–198.

Benbow, C. P. 1986. Physiological correlates of extreme intellectual precocity. *Neuropsychologia* 24:719–725.

Benbow, C. P., and J. C. Stanley. 1981. Mathematical ability: Is sex a factor? *Science* 212:118–119.

Bradshaw, J. L., and N. C. Nettleton. 1983. *Human Cerebral Asymmetry.* Englewood Cliffs, NJ: Prentice-Hall.

Bunge, M. 1967. *Scientific Research: The Search for System.* New York: Springer-Verlag.

Corballis, M. C., and I. L. Beale. 1976. *The Psychology of Left and Right.* Hillsdale, NJ: Lawrence Erlbaum.

Galaburda, A. M., G. F. Sherman, G. D. Rosen, F. Aboitz, and N. Geschwind. 1985. Developmental dyslexia: Four consecutive patients with cortical anomalies. *Annals of Neurology* 18:222–233.

Geffen, G., and K. Quinn. 1984. Hemispheric specialization and ear advantages in processing of speech. *Psychological Bulletin* 96:273–291.

Geschwind, N. 1982. Why Orton was right. *Annals of Dyslexia* 32:13–20.

——— 1984. Cerebral dominance in biological perspective. *Neuropsychologia* 22:675–683.

Geschwind, N., and P. O. Behan. 1982. Lefthandedness: Association with immune disease, migraine and developmental learning disorder. *Proceedings of the National Academy of Sciences* (USA) 79:5097–5100.

——— 1984. Laterality, hormones, and immunity. In *Cerebral Dominance:*

The Biological Foundations, ed. N. Geschwind and A. M. Galaburda. Cambridge, Mass.: Harvard University Press.

Geschwind, N., and A. M. Galaburda. 1985a. Cerebral lateralization: Biological mechanisms, associations, and pathology: I. A hypothesis and a program for research. *Archives of Neurology* 42:429–459.

——— 1985b. Cerebral hypothesis and a program for research. *Archives of Neurology* 42:521–552.

——— 1985c. Cerebral lateralization: Biological mechanisms, associations, and pathology: III. A hypothesis and a program for research. *Archives of Neurology* 42:634–657.

Hansen, O., J. Nerup, and D. Holbek. 1986. A common genetic origin of specific dyslexia and insulin-dependent diabetes mellitus. *Hereditas* 105:165–167.epHellige, J. B. 1990. Hemispheric asymmetry. *Annual Review of Psychology* 41:55–80.

Hugdahl, K. 1987. Lateralization of associative processes: Human conditioning studies. In *Duality and Unity of the Brain,* ed. D. Ottosson. Hampshire, UK: Macmillan Press.

——— ed. 1988. *Handbook of Dichotic Listening: Theory, Methods and Research.* Chichester, UK: Wiley and Sons.

Hugdahl, K., and L. Andersson. 1986. The "forced-attention paradigm" in dichotic listening to CV-syllables: A comparison between adults and children. *Cortex* 22:417–432.

——— 1987. Dichotic listening and reading acquisition in children. A one-year follow-up. *Journal of Clinical and Experimental Neuropsychology* 9:631–649.

——— 1989. Visual half-field tests of lateralized letter presentations: Differences between preliterate and literate children. *International Journal of Neuroscience* 44:215–226.

Hugdahl, K., and C. G. Brobeck. 1986. Hemispheric asymmetry and human electrodermal conditioning: The dichotic extinction paradigm. *Psychophysiology* 23:491–499.

Hugdahl, K., B. Ellertsen, P. E. Waaler, and H. Klove. 1989. Left- and right-handed dyslexic boys: An empirical test of some assumptions of the Geschwind-Behan hypothesis. *Neuropsychologia* 27:223–231.

Hugdahl, K., and M. Franzon. 1985. Visual half-field presentations of incongruent color-words reveal mirror-reversal of language lateralization in dextral and sinistral subjects. *Cortex* 21:359–374.

Hugdahl, K., G. Kvale, H. Nordby, and J. B. Overmier. 1987. Hemispheric asymmetry and human classical conditioning to verbal and non-verbal visual CSs. *Psychophysiology* 24:557–565.

Hugdahl, K., H. Nordby, and G. Kvale. 1990a. Conditional learning and brain asymmetry: Empirical data and a theoretical framework. *Learning and Individual Differences* 1:385–406.

Hugdahl, K., B. Synnevaag, and P. Satz. 1990b. Immune and auto-immune diseases in dyslexic children. *Neuropsychologia* 28:673–680.

Hynd, G. W., and M. Semrud-Clikeman. 1989. Dyslexia and brain morphology. *Psychological Bulletin* 106:447–482.

Kinsbourne, M. 1986. Relationships between nonrighthandedness and diseases of the immune-system. Paper read at the Annual Meeting of International Neuropsychological Society, Denver, CO.

Kolata, G. 1982. Math genius may have hormonal basis. *Science* 222:1312.

Marcel, T., L. Katz, and M. Smith. 1974. Laterality and reading proficiency. *Neuropsychologia* 12:131–140.

Mayes, A. R. 1989. *Human Organic Memory Disorders.* New York: Cambridge University Press.

McKeever, W. F. 1986. Tachistoscopic methods in neuropsychology. In *Experimental Techniques in Human Neuropsychology,* ed. J. Hannay. New York: Oxford University Press.

McKeever, W. F., and A. D. Van Deventer. 1975. Dyslexic adolescents: Evidence of impaired visual and auditory language processing associated with normal lateralization and visual responsivity. *Cortex* 11:361–378.

Orton, S. 1937. *Reading, Writing, and Speech Problems in Children.* New York: Norton.

Parkins, R., R. J. Roberts, S. J. Reinarz, and N. R. Varnes. 1987. CT asymmetries in adult developmental dyslexia. Paper read at the Annual Meeting of the International Neuropsychological Society, Washington, DC.

Pennington, B. F., S. D. Smith, W. J. Kimberling, P. A. Green, and M. M. Haith. 1987. Lefthandedness and immune disorders in familial dyslexics. *Archives of Neurology* 44:634–639.

Rackowski, D., J. W. Kalat, and R. Nebes. 1974. Reliability and validity of some handedness items. *Neuropsychologia* 12:43–47.

Rosen, G. D., A. M. Galaburda, and G. F. Sherman. 1987. Mechanisms of brain asymmetry: New evidence and hypotheses. In *Duality and Unity of the Brain,* ed. D. Ottosson. Hampshire, UK: Macmillan Press.

Salcedo, J. R., B. J. Spiegler, E. Gibson, and D. B. Magilavy. 1985. The autoimmune disease systemic lupus erythematosus is not associated with left-handedness. *Cortex* 21:645–647.

Satz, P., and H. V. Soper. 1986. Lefthandedness, dyslexia and autoimmune disorder: A critique. *Journal of Clinical and Experimental Neuropsychology* 8:453–458.

Satz, P., and S. S. Sparrow. 1970. Specific developmental dyslexia: A theoretical formulation. In *Specific Reading Disability: Advance in Theory and Method,* ed. D. J. Bakker and P. Satz. Rotterdam: Rotterdam University Press.

Schachter, S. C., B. J. Ransil, and N. Geschwind. 1987. Associations of handedness with hair color and learning disabilities. *Neuropsychologia* 25:269–276.

Schur, P. H. 1986. Handedness in systemic lupus erythematosus. *Arthritis and Rheumatism* 29:419–420.

Squire, L. 1987. *Memory and Brain.* New York: Oxford University Press.

Thomson, M. 1975. Laterality and reading attainment. *British Journal of Educational Psychology* 45:317–327.
Urion, D. K. 1988. Nondextrality and autoimmune disorders among relatives of language-disabled boys. *Annals of Neurology* 24:267–269.
Wofsy, D. 1984. Hormones, handedness, and autoimmunity. *Immunology Today* 5:169–170.

8. Fetal Exposure to Maternal Brain Antibodies and Neurological Handicap

Abramsky, O., T. Brenner, and R. P. Lisak. 1979. Significance in neonatal myasthenia gravis of inhibitory effect of amniotic fluid on binding of antibodies to acetylcholine receptor. *Lancet* 2:1333–1335.
Adinolfi, M. 1979. The permeability of the blood-CSF barrier during fetal life in man and rat and the effect of brain antibodies on the development of the CNS. In *Protein Transmission through Living Membranes,* ed. W. A. Hemmings. Amsterdam: Elsevier–North Holland.
——— 1990. Some controversial issues in the field of immunology of reproduction. *Journal of Immunological Research* 2:41–50.
Adinolfi, M., S. E. Beck, S. A. Haddad, and M. J. Seller. 1976. Permeability of blood–cerebrospinal fluid barrier to plasma proteins during fetal and perinatal life. *Nature* 259:140–141.
Adinolfi, M., and S. Haddad. 1977. Levels of plasma proteins in human and rat CSF and the development of the blood-CSF barrier. *Neuropädiatrie* 8:345–353.
Adinolfi, M., J. T. Rick, and S. Liebowitz. 1985. Studies on the biological effects of brain antibodies. In *Immunological Studies of Brain Cells and Functions,* ed. M. Adinolfi and A. Bignami. S.I.M.P. Research Monograph No. 6. London: Spastics International Medical Publications.
Alescio-Lautier, B., D. Metzger, C. Devigne, and B. Soumireu-Mourat. 1989. Microinjection of anti-vasopressin serum into hippocampus in mice: Effects on appetitively reinforced task after intraventricular administration of Arg-vasopressin. *Brain Research* 500:287–294.
Altman, J., and K. Sudarshan. 1975. Postnatal development of locomotion in the laboratory rat. *Animal Behaviour* 23:896–920.
Amtorp, O., and S. C. Sorensen. 1974. The ontogenic development of concentration differences for proteins and ions between plasma and cerebrospinal fluid in rabbits and rats. *Journal of Physiology* 243:387–400.
Arnhold, B. G., and R. Zetterstrom. 1958. Proteins in the cerebrospinal fluid in newborn. An electrophoretic study including hemolytic disease of the newborn. *Pediatrics* 21:279–287.
Auroux, M. 1968. II. Variations des troubles de la capacité d'apprentissage chez le rat en fonction du sexe. *Comptes Rendus des Séances, Société de Biologie* 162:1265–1267.
Auroux, M., M. O. Alnot, and C. Jouvensal. 1967. Perturbations tardives du système nerveux central compatibles avec la vie. I. Baisse de la

capacité d'apprentissage chez le rat par injection d'anticorps hétérospécifiques. *Comptes Rendus des Séances, Société de Biologie* 161:1917–1922.

——— 1968. Perturbations tardives du système nerveux central compatibles avec la vie. II. Etude de la recuperation après alteration chez le rat de la capacité d'apprentissage. *Comptes Rendus des Séances, Société de Biologie* 162:1261–1264.

Bartoccioni, E., A. Evoli, C. Casali, C. Scoppetta, P. Tonali, and C. Provenzano. 1986. Neonatal myasthenia gravis: A clinical and immunological study of seven mothers and their newborn infants. *Journal of Neuroimmunology* 12:155–161.

Behan, P., and N. Geschwind. 1985. Dyslexia, congenital anomalies and immune disorders: The role of the fetal environment. *Annals of the New York Academy of Sciences* 457:13–18.

Behnsen, G. 1926. Farbstoffversuche mit Trypanblau an der Schranke zwischen Blut und Zentralnervensystem der wachsenden Maus. *Münchener Medizinische Wochenschrift* 73:1143–1150.

——— 1927. Uber der Farbstoffspeicherung im Zentralnervensystem der weissen Maus in verschiedenen Alterszustanden. *Zeitschrift für Zellforschung und Mikroskopische Anatomie* 4:515–572.

Bowen, F. P., J. Kosarova, D. Casella, W. J. Nicklas, and S. Berl. 1976. Focal epileptogenic activity induced by topical application of antisera to brain ectomysin-like protein. *Brain Research* 102:363–367.

Brock, D. J. H., and R. G. Sutcliffe. 1972. Alpha-fetoprotein in the antenatal diagnosis of anencephaly and spina bifida. *Lancet* 2:197–199.

Brooks, C. M., and M. E. Peck. 1940. Effects of various cortical lesions on development of placing and hopping reactions in the rat. *Journal of Neurophysiology* 3:66–73.

Cohen, J. J. 1989. Immunity and behaviour. *Journal of Allergy and Clinical Immunology* 79:2–5.

Davson, H. 1967. *Physiology of the Cerebrospinal Fluid.* London: J. and A. Churchill.

——— 1976. The blood-brain barrier. *Journal of Physiology* 255:1–28.

De Felipe, M. C., M. T. Molimero, and J. Del Rio. 1989. Long-lasting neurochemical and functional changes in rats induced by neonatal administration of substance P antiserum. *Brain Research* 485:301–308.

Dobbing, J. 1971. The development of the blood-brain barrier. *Progress in Brain Research* 29:417–425.

Donaldson, J. O., A. S. Penn, R. P. Lisak, O. Abramsky, T. Brenner, and D. L. Scotland. 1981. Anti-acetylcholine receptor antibody in neonatal myasthenia gravis. *American Journal of Diseases in Childhood* 135:222–226.

Evans, C. A. N., J. M. Reynolds, M. L. Reynolds, N. R. Saunders, and M. B. Segal. 1974. The development of a blood-brain barrier mechanism in fetal sheep. *Journal of Physiology* 238:371–386.

Fariello, R. G., G. A. Bubenik, G. M. Brown, and L. J. Grota. 1977. Epileptogenic action of intraventricularly injected antimelatonin antibody. *Neurology* 27:567–570.

Gluecksohn-Waelsch, S. 1957. The effect of maternal immunisation against organ tissues on embryonic differentiation in the mouse. *Journal of Embryology and Experimental Morphology* 5:83–89.

Grazer, F. M., and C. D. Clemente. 1957. Developing blood-brain barrier to trypan blue. *Proceedings of the Society for Experimental Biology, New York* 94:758–760.

Griffin, W. S. T., J. R. Head, D. J. Woodward, and C. Carrol. 1978. Graft-versus-host disease impairs cerebellar growth. *Nature* 275:315–317.

Grontoft, O. 1954. Intracranial haemorrhage and blood-brain barrier problems in the newborn. *Acta Pathologica et Microbiologica Scandinavica, Supplement* 100:1–109.

Harms, D. 1975. Comparative quantitation of immunoglobulin G (IgG) in cerebrospinal fluid and serum of children. *European Neurology* 13:54–64.

Heape, W. 1891. Preliminary note on the transplantation and growth of mammalian ova within a uterine foster mother. *Proceedings of the Royal Society of London (Biology)* 48:457–458.

Heath, R. G., and I. M. Krupp. 1967. Catatonia induced in monkeys by antibrain antibody. *American Journal of Psychiatry* 123:1499–1504.

Heath, R. G., I. M. Krupp, L. W. Byers, and J. I. Liljekvist. 1967. Schizophrenia as an immunological disorder. III. Effects of antimonkey and antihuman antibody on brain function. *Archives of General Psychiatry* 16:24–33.

Hofstein, R., M. Segal, and D. Samuel. 1980. Antibodies to synaptosomal membranes of rat hippocampus and caudate nucleus: Immunological and behavioural characteristics. *Experimental Neurology* 70:307–320.

Johanson, C. E. 1980. Permeability and vascularity of the developing brain: Cerebellum vs. cerebral cortex. *Brain Research* 190:3–16.

Kappers, J. A. 1958. Structural and functional changes in the telencephalic choroid plexus during human ontogenesis. In *The Cerebrospinal Fluid*, ed. G. E. W. Wolstenholme and C. M. O'Connor. London: Churchill.

Karpiak, S. E., and M. M. Rapport. 1975. Behavioural changes in 2-month-old rats following exposure to antibodies against synaptic membranes. *Brain Research* 92:405–413.

Keesey, J., J. Lindstrom, H. Cokely, and C. Herrmann. 1977. Anti-acetylcholine receptor antibody in neonatal myasthenia gravis. *New England Journal of Medicine* 296:55.

Kirman, B. H. 1975. Immune reactions as a possible cause of hitherto unclassified encephalopathy and mental retardation. In *Proceedings of the Third Congress of the International Association for the Scientific Study of Mental Deficiency, The Hague, 1973*, ed. D. A. Primrose. Warsaw: Polish Medical Publishers.

Klingman, G. I., and J. D. Klingman. 1972. Immunosympathectomy as an ontogenic tool. In *Immunosympathectomy,* ed. G. Steiner and E. Schonbaum. Amsterdam: Elsevier.

Kobiler, D., S. Fuchs, and D. Samuel. 1976. The effect of antisynaptosomal plasma membrane antibodies on memory. *Brain Research* 115:129–138.

Lefvert, A. K., and P. O. Osterman. 1983. Newborn infants to myasthenic mothers: A clinical study and an investigation of acetylcholine receptor antibodies in 17 children. *Neurology* 33:133–138.

Leibowitz, S., and R. A. C. Hughes. 1983. *Immunology of the Nervous System.* London: Edward Arnold.

Levi-Montalcini, R. 1964. The nerve growth factor. *Annals of the New York Academy of Sciences* 118:149–170.

——— 1966. The nerve growth factor: Its mode of action on sensory and sympathetic nerve cells. *Harvey Lectures* 60:212–259.

McPherson, C. F. C., and R. P. N. Shek. 1970. Effect of brain isoantibodies on learning and memory in the rat. *Experimental Neurology* 29:1–15.

Mihailovic, L. J., I. Divac, K. Mithovic, D. Milosevic, and B. D. Jankovic. 1969. Effects of intraventricularly injected anti-brain antibodies on delayed alternation and visual discrimination test performance in rhesus monkeys. *Experimental Neurology* 24:325–336.

Mihailovic, L. J., and B. D. Jankovic. 1961. Effects of intraventricularly injected anti-nucleus caudatus antibody on the electrical activity of the cat brain. *Nature* 192:665–666.

Millen, J. W., and A. Hess. 1958. The blood-brain barrier: An experimental study with vital dyes. *Brain* 81:248–257.

Mollison, P. L. 1979. *Blood Transfusion in Medicine.* Oxford: Blackwell Scientific Publishers.

Morel, E., B. Eymard, B. Vernet-der Garabedian, C. Panier, O. Dulac, and J. F. Bach. 1988. Neonatal myasthenia gravis: A new clinical and immunological appraisal on 30 cases. *Neurology* 38:138–142.

Nakao, K., H. Nishitani, M. Susuki, M. Ohta, and K. Hayashi. 1977. Anti-acetylcholine receptor IgG in neonatal myasthenia gravis. *New England Journal of Medicine* 297:169.

Nellhaus, G. 1971. Cerebrospinal fluid immunoglobulin G in childhood. *Archives of Neurology* 24:441–448.

Nolan, P. M., R. Bell, and C. M. Regan. 1987. Acquisition of a brief behavioural experience in the rat is inhibited by the brain-specific monoclonal antibody, F3-87-8. *Neuroscience Letters* 79:346–350.

Ohta, M., M. S. Matsubara, K. Kayashi, K. Nakao, and H. Nishitani. 1981. Acetylcholine receptor antibodies in infants of mothers with myasthenia gravis. *Neurology* 31:1019–1022.

Otila, E. 1948. Studies on the cerebrospinal fluid in premature infants. *Acta Paediatrica* 35 (Supplement 8):3–100.

Plioplys, A. V., A. Greaves, and W. Yoshida. 1989. Anti-CNS antibodies in childhood neurologic diseases. *Neuropediatrics* 20:93–102.

Rick, J. T., A. N. Gregson, S. Leibowitz, and M. Adinolfi. 1980. Behavioral changes in adult rats following administration of antibodies against brain gangliosides. *Developmental Medicine and Child Neurology* 22:719–724.

Rick, J. T., A. N. Gregson, M. Adinolfi, and S. Leibowitz. 1981. The behaviour of immature and mature rats, exposed prenatally to antiganglioside antibodies. *Journal of Neuroimmunology* 1:413–419.

Saunders, N. R., K. M. Dziegelewska, D. H. Malinowska, M. L. Reynolds, J. M. Reynolds, and K. Mollgard. 1976. Permeability of blood–cerebrospinal fluid barrier during fetal and perinatal life. *Nature* 262:156.

Sazonova, L. V., and V. M. Yavkin. 1977. A study of anti-brain autoimmune antibodies in mental retardation of parasyphilitic origin. In *Research to Practice in Mental Retardation,* vol. 3, *Biomedical Aspects,* ed. P. Mittler.

Scott, J. S. 1976. Immunological diseases in pregnancy. In *Immunology of Human Reproduction,* ed. J. S. Scott and W. R. Jones. London: Academic Press.

Scott, J. S., P. J. Maddison, P. V. Taylor, E. Esscher, O. Scott, and R. P. Skinner. 1983. Connective-tissue disease, antibodies to ribonucleoprotein and congenital heart block. *New England Journal of Medicine* 309:209–212.

Seller, M. J., and M. Adinolfi. 1975. Levels of albumin, alpha fetoprotein and IgG in human fetal cerebrospinal fluid. *Archives of Disease in Childhood* 50:484–485.

Semenov, S. F., A. P. Chuprikov, and V. P. Kokhanov. 1968. Anti-brain antibodies in women with brain abnormalities and in their newborn babies. *Zhurnal Nevropathologhi i Psikhiatrii* 68:1819–1926.

Semenov, S. F., M. S. Pevsner, M. G. Reidiboim, and R. D. Kogan. 1972. Comparative study of anti-brain antibodies and brain antigens in the blood serum of children with some developmental abnormalities. *Pediatriya* 5:49–54.

Semenov, S. F., K. N. Nazarov, and A. P. Chuprikov. 1973. *Autoimmunniye protsessy pri vrozhdennykh entsefalopatiyakh, epilepsi i shizofrenii* [Autoimmune processes in congenital encephalopathy, epilepsy and schizophrenia]. Moscow: Meditsina Publishing House.

Statz, A., and K. Felgenhauer. 1983. Development of the blood-CSF barrier. *Developmental Medicine and Child Neurology* 25:152–161.

Stern, L., and R. Peyrot. 1927. Le fonctionnement de la barrière hématoencéphalique aux divers stades de développement chez diverses espèces animals. *Comptes Rendus des Séances, Société de Biologie* 96:1124–1126.

Stubbs, F. G., and M. L. Crawford. 1977. Depressed lymphocyte responsiveness in autistic children. *Journal of Autism and Childhood Schizophrenia* 7:49–55.

Tang, Y. 1935. On the development of different placing reactions in the albino rat. *Chinese Journal of Physiology* 9:339–344.

Thorley, J. D., J. M. Kaplan, R. K. Holmes, G. H. McCracken, and J. P. Sanford. 1975. Passive transfer of antibodies of maternal origin from blood to cerebrospinal fluid in infants. *Lancet* 1:651–653.

Todd, R. D., and R. D. Ciaranello. 1985. Demonstration of inter- and intraspecies differences in serotonin binding sites by antibodies from an autistic child. *Proceedings of the National Academy of Sciences* (USA) 82:612–616.

Todd, R. D., J. M. Hickok, G. M. Anderson, and D. J. Cohen. 1988. Antibrain antibodies in infantile autism. *Biology and Psychiatry* 23:644–647.

Vernet-der Garabedian, B., B. Eymard, J. F. Bach, and E. Morel. 1989. Alpha-bungarotoxin blocking antibodies in neonatal myasthenia gravis: Frequency and selectivity. *Journal of Neuroimmunology* 21:41–47.

Warren, R. P., A. Foster, and N. C. Margaretten. 1987. Reduced natural killer cell activity in autism. *American Journal of Clinical Adolescent Psychiatry* 26:333–335.

Warren, R. P., A. Foster, N. C. Margaretten, and N. C. Pace. 1986. Immune abnormalities in patients with autism. *Journal of Autism and Developmental Disease* 16:189–197.

Weizmann, A., R. Weizmann, G. A. Szekely, H. Wejsenbeek, and E. Livni. 1982. Abnormal immune response to brain tissue antigen in the syndrome of autism. *American Journal of Psychiatry* 139:1462–1465.

Wenzel, D., and K. Felgenhauer. 1976. The development of the blood-CSF barrier after birth. *Neuropädiatrie* 7:175–181.

Widell, S. 1958. On the cerebrospinal fluid in normal children and patients with acute abacterial meningoencephalitis. *Acta Paediatrica* 47 (Supplement 115):1–102.

9. Hormones and Cerebral Organization

Baker, M. A., ed. 1987. *Sex Differences in Human Performance.* New York: Wiley Press.

Bauer, R. 1978. Ontogeny of two-way avoidance in male and female rats. *Developmental Psychobiology* 11:103–116.

Beatty, W. 1979. Gonadal hormones and sex differences in nonreproductive behaviors in rodents: Organizational and activational influences. *Hormones and Behavior* 12:112–163.

Beatty, W. 1984. Hormonal organization of sex differences in play fighting and spatial behavior. In *Progress in Brain Research,* vol. 61, ed. G. J. De Vries, J. P. C. De Bruin, H. B. M. Uylings, and M. A. Corner. Amsterdam: Elsevier Science Publishers.

Beatty, W. W., K. C. Gregoire, and L. L. Parmiter. 1973. Sex differences in retention of passive avoidance behavior in rats. *Bulletin of the Psychonomic Society* 2:99–100.

Bell, A. D., and S. Variend. 1985. Failure to demonstrate sexual dimorphism of the corpus callosum in childhood. *Journal of Anatomy* 143:143–147.

Benbow, C. P., and J. C. Stanley. 1981. Mathematical ability: Is sex a factor? *Science* 212:118–119.

——— 1983. Sex differences in math reasoning ability: More facts. *Science* 222:1029–1031.

Bernstein, M. E. 1954. Studies in the human sex ratio. Evidence of genetic variations in the primary sex ratio in man. *Journal of Heredity* 45:59–64.

Berrebi, A. S., R. H. Fitch, D. L. Ralphe, J. O. Denenberg, V. L. Friedrich, Jr., and V. H. Denenberg. 1988. Corpus callosum: Region-specific effects of sex, early experience, and age. *Brain Research* 438:216–224.

Bleier, R., L. Houston, and W. Byne. 1986. Can the corpus callosum predict gender, age, handedness, or cognitive differences? *Trends in Neuroscience* 9:391–394.

Bryden, M. P. 1982. *Laterality: Functional Asymmetry in the Brain.* New York: Academic Press.

Buffery, A. W. H., and J. A. Gray. 1972. Sex differences in the development of spatial and linguistic skills. In *Gender Differences: Their Ontogeny and Significance,* ed. C. Ounsted and D. C. Taylor, 123–157. Edinburgh: Churchill Livingstone.

Byne, W., R. Bleier, and L. Houston. 1986. Individual, age and sex related differences in human corpus callosum: A study using magnetic resonance imaging. *Society for Neuroscience Abstracts* 12:719.

Camp, D., T. Robinson, and J. Becker. 1984. Sex differences in the effect of early experience on the development of behavioral and brain asymmetries in rats. *Physiology and Behavior* 33:433–439.

Chambers, K. C. 1985. Sexual dimorphism as an index of hormonal influences on conditioned food aversions. *Annals of the New York Academy of Science* 443:110–125.

Clark, A. S., and P. S. Goldman-Rakic. 1989. Gonadal hormones influence the emergence of cortical function in nonhuman primates. *Behavioral Neuroscience* 103:1287–1295.

Clarke, S., R. Kraftsik, H. van der Loos, and G. M. Innocenti. 1989. Forms and measures of adult and developing human corpus callosum: Is there sexual dimorphism? *Journal of Comparative Neurology* 280:213–230.

Dawson, J., Y. M. Cheung, and R. T. S. Lau. 1975. Developmental effects of neonatal sex hormones on spatial and activity skills in the white rat. *Biological Psychology* 3:213–229.

deLacoste-Utamsing, C. M., and R. L. Holloway. 1982. Sexual dimorphism in the human corpus callosum. *Science* 216:1431–1432.

Demeter, S., J. Ringo, and R. W. Doty. 1985. Sexual dimorphism in the human corpus callosum? *Society for Neuroscience Abstracts* 11:868.

Denenberg, V. H., A. S. Berrebi, and R. H. Fitch. 1989. A factor analysis of the rat's corpus callosum. *Brain Research* 497:271–279.

Denenberg, V. H., J. T. Brumaghim, G. C. Haltmeyer, and M. X. Zarrow. 1967. Increased adrenocortical activity in the neonatal rat following handling. *Endocrinology* 81:1047–1052.

Denenberg, V. H., R. H. Fitch, L. M. Schrott, P. E. Cowell, and N. S. Waters. 1991a. Corpus callosum: Interactive effects of testosterone and handling in the rat. *Behavioral Neuroscience* 105:562–566.

Denenberg, V. H., A. Kertesz, and P. E. Cowell. 1991b. Corpus callosum: Sex-handedness interactions with human MRIs. In press.

Denenberg, V. H., and D. A. Yutzey. 1985. Hemispheric laterality, behavioral asymmetry, and the effects of early experience in rats. In *Cerebral Lateralization in Nonhuman Species,* ed. S. D. Glick, pp. 109–133. Orlando: Academic Press.

Denenberg, V. H., and M. X. Zarrow. 1971. Effects of handling in infancy upon adult behavior and adrenocortical activity: Suggestions for a neuroendocrine mechanism. In *Early Childhood: The Development of Self-Regulatory Mechanisms,* ed. D. N. Walcher and D. L. Peters, pp. 39–64. New York: Academic Press.

Denti, A., and A. Epstein. 1972. Sex differences in the acquisition of two kinds of avoidance behavior in rats. *Physiology and Behavior* 8:611–615.

Denti, A., and J. Negroni. 1975. Activity and learning in neonatally hormone treated rats. *Acta Physiologica Latinoamerica* 25:99–106.

Diamond, M. C. 1984. Age, sex and environmental influences. In *Cerebral Dominance: The Biological Foundations,* ed. N. Geschwind and A. M. Galaburda, pp. 134–146. Cambridge: Harvard University Press.

Diamond, M., G. Dowling, and R. Johnson. 1981. Morphologic cerebral cortical asymmetry in male and female rats. *Experimental Neurology* 71:261–268.

Diamond, M., R. E. Johnson, and J. Ehlert. 1979. A comparison of cortical thickness in male and female rats—normal and gonadectomized, young and adult. *Behavioral Neurology* 26:485–491.

Earley, C. J., and B. E. Leonard. 1978. Androgenic involvement in conditioned taste aversion. *Hormones and Behavior* 11:1–11.

Fitch, R. H., A. S. Berrebi, P. E. Cowell, L. M. Schrott, and V. H. Denenberg. 1990a. Corpus callosum: Effects of neonatal hormones on sexual dimorphism in the rat. *Brain Research* 515:111–116.

Fitch, R. H., P. E. Cowell, L. M. Schrott, and V. H. Denenberg. 1990b. Corpus callosum: Perinatal anti-androgen and callosal demasculinization. *International Journal of Developmental Neuroscience* 9:35–38.

——— 1991a. Corpus callosum: Ovarian hormones and feminization. *Brain Research* 542:313–317.

Fitch, R. H., R. F. McGivern, E. Redei, L. M. Schrott, P. E. Cowell, and V. H. Denenberg. 1992. Neonatal ovariectomy and pituitary-adrenal responsiveness in the adult rat. *Acta Endocrinology* 126:44–48.

Fleming, D., R. H. Anderson, R. Rhees, E. Kinghorn, and J. Bakaitis. 1986. Effects of prenatal stress on sexually dimorphic asymmetries in the cerebral cortex of the male rat. *Brain Research Bulletin* 16:395–398.

Frankfurt, M., E. Gould, C. S. Woolley, and B. S. McEwen. 1990. Gonadal steroids modify dendritic spine density in ventromedial hypothalamic neurons: A Golgi study in the adult rat. *Neuroendocrinology* 51:530–535.

Galaburda, A. M., J. Corsiglia, G. D. Rosen, and G. F. Sherman. 1987. Planum temporale asymmetry: Reappraisal since Geschwind and Levitsky. *Neuropsychologia* 25:853–868.

Galaburda, A. M., and T. M. Kemper. 1979. Cytoarchitectonic abnormalities in developmental dyslexia: A case study. *Annals of Neurology* 6:94–100.

Garron, D. 1977. Sex linked recessive inheritance of spatial and numerical abilities and Turner's syndrome. *Psychological Review* 77:147–152.

Geschwind, N., and P. O. Behan. 1982. Left-handedness: Association with immune disease, migraine and developmental learning disorders. *Proceedings of the National Academy of Sciences* (USA) 79:5097–5100.

Geschwind, N., and A. M. Galaburda. 1985. Cerebral lateralization: Biological mechanisms, associations, and pathology (parts I, II, and III). *Archives of Neurology* 42:428–654.

Gould, F., C. S. Woolley, M. Frankfurt, and B. S. McEwen. 1990. Gonadal steroids regulate dendritic spine density in hippocampal pyrimidal cells in adulthood. *Journal of Neuroscience* 10:1286–1291.

Graham, J. M., Jr., A. S. Bashir, R. E. Stark, A. Silbert, S. Walzer. 1990. Oral and written language abilities of XXY boys: Implications for anticipatory guidance. *Pediatrics* 8:795–806.

Graham, J. M., Jr., A. S. Bashir, R. E. Stark, and P. Tallal. 1982. Auditory processing in unselected XXY boys. *Clinical Research* 30:118A.

Habib, M. 1989. Anatomical asymmetries of the human cerebral cortex. *International Journal of Neuroscience* 47:67–79.

Halpern, D. F. 1986. *Sex Differences in Cognitive Abilities.* London: Erlbaum.

Harshman, R. A., E. Hampson, and S. A. Berenbaum. 1983. Individual differences in cognitive abilities and brain organization: Part 1. Sex and handedness differences in ability. *Canadian Journal of Psychology* 37:144–192.

Haslam, R. H. A., J. T. Dalby, R. D. Johns, and A. W. Rademaker. 1981. Cerebral asymmetry in developmental dyslexia. *Archives of Neurology* 38:679–682.

Heisler, L. E., R. L. Reid, and D. A. Van Vugt. 1990. Ovarian steroid modulation of CRH stimulation of cortisol in rhesus monkeys. *Society for Neuroscience Abstracts* 16:7.

Hier, D. B., M. LeMay, P. B. Rosenberger, and V. P. Peblo. 1978. Developmental dyslexia. *Archives of Neurology* 35:90–92.

Hugdahl, K., B. Synnevag, and P. Satz. 1990. Immune and autoimmune diseases in dyslexic children. *Neuropsychologia* 28:673–679.

Hynd, G. W., and M. Semrud-Clikeman. 1989. Dyslexia and brain morphology. *Psychological Bulletin* 3:447–482.

James, W. H. 1980. Gonadotropins and the human secondary sex ratio. *British Medical Journal* 281:711–712.

——— 1983. Timing of fertilization and the sex ratio of offspring. In *Sex Selection of Children,* ed. N. G. Bennett, 73–99. New York: Academic Press.

——— 1986. Hormonal control of sex ratio. *Journal of Theoretical Biology* 118:427–441.

——— 1988. Testosterone levels, handedness, and sex ratio at birth. *Journal of Theoretical Biology* 133:261–266.

Jernigan, T. L., J. R. Hesselink, E. Sowell, and P. Tallal. 1991. Cerebral morphology and MRI on language and learning impaired children. *Archives of Neurology* 48:539–545.

Jernigan, T. L., P. Tallal, and J. R. Hesselink. 1987. Cerebral morphology on magnetic resonance imaging in developmental dysphasia. *Society for Neuroscience Abstracts* 13:651.

Joseph, R., S. Hess, and E. Birecree. 1978. Effects of hormone manipulations and exploration on sex differences in maze learning. *Behavioral Biology* 24:364–377.

Juraska, J. M., and J. R. Kopcik. 1988. Sex and environmental influences on the size and ultrastructure of the rat corpus callosum. *Brain Research* 450:1–8.

Kertesz, A., M. Polk, J. Howell, and S. Black. 1987. Cerebral dominance, sex, and callosal size in MRI. *Neurology* 37:1385–1388.

Kime, D. E., G. P. Vinson, P. W. Major, and R. Kilpatrick. 1980. Adrenal-gonad relationships. In *General, Comparative and Clinical Endocrinology of the Adrenal Cortex,* vol. 3, ed. I. C. Jones and I. W. Henderson. New York: Academic Press.

Kimura, D. 1987. Sex differences, human brain organization. In *Encyclopedia of Neuroscience.* Boston: Birkhauser Press.

Kimura, D., and R. Harshman. 1984. Sex differences in brain organization for verbal and non-verbal functions. In *Progress in Brain Research,* vol. 61, ed. G. J. De Vries, J. P. C. De Bruin, H. B. M. Uylings, and M. A. Corner. Amsterdam: Elsevier Science Publishers.

Krasnoff, A., and L. Weston. 1976. Puberal status and sex differences: Activity and maze behavior in rats. *Developmental Psychobiology* 9:261–269.

Kreuz, L. E., R. M. Rosew, and J. R. Jennings. 1972. Suppression of plasma testosterone levels and psychological stress. *Archives of General Psychiatry* 26:479–482.

Larsen, J. P., T. Hoien, I. Lundberg, and H. Odegaard. 1990. MRI evaluation of the size and symmetry of the planum temporale in adolescents with developmental dyslexia. *Brain and Language* 39:289–301.

Levy, J. 1976. Cerebral lateralization and spatial ability. *Behavioral Genetics* 6:171–188.

Levy, J., and T. Nagylaki. 1972. A model for the genetics of handedness. *Genetics* 72:117–128.

Levy, J., and M. Reid. 1978. Variations in cerebral organization as a function of handedness, hand-posture in writing, and sex. *Journal of Experimental Psychology: General* 107:119–144.

Levy-Agresti, J., and R. W. Sperry. 1968. Differential perceptual capacities in major and minor hemispheres. *Proceedings of the National Academy of Science* (USA) 61:1151.

Maccoby, E. E., and C. N. Jacklin. 1974. *The Psychology of Sex Differences.* Stanford: Stanford University Press.

Masica, D. N., J. Money, A. A. Ehrhardt, and V. G. Lewis. 1969. IQ, fetal sex hormones, and cognitive patterns: Studies of the testicular feminizing syndrome of androgen insensitivity. *Johns Hopkins Medical Journal* 124:34–43.

Matsumoto, A., and Y. Arai. 1976. Effect of estrogen on early postnatal development of synaptic formation in the hypothalamic arcuate nucleus of female rats. *Neuroscience Letters* 2:79–82.

——— 1981. Effect of androgen on sexual differentiation of synaptic organization in the hypothalamic arcuate nucleus: An ontogenetic study. *Neuroendocrinology* 33:166–169.

McGlone, J. 1978. Sex differences in functional brain asymmetry. *Cortex* 14:122–128.

——— 1980. Sex differences in human brain asymmetry: A critical survey. *Behavioral and Brain Sciences* 3:215–263.

McGlone, J., and A. Kertesz. 1973. Sex differences in cerebral processing of visuo-spatial tasks. *Cortex* 9:313–320.

Meaney, M. J., D. H. Aitken, S. Bhatnagar, C. Van Berkel, and R. M. Sapolsky. 1988. Postnatal handling attenuates neuroendocrine, anatomical, and cognitive impairments related to the aged hippocampus. *Science* 238:766–768.

Meaney, M. J., D. H. Aitken, S. R. Bodnoff, L. J. Iny, J. E. Tatarewicz, and R. M. Sapolsky. 1985. Early postnatal handling alters glucocorticoid receptor concentrations in selected brain regions. *Behavioral Neuroscience* 99:765–770.

Meyer, J. S. 1983a. Prevention of adrenalectomy induced brain growth stimulation by corticosterone treatment. *Physiology and Behavior* 41:391–395.

——— 1983b. Early adrenalectomy stimulates subsequent growth and development of the rat brain. *Experimental Neurology* 82:432–446.

Meyer, J. S., and K. R. Fairman. 1985. Early adrenalectomy increases myelin content of the brain. *Developmental Brain Research* 17:1–9.

Miyakawa, M., and Y. Arai. 1987. Synaptic plasticity to estrogen in the lateral septum of adult male and female rats. *Brain Research* 436:184–188.

Moore, T. 1967. Language and intelligence: A longitudinal study of the first eight years. *Human Development* 10:88–106.

Nielson, J., A. M. Sorenson, and K. Sorenson. 1981. Mental development of unselected children with sex-chromosomal abnormalities. *Human Genetics* 59:324–332.

Nordeen, E. J., K. W. Nordeen, D. R. Sengelaub, and A. P. Arnold. 1985. Androgens prevent normally occurring cell death in a sexually dimorphic spinal nucleus. *Science* 229:671–673.

Pappas, C. T. E., M. C. Diamond, and R. E. Johnson. 1978. Effects of ovariectomy and differential experience in the rat cerebral cortical morphology. *Brain Research* 154:53–60.

——— 1979. Morphological changes in the cerebral cortex of rats with altered levels of ovarian hormones. *Behavioral and Neural Biology* 26:298–310.

Plante, E., L. Swisher, and R. Vance. 1989. Anatomical correlates of normal and impaired language in a set of dizygotic twins. *Brain and Language* 37:643–655.

Puck, M., K. Tennes, W. Frankenburg, K. Bryant, and A. Robinson. 1975. Early childhood development of four boys with 47 XXY karyotype. *Clinical Genetics* 7:8–20.

Ray, W. J., S. Georgiou, and R. Ravizza. 1979. Spatial abilities, sex differences, and lateral eye movements. *Developmental Psychology* 15:455–457.

Ray, W. J., N. Newcombe, J. Semon, and P. M. Cole. 1981. Spatial abilities, sex differences and EEG functioning. *Neuropsychologica* 19:719–722.

Resnick, S., and S. A. Berenbaum. 1982. Cognitive functioning in individuals with congenital hyperplasia. *Behavioral Genetics* 12:594–595.

Robinson, A., H. Lubs, J. Nielson, and K. Sorenson. 1979. Summary of clinical findings: Profiles of children with 47XXX, 47XXY, and 47XYY karyotypes. *Birth Defects: Original Article Series* 15:261–266.

Rosenberger, P. B., and D. B. Hier. 1980. Cerebral asymmetry and verbal intellectual deficits. *Annals of Neurology* 8:300–304.

Rovet, J., and C. Netley. 1982. Processing deficits in Turner's syndrome. *Developmental Psychology* 18:77–94.

Rumsey, J. M., R. Dorwart, M. Vermes, M. B. Denckla, M. J. P. Kruesi, and J. L. Rapaport. 1986. Magnetic resonance imaging of brain anatomy in severe developmental dyslexia. *Archives of Neurology* 43:1045–1046.

Sanders, B., and M. Soares. 1986. Sexual maturation and spatial ability in college students. *Developmental Psychology* 22:199–203.

Sas, M., and J. Szollosi. 1980. The sex ratio of children of fathers with spermatic disorders following hormone therapy. *Orvosi Hetilap* 121:2807–2808.

Shaywitz, S. E., B. A. Shaywitz, J. M. Fletcher, and M. D. Escobar. 1990. Prevalence of reading disabilities in boys and girls. Results of the Connecticut Longitudinal Study. *Journal of the American Medical Association* 264:998–1002.

Shaywitz, S. E., V. A. Towle, D. K. Kessne, and B. A. Shaywitz. 1988. Prevalence of dyslexia in boys as compared to girls in an epidemiological sample: Contrasts between children identified by research criteria and by the school system. *Annals of Neurology* 24:313–314.

Stewart, J., and B. Kolb. 1988. Asymmetry in the cerebral cortex of the rat: An analysis of the effects of neonatal gonadectomy on cortical thickness in three strains of rats. *Behavioral and Neural Biology* 49:344–360.

Stewart, J., A. Skavarenina, and J. Pottier. 1975. Effects of neonatal androgens on open-field behavior and maze-learning in the prepubescent and adult rat. *Physiology and Behavior* 14:291–295.

Tallal, P. 1980. Auditory temporal perception, phonics, and reading disabilities in children. *Brain and Language* 9:182–198.

——— 1988. Developmental language disorders. *Learning Disabilities: Proceedings of the National Conference,* ed. J. F. Kavanagh and T. J. Truss, Jr., 181–272. Parkton, MD: York Press.

——— 1989. Unexpected sex-ratios in families of language/learning impaired children. *Neuropsychologia* 27:987–998.

Tallal, P., and F. Newcombe. 1978. Impairment of auditory perception and language comprehension in dysphasia. *Brain and Language* 5:13–24.

Tallal, P., and M. Piercy. 1974. Developmental aphasia: Rate of auditory processing and selective impairment of consonant perception. *Neuropsychologia* 12:83–93.

Tallal, P., R. Ross, and S. Curtiss. 1989. Familial aggregation in specific language impairment. *Journal of Speech and Hearing Disorders* 54:167–173.

Tallal, P., R. E. Stark, and E. D. Mellits. 1985a. The relationship between auditory temporal analysis and receptive language development: Evidence from studies of developmental language disorder. *Neuropsychologica* 23:527–534.

——— 1985b. Identification of language-impaired children on the basis of rapid perception and production skills. *Brain and Language* 25:314–322.

Toran-Allerand, C. 1984. On the genesis of sexual differentiation of the central nervous system: Morphogenetic consequences of steroidal exposure and possible role of alpha-feta-protein. In *Progress in Brain Research* vol. 61, ed. G. J. De Vries, J. P. C. De Bruin, H. B. M. Uylings, and M. A. Corner. Amsterdam: Elsevier Science Publishers.

Waber, D. P. 1976. Sex differences in cognition: A function of maturation rate? *Science* 192:572–574.

Waber, D. P., M. B. Mann, J. Merola, and D. M. Moylan. 1985. Physical maturation rate and cognitive performance in early adolescence: A longitudinal evaluation. *Developmental Psychology* 21:666–681.

Ward, I. 1972. Prenatal stress feminizes and demasculinizes the behavior of males. *Science* 175:82–84.

——— 1983. Reproductive functioning in the prenatally stressed female rat. *Developmental Psychobiology* 16:111–118.

Weinberg, J., E. A. Krahn, and S. Levine. 1978. Differential effects of handling on exploration in male and female rats. *Developmental Psychobiology* 11:251–259.

Witelson, S. F. 1985. The brain connection: The corpus callosum is larger in left-handers. *Science* 229:665–667.

——— 1989. Hand and sex differences in the isthmus and genu of the human corpus callosum. *Brain* 112:799–835.

——— 1991. Sexual differentiation of the human tempero-parietal region for functional asymmetry: Neuroanatomical evidence. *Psychoneuroendocrinology* 16:131–153.

Witelson, S. F., and D. L. Kigar. 1988. Anatomical development of the corpus callosum in humans: A review with reference to sex and cognition. In *Brain Lateralization in Children: Developmental Implications*, ed. D. L. Molfese, and S. J. Segalowitz. New York: Guilford Press.

Wittig, M. A., and A. C. Peterson, eds. 1979. *Sex-Related Differences in Cognitive Functioning: Developmental Issues*. New York: Academic Press.

Zimmerberg, B., and L. A. Mickus. 1990. Sex differences in corpus callosum: Influence of prenatal alcohol exposure and maternal undernutrition. *Brain Research* 537:115–22.

Zimmerberg, B., and L. V. Scalzi. 1989. Commissural size in neonatal rats:

Effects of sex and prenatal alcohol exposure. *International Journal of Developmental Neuroscience* 7:81–86.

10. Genes and Genders

Ansara, A., N. Geschwind, A. M. Galaburda, M. Albert, and N. Gartrell. 1981. *Sex Differences in Dyslexia.* Towson, MD: Orton Dyslexia Society.

Bakwin, H. 1973. Reading disability in twins. *Developmental Medicine and Child Neurology* 15:184–187.

Bisgaard, M. L., H. Eiberg, N. Moller, E. Niebuhr, and J. Mohr. 1987. Dyslexia and chromosome 15 heteromorphism: Negative LOD score in a Danish material. *Clinical Genetics* 32:118–119.

Carter, C. O. 1969. Genetics of common disorders. *British Medical Bulletin* 25:52–57.

——— 1973. Multifactorial genetic disease. In *Medical Genetics,* ed. V. A. McKusick and R. Clairborne, 199–208. New York: HP Publishing Company.

Decker, S. N., and S. G. Vanderberg. 1985. Colorado twin study of reading disability. In *Biobehavioral Measures of Dyslexia,* ed. D. B. Gray and J. F. Kavanagh, 123–135. Parkton, MD: York Press.

DeFries, J. C. 1985. Colorado reading project. In *Biobehavioral Measures of Dyslexia,* ed. D. B. Gray and J. F. Kavanagh, 107–122. Parkton, MD: York Press.

DeFries, J. C., and S. N. Decker. 1982. Genetic aspects of reading disability: The Colorado Family Reading Study. In *Reading Disability: Varieties and Treatments,* ed. R. N. Malatesha and P. G. Aaron, 255–279. New York: Academic Press.

DeFries, J. C., and D. W. Fulker. 1985. Multiple regression analysis of twin data. *Behavior Genetics* 15:467–473.

——— 1988. Multiple regression analysis of twin data: Etiology of deviant scores versus individual differences. *Acta Geneticae Medicae et Gemellologiae* 37:205–216.

DeFries, J. C., D. W. Fulker, and M. C. LaBuda. 1987. Evidence for a genetic aetiology in reading disability of twins. *Nature* 329:537–539.

DeFries, J. C., and J. J. Gillis. 1991. Etiology of reading deficits in learning disabilities. In *Neuropsychological Foundations of Learning Disabilities,* ed. J. E. Obrzut and G. W. Hynd, 29–47. Orlando: Academic Press.

DeFries, J. C., G. P. Vogler, and M. C. LaBuda. 1986. Colorado family reading study: An overview. In *Behavior Genetics: Principles and Applications II,* ed. J. L. Fuller and E. C. Simmel, 29–56. Hillsdale, NJ: Lawrence Erlbaum Associates.

Dunn, L. M., and F. C. Markwardt. 1970. *Examiner's Manual: Peabody Individual Achievement Test.* Circle Pines, MN: American Guidance Service.

Elston, R. C., and K. C. Yelverton. 1975. General models for segregation analysis. *American Journal of Human Genetics* 27:31–45.

Finucci, J. M., and B. Childs. 1981. Are there really more dyslexic boys than girls? In Ansara et al. (1981), 1–9.

——— 1983. Dyslexia: Family studies. In *Genetic Aspects of Speech and Language Disorders,* ed. C. Ludlow and J. A. Cooper, 157–167. New York: Academic Press.

Geschwind, N. 1981. A reaction to the conference on sex differences and dyslexia. In Ansara et al. (1981), xiii–xviii.

Harris, E. L. 1986. The contribution of twin research to the study of the etiology of reading disability. In *Genetics and Learning Disabilities,* ed. S. D. Smith, 3–19. San Diego: College-Hill Press.

Hay, D. A., and P. J. O'Brien. 1982. Problems of twins in developmental behavior genetics. *Behavior Genetics* 12:587 (abstract).

LaBuda, M. C., J. C. DeFries, and D. W. Fulker. 1986. Multiple regression analysis of twin data obtained from selected samples. *Genetic Epidemiology* 3:425–433.

LaBuda, M. C., J. C. DeFries, and B. F. Pennington. 1990. Reading disability: A model for the genetic analysis of complex behavioral disorders. *Journal of Counseling and Development* 68:645–651.

Lander, E. S., and D. Botstein. 1989. Mapping Mendelian factors underlying quantitative traits using RFLP linkage maps. *Genetics* 121:185–199.

Lewitter, F. I., J. C. DeFries, and R. C. Elston. 1980. Genetic models of reading disability. *Behavior Genetics* 10:9–30.

Lykken, D. T., A. Tellegen, and R. DeRubeis. 1978. Volunteer bias in twin research: The rule of two-thirds. *Social Biology* 25:1–9.

Nichols, R. C., and W. C. Bilbro. 1966. The diagnosis of twin zygosity. *Acta Genetica* 16:265–275.

Olson, R., B. Wise, F. Connors, J. Rack, and D. Fulker. 1989. Specific deficits in component reading and language skills: Genetic and environmental influences. *Journal of Learning Disabilities* 22:339–348.

Plomin, R., J. C. DeFries, and G. E. McClearn. 1990. *Behavioral Genetics: A Primer,* 2d ed. New York: Freeman.

Shaywitz, S. E., B. A. Shaywitz, J. M. Fletcher, and M. D. Escobar. 1990. Prevalence of reading disability in boys and girls. *Journal of the American Medical Association* 264:998–1002.

Smith, S. D., and D. E. Goldgar. 1986. Single gene analyses and their application to learning disabilities. In *Genetics and Learning Disabilities,* ed. S. D. Smith. San Diego: College-Hill Press.

Smith, S. D., W. J. Kimberling, B. F. Pennington, and H. A. Lubs. 1983. Specific reading disability: Identification of an inherited form through linkage analysis. *Science* 219:1345–1347.

Smith, S. D., B. F. Pennington, W. J. Kimberling, and P. S. Ing. 1990. Familial dyslexia: Use of genetic linkage data to define subtypes. *Journal of the American Academy of Child and Adolescent Psychiatry* 29:204–213.

SPSS-X. 1990. *Statistical Package for the Social Sciences,* 3d ed. [Computer Program]. Chicago: SPSS Inc.

Stevenson, J., P. Graham, H. Fredman, and V. McLoughlin. 1984. The genetics of reading disability. In *The Biology of Human Intelligence,* ed. C. J. Turner and H. B. Miles. Nafferton: Nafferton Books Limited.

——— 1987. A twin study of genetic influences on reading and spelling ability and disability. *Journal of Child Psychology and Psychiatry* 28:229–247.

Symmes, J. S., and J. L. Rapoport. 1972. Unexpected reading failure. *American Journal of Orthopsychiatry* 42:82–91.

Thompson, J. S., and M. W. Thompson. 1986. *Genetics in Medicine,* 4th ed. Philadelphia: Saunders.

Vogel, S. A. 1990. Gender differences in intelligence, language, visual-motor abilities, and academic achievement in males and females with learning disabilities: A review of the literature. *Journal of Learning Disabilities* 23:44–52.

Vogler, G. P., J. C. DeFries, and S. N. Decker. 1985. Family history as an indicator of risk for reading disability. *Journal of Learning Disabilities* 18:419–421.

Wechsler, D. 1974, 1981. *Examiner's Manual: Wechsler Adult Intelligence Scale—Revised.* New York: Psychological Corporation.

Wong, B. 1986. Problems and issues in the definition of learning disabilities. In *Psychological and Educational Perspectives on Learning Disabilities,* ed. J. K. Torgesen and B. Wong, 3–25. New York: Academic Press.

Zerbin-Rüdin, E. 1967. Kongenitale Wortblindheit oder Spezifische Dyslexie [Congenital word-blindness]. *Bulletin of the Orton Society* 17:47–56.

11. Neurological Arguments for a Joint Developmental Dysphasia-Dyslexia Syndrome

Aimard, P. 1982. *L'enfant et son langage.* Villeurbanne, France: SIMEP.

Ajuriaguerra, J. de, A. Jaeggi, F. Guignard, F. Kocher, M. Maquard, S. Roth, and E. Schmid. 1976. The development and prognosis of dysphasia in children. In *Normal and Deficient Language,* ed. D. M. Morehead and A. E. Morehead, 345–385. Baltimore: University Park Press.

Amorosa, H. 1981. The timing of speech and hand motor coordination in language delayed children. *Journal of the Acoustics Society of America* 71:22.

Annett, M. 1985. *Left, Right, Hand and Brain: The Right Shift Theory.* London: Lawrence Erlbaum.

Aram, D. M., and J. E. Nation. 1975. Patterns of language behaviour in children with developmental language disorders. *Journal of Speech and Hearing Research* 18:229–241.

Benton, A. L. 1964. Developmental asphasia and brain damage. *Cortex* 1:40–52.

Bertoncini, J., J. Morais, R. Bijeljac-Babic, S. McAdams, I. Peretz, and J. Mehler. 1989. Dichotic perception and laterality in neonates. *Brain and Language* 37:591–605.

Bever, T. G., C. Carrithers, W. Cowart, and D. J. Townsend. 1989. Language processing and familial handedness. In *From Reading to Neurons,* ed. A. M. Galaburda, 331–360. London: MIT Press.

Birch, H. G., and L. Belmont. 1964. Auditory-visual integration in normal and retarded readers. *American Journal of Orthopsychiatry* 34:852–861.

Bishop, D. V. M. 1988. Otitis media and developmental language disorder: A reply to Gordon. *Journal of Child Psychology and Psychiatry* 29:365–368.

Bishop, D. V. M., and A. Edmundson. 1987. Specific language impairment as a maturational lag: Evidence from longitudinal data on language and motor development. *Developmental Medicine and Child Neurology* 29:442–459.

Boder, E. 1973. Developmental dyslexia: A diagnostic approach based on three atypical reading-spelling patterns. *Developmental Medicine and Child Neurology* 15:663–687.

Bradley, L., and P. E. Bryant. 1981. Categorizing sounds and learning to read: A causal connection. *Nature* 271:746–747.

Bruner, J. 1983. *Savoir Faire, Savoir Dire.* Paris: PUF.

Changeux, J.-P. 1985. *Neuronal Man. The Biology of Mind.* New York: Oxford University Press.

Chavance, M., G. Dellatolas, M. G. Bousser, R. Amor, B. Grardel, A. Kahan, M. F. Kahn, J. P. Le Floch, and G. Tchobroutsky. 1990. Handedness, immune disorders and information bias. *Neuropsychologia* 28:429–441.

Chi, J. G., E. C. Dooling, and F. H. Gilles. 1977. Left-right asymmetries of the temporal speech areas of the human fetus. *Archives of Neurology* 34:346–348.

Cohen, M., R. Campbell, and F. Yaghmai. 1989. Neuropathological abnormalities in developmental dysphasia. *Annals of Neurology* 25:567–570.

Cohen, D. J., B. Caparulo, and B. Shaywitz. 1976. Primary childhood aphasia and childhood autism. *Journal of the American Academy of Child Psychiatry* 15:604–645.

Cossu, G., and J. C. Marshall. 1986. Theoretical implications of the hyperlexia syndrome: Two new Italian cases. *Cortex* 4:579–589.

Crain, S. 1989. Why poor readers misunderstand spoken sentences. In *Phonology and reading disability. Solving the reading puzzle,* ed. D. Shankweiler and I. Y. Liberman, 133–165. IARLD Monograph Series no. 6. Ann Arbor: University of Michigan Press.

Critchley, M. 1964, 1970. *Developmental Dyslexia.* London: Heinemann.

Dalby, M. A. 1977. Aetiological studies in language retarded children. *Neuropaediatrie* 8:499–500.

Damasio, A. R. 1989. Reflections on visual recognition. In *From Reading to Neurons,* ed. A. M. Galaburda, 361–376. Cambridge, Mass.: MIT Press.

Dawson, G., C. Finley, S. Phillips, and A. Lewy. 1989. A comparison of hemispheric asymmetries in speech-related brain potentials of autistic and dysphasic children. *Brain and Language* 37:26–41.

De Grauw, A., C. Njiokiktjien, and L. de Sonneville. 1989. Morphometry of the cerebral hemispheres in developmental dysphasia. *Annals of Neurology Abstract* 26:482.

Denckla, M. B. 1985. Motor coordination in dyslexic children. Theoretical and clinical implications. In *Dyslexia,* ed. F. Duffy and N. Geschwind, 187–195. Boston: Little, Brown.

Duffy, F. H., M. B. Denckla, P. H. Bartels, and G. Sandini. 1980. Dyslexia: Regional differences in brain electrical activity by topographic mapping. *Annals of Neurology* 7:412–420.

Eimas, P. D. 1985. The perception of speech in early infancy. *Scientific American* 252:34–40.

Eisenson, J. 1968. Developmental aphasia (dyslogia): A postulation of a unitary concept of the disorder. *Cortex* 3:184–200.

Elliott, D. E., and R. M. Needleman. 1976. The syndrome of hyperlexia. *Brain and Language* 3:339–349.

Entus, A. K. 1977. Hemispheric asymmetry in processing of dichotically presented speech and nonspeech stimuli by infants. In *Language Development and Neurological Theory,* ed. S. J. Segalowitz and F. A. Gruber, 63–73. New York: Academic Press.

Flynn, J. M., and W. M. Deering. 1989. Subtypes of dyslexia: Investigation of Boder's system using quantitative neurophysiology. *Developmental Medicine and Child Neurology* 31:215–223.

Frith, U. 1986. A developmental framework for developmental dyslexia. *Annals of Dyslexia* 36:69–81.

Frumkin, B., and I. Rapin. 1980. Perception of vowels and consonant-vowels of varying duration in language impaired children. *Neuropsychologia* 18:443–454.

Galaburda, A. M. 1983. Developmental dyslexia, current anatomical research. *Annals of Dyslexia* 33:41–48.

——— 1988. The pathogenesis of childhood dyslexia. In *Language, Communication and the Brain,* ed. F. Plum, 127–137. New York: Raven Press.

Galaburda, A. M., J. Corsiglia, G. D. Rosen, and G. F. Sherman. 1987. Planum temporale asymmetry: Reappraisal since Geschwind and Levitsky. *Neuropsychologia* 25:853–868.

Galaburda, A. M., and T. Kemper. 1979. Cytoarchitectonic abnormalities in developmental dyslexia. A case study. *Annals of Neurology* 6:94–100.

Galaburda, A. M., F. Sanides, and N. Geschwind. 1978. Human brain: Cytoarchitectonic left-right asymmetries in the temporal speech region. *Archives of Neurology* 35:812–817.

Galaburda, A. M., G. F. Sherman, G. D. Rosen, F. Aboitiz, and N. Geschwind. 1985. Developmental dyslexia: Four consecutive patients with cortical anomalies. *Annals of Neurology* 18:222–233.

Gathercole, S. E., and A. D. Baddeley. 1989. The role of phonological memory in normal and disordered language development. In *Brain and Reading,* ed. C. von Euler, I. Lundberg, and G. Lennerstrand, 245–256. London: Macmillan Press.

Geschwind, N., and P. Behan. 1982. Lefthandedness: Association with immune disease, migraine, and developmental learning disorder. *Proceedings of the National Academy of Sciences* (USA) 79:5097–5100.

——— 1983. Laterality, hormones and immunity. In *Cerebral Dominance:*

The Biological Foundations, ed. N. Geschwind and A. M. Galaburda, 211–224. Cambridge, Mass: Harvard University Press.

Gilles, F. H., A. Leviton, and E. C. Dooling. 1983. *The Developing Human Brain.* Boston: John Wright.

Gordon, A. G. 1988. Some comments on Bishop's annotation "Developmental dysphasia and otitis media." *Journal of Child Psychology and Psychiatry* 39:361–363.

Gould, J. H., and D. J. Glencross. 1990. Do children with a specific reading disability have a general serial-ordering deficit? *Neuropsychologia* 28:271–278.

Grohnfeldt, M. 1986. *Störungen der Sprachentwicklung.* Berlin: Marhold.

Hall, D. 1989. Delayed speech in children. *British Medical Journal* 49:7–8.

Hanson, V. L. 1989. Phonology and reading: Evidence from profoundly deaf readers. In *Phonology and Reading Disability: Solving the Reading Puzzle,* ed. D. Shankweiler and I. Y. Liberman, 69–89. IARLD Monograph Series no. 6. Ann Arbor: University of Michigan Press.

Haslam, R. H. A., J. T. Dalby, R. D. Johns, and A. W. Rademaker. 1981. Cerebral asymmetry in developmental dyslexia. *Archives of Neurology* 38:679–682.

Hecox, K. E. 1988. Language development: A sensory development and signal processing perspective. In *Language, Communication, and the Brain,* ed. F. Plum, 77–86. New York: Raven Press.

Hier, D. B., M. LeMay, P. B. Rosenberger, and V. P. Perlo. 1978. Developmental dyslexia: Evidence for a subgroup with a reversal cerebral asymmetry. *Archives of Neurology* 35:90–92.

Hillebrandt, J. 1983. Perceptual organization of speech sounds by infants. *Journal of Speech and Hearing Disabilities* 26:268–282.

Holtz, A. 1987. Die Entwicklungsdysphasie—Sprachpathologisches Konzept oder psycholinguistisches Chaos? *Sprache—Stimme—Gehor* 11:21–26.

Ingram, T. T. S. 1971. Developmental disorders of speech. In *Handbook of Clinical Neurology,* vol. 4, ed. P. J. Vinken and G. W. Bruyn, 427–434. Amsterdam: Elsevier.

Johnston, R. B., R. E. Stark, E. D. Mellits, and P. Tallal. 1981. Neurological status of language-impaired and normal children. *Annals of Neurology* 10:159–163.

Katz, R. B. 1986. Phonological deficiencies in children with reading disability: Evidence from an object naming task. *Cognition* 22:225–57.

Kelly, M. S., C. T. Best, and U. Kirk. 1989. Cognitive processing deficits in reading disabilities: A prefrontal cortical hypothesis. *Brain and Cognition* 11:275–293.

Kertesz, A., S. E. Black, M. Polk, J. Howell. 1986. Cerebral asymmetries on magnetic resonance imaging. *Cortex* 22:117–127.

Kinsbourne, M. 1989. Neurological theories of dyslexia. In *Child Neurology and Developmental Disabilities,* ed. J. H. French, J. Harel, and P. Casaer, 279–288. Baltimore: P. H. Brookes.

Klicpera, C. 1984. Der neuropsychologische Beitrag zur Legasthenieforschung. Eine Uebersicht uber wichtige Erklarungsmodelle und Befunde. *Fortschritte der Neurologie und Psychiatrie* 52:93–103.

Knox, C., and D. Kimura. 1970. Cerebral processing of nonverbal sounds in boys and girls. *Neuropsychologia* 8:227–237.

Leboyer, M., D. N. Osherson, M. Nosten, and P. Roubertoux. 1988. Is autism associated with anomalous dominance? *Journal of Autism and Developmental Disorders* 18:539–551.

LeMay, M. 1976. Morphological cerebral asymmetries of modern man, fossil man and nonhuman primates. *Annals of the New York Academy of Sciences* 280:349–366.

Leviton, A. 1987. Single-cause attribution. *Developmental Medicine and Child Neurology* 29:805–814.

Licht, R., D. J. Bakker, A. Kok, and A. Bouma. 1988. The development of lateral event-related potentials (ERP's) related to wordnaming: A four year longitudinal study. *Neuropsychologia* 36:327–340.

Lindgren, S. D., and L. C. Richman. 1984. Immediate memory functions of verbally deficient reading-disabled children. *Journal of Learning Disabilities* 17:222–225.

Lindgren, S. D., L. C. Richman, and M. J. Eliason. 1986. Memory processes in reading disability subtypes. *Developmental Neuropsychology* 2:173–181.

Lovett, M. W., M. J. Ransby, N. Hardwyck, M. S. Johns, and S. A. Donaldson. 1989. Can dyslexia be treated? Treatment-specific and generalised treatment effects in dyslexic children's response to remediation. *Brain and Language* 37:90–121.

Ludlow, C. L. 1980. Children's language disorders, recent research advances. *Annals of Neurology* 7:497–507.

Lundberg, I. 1989. Lack of phonological awareness: A critical factor in dyslexia. In *Brain and Reading,* ed. C. von Euler, I. Lundberg, and G. Lennerstrand, 221–232. Wenner-Gren International Symposium Series, vol. 54. New York: Stockton Press.

Maccario, M., S. J. Hefferen, S. J. Keblusek, and K. A. Lipinski. 1982. Developmental dysphasia and electroencephalographic abnormalities. *Developmental Medicine and Child Neurology* 24:141–155.

Mattis, S., J. H. French, I. Rapin. 1975. Dyslexia in children and young adults: Three independent neuropsychological syndromes. *Developmental Medicine and Child Neurology* 17:150–163.

McKeefer, W. F., and D. A. Rich. 1990. Left handedness and immune disorders. *Cortex* 26:33–40.

Mehler, J., P. Jusczyk, G. Lambertz, N. Halsted, J. Bertoncini, and C. Amiel-Tison. 1988. A precursor of language acquisition in young infants. *Cognition* 29:143–178.

Molfese, D. L., R. B. Freeman, and D. S. Palermo. 1975. The ontogeny of brain lateralization for speech and non-speech stimuli. *Brain and Language* 2:356–368.

Morse, P. A. 1972. The discrimination of speech and nonspeech stimuli in early infancy. *Journal of Experimental Child Psychology* 14:477–492.

Murphy, L. A., A. Pollatsek, and A. D. Well. 1988. Developmental dyslexia and word retrieval deficits. *Brain and Language* 35:1–23.

Myklebust, J. B., ed. 1978. *Progress in Learning Disabilities*, vol. 4, 1–39. New York: Grune and Stratton.

Njiokiktjien, C. 1983. Callosal dysfunction as a possible pathogenetic factor in developmental dysphasia. *Neuropadiatrie (Abstract)* 14:123.

——— 1988. Pediatric behavioural neurology. *Clinical Principles*, vol. 1. Amsterdam: Suyi Publications.

——— 1984. Zur Pathogenese von Sprachentwicklungsstörungen. In *Entwicklungsneurologie*, ed. R. Michaelis, R. Nolte, M. Buchwald-Saal, and G. H. Haas, 83–85. Stuttgart: Kohlhammer.

——— 1990. Developmental dysphasia: Clinical importance and underlying neurological causes. Review I. *Acta Paedopsychiatrica* 53:126–37.

Njiokiktjien, C., C. Amiel-Tison, T. J. De Grauw, G. Ramaekers, F. Visscher, and R. Wennekes. 1991. Interhemispheric dysfunction as a contributor to learning disabilities. In *Modern Perspectives of Child Neurology*, ed. Y. Fukuyama, S. Kamoshita, C. Ohtsuka, and Y. Suzuki, 249–251. Tokyo: Japanese Society of Child Neurology.

Njiokiktjien, C., J. Valk, and G. Ramaekers. 1988. Malformation or damage of the corpus callosum? A clinical and MRI study. *Brain and Development* 10:92–99.

Ojemann, G. A. 1988. Some brain mechanisms for reading. In *Brain and Reading*. ed. I. Wennergren, C. von Euler, I. Lundberg, and G. Lennerstrand, 47–59. London: Macmillan Press.

Patten, B. M. 1973. Visually mediated thinking. A report of the case of Albert Einstein. *Journal of Learning Disabilities* 6:15–20.

Pennington, B. F. 1990. Annotation: The genetic of dyslexia. *Journal of Child Psychology and Psychiatry* 31:193–201.

Piazza, D. M. 1977. Cerebral lateralization in young children as measured by dichotic listening and finger-tapping tasks. *Neuropsychologia* 15:417–425.

Pinkerton, F., D. R. Watson, and R. J. McClelland. 1989. A neuropsychological study of children with reading, writing and spelling difficulties. *Developmental Medicine and Child Neurology* 31:569–581.

Pribram, K. H. 1971. *Languages of the Brain*. Englewood Cliffs, N.J.: Prentice-Hall.

Rapin, I. D., and D. A. Allen. 1982. Developmental language disorders, nosologic considerations. In *Neuropsychology of Language, Reading and Spelling*, ed. I. D. Rapin and A. Allen, 157–186. New York: Academic Press.

——— 1986. The physician's assessment and management of young children with developmental language disorders. In *Neuropaediatrie II. Padiatrische Fortbildungskurse fur die Praxis*, ed. E. Rossi, 1–12. Basel: Karger.

——— 1988. Syndromes in developmental dysphasia and adult aphasia. In *Language, Communication and the Brain*, ed. F. Plum, 57–75. New York: Raven Press.

Rentz, R., G. Niebergall, and D. Gobel. 1986. Feinneurologische Befunde bei sprachgestorten Schulkindern. *Klinische Paediatrie* 198:107–113.

Rich, D. A., and W. F. McKeever. 1990. An investigation of immune system disorder as a "marker" for anomalous dominance. *Brain and Cognition* 12:55–72.

Rosenberger, P. B., and D. B. Hier. 1980. Cerebral asymmetry and verbal intellectual deficits. *Annals of Neurology* 8:300–304.

Rumsey, J. M., R. Dorwart, M. Vermess, M. B. Denckla, M. J. P. Kruesi, and J. L. Rapoport. 1986. Magnetic resonance imaging of brain anatomy in severe developmental dyslexia. *Archives of Neurology* 43:1045–1046.

Seidenberg, M. S., and J. L. McClelland. 1989. Visual word recognition and pronunciation: A computational model of acquisition, skilled performance and dyslexia. In *From Reading to Neurons*, ed. A. M. Galaburda, 255–305. Cambridge, Mass.: MIT Press.

Selnes, O. A., D. S. Knopman, N. Niccum, and A. B. Rubens. 1985. The critical role of Wernicke's area in sentence repetition. *Annals of Neurology* 17:549–557.

Shalev, R. S., R. Weirtman, and N. Amir. 1988. Developmental dyscalculia. *Cortex* 24:555–561.

Shankweiler, D., and I. Y. Liberman, ed. 1990. *Phonology and Reading Disability. Solving the Reading Puzzle.* IARLD Monograph Series no. 6. Ann Arbor: University of Michigan Press.

Smith, I. M., and S. E. Bryson. 1988. Monozygotic twins concordant for autism and hyperlexia. *Developmental Medicine and Child Neurology* 30:527–535.

Söderbergh, R. 1984. Teaching Swedish deaf preschool children to read. In *Child Language: An International Perspective*, ed. P. S. Dale and D. Ingram, 373–386. Baltimore: University Park Press.

Sonneville L. de, and C. Njiokiktjien. 1988. Pediatric behavioural neurology. In *Aspects of Information Processing: A Computer-Based Approach of Development and Disorders*, vol. 2, 208. Amsterdam: Suyi Publications.

Tallal, P. 1981. Language disabilities in children: Perceptual correlates. *International Journal of Pediatric Otolaryngology* 3:1–13.

Tallal, P., and W. Katz. 1989. Neuropsychological and neuroanatomical studies of developmental language/reading disorders: Recent advances. In *Brain and Reading*, ed. C. von Euler, I. Lundberg, and G. Lennerstrand, 183–196. Wenner-Gren International Symposium Series, vol. 54. New York: Stockton Press.

Tallal, P., R. E. Stark, C. Kallman, and D. Mellits. 1982. Developmental dysphasia, relation between acoustic processing deficits and verbal processing. *Neuropsychologia* 18:273–284.

Tan, X. S. T. 1990. Developmental dysphasia [Dutch]. In *Omtrent Logopedie* 6:145–166.

Trevarthen, C. 1986. Form, significance and psychological potential of hand gestures of infants. In *The Biological Foundations of Gestures: Motor and Semiotic Aspects*, ed. J.-L. Nespoulous, P. Perron, and A. Roch Lecours, 149–202. New Jersey: LEA.

Tunmer, W. E. 1989. The role of language related factors in reading disability. In *Phonology and Reading Disability. Solving the Reading Puzzle,* ed. D. Shankweiler and I. Y. Liberman, 91–131. IARLD Monograph Series no. 6. Ann Arbor: University of Michigan Press.

Urion, D. K. 1988. Nondextrality and autoimmune disorders among relatives of language-disabled boys. *Annals of Neurology* 24:267–269.

Van Dongen, H. R., M. C. B. Loonen, and K. J. Van Dongen. 1985. Anatomical basis for acquired fluent asphasia in children. *Annals of Neurology* 17:306–309.

Vellutino, F. 1987. Dyslexia. *Scientific American* 256:20–27.

Wada, J. A., R. Clarke, and A. Hamm. 1975. Cerebral hemisphere asymmetry in humans: Cortical speech zones in 100 adult and 100 infant brains. *Archives of Neurology* 32:239–246.

Warrington, E. K. 1967. The incidence of verbal disability associated with retardation reading. *Neuropsychologia* 5:175–179.

Witelson, S. 1977. Developmental dyslexia: Two right hemispheres and none left. *Science* 195:309–311.

Witelson, S. F., and W. Pallie. 1973. Left hemisphere specialisation for language in the newborn: Neuroanatomical evidence of asymmetry. *Brain* 96:671–676.

Wolf, M. 1986. Rapid alternating stimulus naming in the developmental dyslexias. *Brain and Language* 27:360–379.

Wolf, M., and H. Goodglass. 1986. Dyslexia, dysnomia, and lexical retrieval: A longitudinal investigation. *Brain and Language* 28:154–168.

Wolfus, B., M. Moscovitch, and M. Kinsbourne. 1980. Subgroups of developmental language impairment. *Brain and Language* 10:152–171.

Wood, K. M., L. C. Richman, and M. J. Eliason. 1989. Immediate memory functions in reading disability subtypes. *Brain and Language* 36:181–192.

Woods, B. T. 1985. Developmental dysphasia. In *Handbook of Clinical Neurology,* vol. 2, *Neurobehavioural Disorders,* ed. A. M. Frederiks, 139–145. Amsterdam: Elsevier Science Publications.

Wyke, M., ed. 1978. *Developmental Dysphasia.* London: Academic Press.

Yakovlev, P. I., and A. Roch-Lecours. 1967. The myelogenetic cycles of regional maturation of the brain. In *Regional Maturation of the Brain in Early Life,* ed. A. Minkovski, 3–70. Oxford: Blackwell.

Yule, W., and M. Rutter, ed. 1987. Language development and disorders. Oxford: Mac Keith Press.

Zaidel, E., and A. M. Peters. 1981. Phonological encoding and ideographic reading by the disconnected right hemisphere: Two case studies. *Brain and Language* 14:205–234.

12. Parallel Processing in the Visual System and the Brain

Abeles, M., and M. H. Goldstein, Jr. 1970. Functional architecture in cat primary auditory cortex: Columnar organization and organization according to depth. *Journal of Neurophysiology* 33:172–187.

Barlow, H. B. 1981. Critical limiting factors in the design of the eye and visual cortex. The Ferrier Lecture, 1980. *Proceedings of the Royal Society (London)* 212:1–34.

Bassi, C. J., and J. C. Galanis. 1990. Impairment of binocular vision may indicate early glaucomatous damage. *Investigative Ophthalmology and Visual Science (Supplement)* 31:230.

Bodamer, J. 1947. Die Prosop-Agnosie. *Archiv für Psychiatrie Nervenkrankheiten* 179:6–54.

Burkhalter, A., and D. C. Van Essen. 1986. Processing of color, form and disparity information in visual areas VP and V2 of ventral extrastriate cortex in the macaque monkey. *Journal of Neuroscience* 6:2327–2351.

Campbell, F. W., and L. Maffei. 1981. The influence of spatial frequency and contrast on the perception of moving patterns. *Vision Research* 21:713–721.

Casagrande, V. A., P. D. Beck, G. J. Condo, and E. A. Lachica. 1990. Intrinsic connections of CO blobs in striate cortex of primates. *Investigative Ophthalmology and Visual Science* 31:1945.

Cavanagh, P. 1988. Pathways in early vision. In *Computational Processes in Human Vision: An Interdisciplinary Perspective,* ed. Zenon Pylyshyn. Norwood, NJ: Ablex.

Cavanagh, P., J. Boeglin, and O. E. Favreau. 1985. Perception of motion in equiluminous kinematograms. *Perception* 14:151–162.

Cavanagh, P., and Y. Leclerc. 1985. Shadow constraints. *Investigative Ophthalmology and Visual Science.* 26:282.

Cusick, C. G., and J. H. Kaas. 1988. Cortical connections of area 18 and dorsolateral visual cortex in squirrel monkeys. *Visual Neuroscience* 1:211–237.

Damasio, A., T. Yamada, H. Damasio, J. Corbett, and J. McKee. 1980. Central achromatopsia: Behavioral, anatomic, and physiologic aspects. *Neurology* 30:1064–1071.

De Monasterio, F. M., and P. Gouras. 1975. Functional properties of ganglion cells of the rhesus monkey retina. *Journal of Physiology* (London) 251:167–195.

Derrington, A. M., J. Krauskopf, and P. Lennie. 1984. Chromatic mechanisms in lateral geniculate nucleus of macaque. *Journal of Physiology* (London) 357:241–265.

Derrington, A. M., and P. Lennie. 1984. Spatial and temporal contrast sensitivities of neurones in lateral geniculate nucleus of macaque. *Journal of Physiology* (London) 357:219–240.

DeValois, R. L., I. Abramov, and G. H. Jacobs. 1966. Analysis of response patterns of LGN cells. *Journal of the Optical Society of America* 56:966–977.

DeValois, R. L., D. M. Snodderly, E. W. Yund, and N. K. Hepler. 1977. Responses of macaque lateral geniculate cells to luminance and color figures. *Sensory Processes* 1:244–259.

DeYoe, E. A., and D. C. Van Essen. 1985. Segregation of efferent connections and receptive field properties in visual area V2 of the macaque. *Nature* 317:58–61.

Dubner, R., and S. M. Zeki. 1971. Response properties and receptive fields of cells in an anatomically defined region of the superior temporal sulcus. *Brain Research* 35:528–531.

Fitzpatrick, D., K. Itoh, and I. T. Diamond. 1983. The laminar organization of the lateral geniculate body and the striate cortex in the squirrel monkey (*Saimiri sciureus*). *Journal of Neuroscience* 3:673–702.

Fitzpatrick, D., J. S. Lund, and G. G. Blasdel. 1985. Intrinsic connections of macaque striate cortex: Afferent and efferent connections of lamina 4C. *Journal of Neuroscience* 5:3329–3349.

Goldman, P., and W. J. H. Nauta. 1977. Columnar distribution of cortico-cortical fibers in the frontal association, limbic, and motor cortex of the developing rhesus monkey. *Brain Research* 122:393–413.

Gouras, P. 1968. Identification of cone mechanisms in monkey ganglion cells. *Journal of Physiology* (London) 199:533–547.

——— 1969. Antidromic responses of orthodromically identified ganglion cells in monkey retina. *Journal of Physiology* (London) 204:407–419.

Gregory, R. L. 1972. Cognitive contours. *Nature* 238:51.

——— 1977. Vision with isoluminant colour contrast: 1. A projection technique and observations. *Perception* 6:113–119.

Guillery, R. W. 1979. A speculative essay on geniculate lamination and its development. *Progress in Brain Research* 51:403–418.

Hubel, D. H., and M. S. Livingstone. 1990. Color and contrast sensitivity in the lateral geniculate body and primary visual cortex of the macaque monkey. *Journal of Neuroscience* 10:2223–2237.

Hubel, D. H., and T. N. Wiesel. 1972. Laminar and columnar distribution of geniculo-cortical fibers in the macaque monkey. *Journal of Comparative Neurology* 146:421–450.

Ives, H. E. 1923. A chart of the flicker photometer. *Journal of the Optical Society of America, Review Science Instructions* 7:363–365.

Kanizsa, G. 1979. Essays on Gestalt perception. In *Organization in Vision,* New York: Praeger.

Kaplan, E., and R. M. Shapley. 1982. X and Y cells in the lateral geniculate nucleus of the macaque monkey. *Journal of Physiology* (London) 330:125–143.

——— 1986. The primate retina contains two types of ganglion cells, with high and low contrast sensitivity. *Proceedings of the National Academy of Science* (USA) 83:2755–2757.

Kofka, K. 1935. *Principles of Gestalt Psychology.* New York: Harcourt, Brace.

Krüger, J. 1979. Responses to wavelength contrast in the afferent visual systems of the cat and the rhesus monkey. *Vision Research* 19:1351–1358.

Leventhal, A. G., R. W. Rodieck, and B. Dreher. 1981. Retinal ganglion cell classes in the Old World monkey: Morphology and central projections. *Science* 213:1139–1142.

Levitt, J. B., and J. A. Movshon. 1990. Homogeneity of response properties of neurons in macaque V2. *Investigative Ophthalmology and Visual Science (Supplement)* 31:89.

Liebergall, D., S. Solomon, and J. Schultz. 1990. The effect of chromic open angle glaucoma on stereoacuity. *Investigative Ophthalmology and Visual Science (Supplement)* 31:230.

Liebmann, S. 1926. Uber das Verhalten farbiger Formen bei Helligkeitsgleichheit von Figur und Grund. *Psychologische Forschung* 9:300–353.

Livingstone, M. S., and D. H. Hubel. 1982. Thalamic inputs to cytochrome oxidase-rich regions in monkey visual cortex. *Proceedings of the National Academy of Science* (USA) 79:6098–6101.

——— 1984a. Anatomy and physiology of a color system in the primate visual cortex. *Journal of Neuroscience* 4:309–356.

——— 1984b. Specificity of intrinsic connections in primate primary visual cortex. *Journal of Neuroscience* 4:2830–2835.

——— 1987a. Connections between layer 4B of area 17 and thick cytochrome oxidase stripes of area 18 in the squirrel monkey. *Journal of Neuroscience* 7:3371–3377.

——— 1987b. Psychophysical evidence for separate channels for the perception of form, color, movement, and depth. *Journal of Neuroscience* 7:3416–3468.

Logothetis, N. K., P. H. Schiller, E. R. Charles, and A. C. Hurlbert. 1990. Perceptual deficits and the activity of the color-opponent and broadband pathways at isoluminance. *Science* 247:214–217.

Lovegrove, W., R. Martin, and W. Slaghuis. 1986. A theoretical and experimental case for a visual deficit in specific reading disability. *Cognitive Neuropsychology* 3:225–267.

Lu, C., and D. H. Fender. 1972. The interaction of color and luminance in stereoscopic vision. *Investigative Ophthalmology* 11:482–490.

Lund, J. S. 1973. Organization of neurons in the visual cortex, area 17, of the monkey (*Macaca mulatta*). *Journal of Comparative Neurology* 147:455–475.

——— 1987. Local circuit neurons of macaque monkey striate cortex: I. Neurons of laminae 4C and 5A. *Journal of Comparative Neurology* 257:60–92.

Lund, J. S., and R. G. Boothe. 1975. Interlaminar connections and pyramidal neuron organization in the visual cortex, area 17, of the macaque monkey. *Journal of Comparative Neurology* 159:305–334.

Lund, J. S., R. D. Lund, A. E. Hendrickson, A. H. Bunt, and A. F. Fuchs. 1975. The origin of efferent pathways from the primary visual cortex, area 17, of the macaque monkey as shown by retrograde transport of horseradish peroxidase. *Journal of Comparative Neurology* 164:287–304.

Martin, F., and W. Lovegrove. 1987. Flicker contrast sensitivity in normal and specifically disabled readers. *Perception* 16:215–221.

Maunsell, J. H. R., and D. C. Van Essen. 1983. Functional properties of neurons in middle temporal visual area of the macaque monkey. II. Binocular interactions and sensitivity to binocular disparity. *Journal of Neurophysiology* 49:1148–1167.

McGuire, P. K., S. Hockfield, and P. S. Goldman-Rakic. 1989. Distribution of Cat-301 immunoreactivity in the frontal and parietal lobes of the macaque monkey. *Journal of Comparative Neurology* 288:280–296.

Mountcastle, V. B. J. 1957. Modality and topographic properties of single neurons of cat's somatic sensory cortex. *Journal of Neurophysiology* 20:408–434.

Pearlman, A. L., J. Birch, and J. C. Meadows. 1979. Cerebral colorblindness: An acquired defect in hue discrimination. *Annals of Neurology* 5:253–261.

Pfeiffer, N., D. Birkner-Binder, and M. Bach. 1990. Pattern-electroretinogram (PERG) in early glaucoma: Effects of check size, contrast and retinal eccentricity. *Investigative Ophthalmology and Visual Science (Supplement)* 31:231.

Rubin, E. 1915. *Synsoplevede Figurer.* Copenhagen: Glydendalska.

Schiller, P. H., N. K. Logothetis, and E. R. Charles. 1990. Functions of the colour-opponent and broad-band channels of the visual system. *Nature* 343:68–70.

Schiller, P. H., and J. G. Malpeli. 1978. Functional specificity of lateral geniculate nucleus laminae of the rhesus monkey. *Journal of Neurophysiology* 41:788–797.

Shapley, R. M., E. Kaplan, and R. Soodak. 1981. Spatial summation and contrast sensitivity of X and Y cells in the lateral geniculate nucleus of the macaque. *Nature* 292:543–545.

Sherman, S. M. 1985. In *Progress in Psychobiology and Physiological Psychology* 2:233–3224.

Stein, J., P. Riddell, and S. Fowler. 1987. Fine binocular control in dyslexic children. In *Aphasia,* ed. C. Rose. London: Whurr Wyke.

——— 1989. Disordered right hemisphere function in developmental dyslexia. In *Brain and Reading,* ed. C. von Euler, I. Lundberg, and G. Lennerstrand. New York: Stockton Press.

Tallal, P., R. E. Stark, and E. D. Mellits. 1985. Identification of language-impaired children on the basis of rapid perception and production skills. *Brain Language* 25:314–322.

Ts'o, D. Y., and C. D. Gilbert. 1988. The organization of chromatic and spatial interactions in the primate striate cortex. *Journal of Neuroscience* 8:1712–1727.

Wiesel, T. N., and D. H. Hubel. 1966. Spatial and chromatic interactions in the lateral geniculate body of the rhesus monkey. *Journal of Neurophysiology* 29:1115–1156.

Williams, M. C., and K. LeCluyse. 1989. *The perceptual consequences of a temporal processing deficit in reading disabled children.* in press.

Woolsey, T. A., and H. Van der Loos. 1970. The structural organization of layer IV in the somatosensory region (SI) of mouse cerebral cortex. The description of a cortical field composed of discrete cytoarchitectonic units. *Brain Research* 17:205–242.

Zihl, J., D. Von Cramon, and N. Mai. 1983. Selective disturbance of movement vision after bilateral brain damage. *Brain* 106:313–340.

13. The Neurobiology of Learning Disabilities

Aram, D. M. 1988. Language sequelae of unilateral brain lesions in children. In *Language, Communication and the Brain,* ed. F. Plum, 171–197. New York: Raven.

Bailey, P., and G. Von Bonin. 1951. *The Isocortex of Man.* Urbana, IL: University of Illinois Press.

Bear, D., D. Schiff, J. Saver, M. Greenberg, and R. Freeman. 1986. Quantitative analysis of cerebral asymmetries: Fronto-occipital correlation, sexual dimorphism, and association with handedness. *Archives of Neurology* 43:598–603.

Budinger, T. F. 1981. Nuclear magnetic resonance (NMR) in vivo studies: Known thresholds for health effects. *Journal of Computer Assisted Tomography* 5:800–811.

Caviness, V. S., Jr., P. A. Filipek, and D. N. Kennedy. 1989. Magnetic resonance technology in human brain science; blueprint for a program based upon morphometry. *Brain and Development* 1:1–13.

Creasey, H., J. M. Rumsey, M. Schwartz, R. Duara, J. L. Rapoport, and S. I. Rapoport. 1986. Brain morphometry in autistic men as measured by volumetric computed tomography. *Archives of Neurology* 43:669–672.

Filipek, P. A., D. N. Kennedy, V. S. Caviness, Jr., S. Klein, and I. Rapin. 1987. In vivo MRI-based volumetric brain analysis in subjects with verbal auditory agnosia. *Annals of Neurology* 22:410.

Filipek, P. A., D. N. Kennedy, S. K. Kennedy, and V. S. Caviness, Jr. 1988. Shape analysis of the brain of two siblings based upon magnetic resonance imaging. *Annals of Neurology* 24:355.

Filipek, P. A., D. N. Kennedy, V. S. Caviness, Jr., S. L. Rossnick, T. A. Spraggins, and P. M. Starewicz. 1989. Magnetic resonance imaging-based brain morphometry: Development and application to normal subjects. *Annals of Neurology* 5:61–67.

Filipek, P. A., D. N. Kennedy, and V. S. Caviness, Jr. 1990a. Neuroimaging in child neuropsychology. In *Handbook of Neuropsychology,* vol. 6, ed. I. Rapin and S. Segalowitz. Amsterdam: Elsevier.

Filipek, P. A., D. N. Kennedy, J. Rademacher, and V. S. Caviness, Jr. 1990b. Error and variability incurred with MRI-based morphometry. *Annals of Neurology* 28:459.

Galaburda, A. M., G. F. Sherman, G. D. Rosen, F. Aboitiz, and N. Geschwind. 1985. Developmental dyslexia: Four consecutive patients with cortical anomalies. *Annals of Neurology* 18:222–223.

Geschwind, N., and P. Behan. 1982. Left-handedness: Association with immune disease, migraine, and developmental learning disorder. *Proceedings of the National Academy of Science* (USA) 79:5097–5100.

Haslam, R. H. A., J. T. Dalby, R. D. Johns, and A. W. Rademaker. 1981. Cerebral asymmetry in developmental dyslexia. *Archives of Neurology* 38:679–682.

Hier, D. B., M. LeMay, P. B. Rosenberger, and V. P. Perlo. 1978. Developmental dyslexia. *Archives of Neurology* 35:90–92.

Jernigan, T. L., and P. A. Tallal. 1990. Late childhood changes in brain morphology observable with MRI. *Developmental Medicine and Child Neurology* 32:379–385.

Jernigan, T. L., P. Tallal, and J. R. Hesselink. 1989. Cerebral morphology on MR in language/reading-impaired children. *Neurology* 39:138.

Jouandet, M. L., M. J. Tramo, D. M. Herron, A. Hermann, W. C. Loftus, J. Bazell, and M. S. Gazzaniga. 1989. Brainprints: Computer-generated two-dimensional maps of the human cerebral cortex in vivo. *Journal of Cognitive Neuroscience* 1:88–117.

Kennedy, D. N., P. A. Filipek, and V. S. Caviness, Jr. 1990. Fourier shape analysis of anatomic structures. In *Recent Advances in Fourier Analysis and Its Applications,* ed. J. S. Byrnes and J. L. Byrnes, 17–28. Dordrecht, The Netherlands: Kluwer Academic Publishers.

Kennedy, D. N., S. K. Kennedy, P. A. Filipek, and V. S. Caviness, Jr. 1988. Morphometric characterization of brain shape by Fourier analysis. *Magnetic Resonance in Medicine* 7:993.

Kosslyn, S. M. 1988. Aspects of cognitive neuroscience of mental imagery. *Science* 240:1621–1626.

Kretschmann, H. J., A. Schleicher, J. F. Grottschreiber, and W. Kullman. 1979. The Yakolev collection: A pilot study of its suitability for the morphometric documentation of the human brain. *Journal of Neurological Science* 43:111–126.

Krieg, W. J. S. 1973. *Architectonics of Human Cerebral Fiber Systems.* Evanston, IL: Brain Books.

Mountcastle, V. B. 1978. An organizing principle for cerebral function: The unit module and the distributed system. In *The Mindful Brain: Cortical Organization and the Group-Selective Theory of Higher Brain Function,* ed. G. M. Edelman and V. B. Mountcastle, 7–50. Cambridge: MIT Press.

Nass, R., and H. D. Peterson. 1989. Differential effects of congenital left and right brain injury. *Brain and Cognition* 9:258–266.

Pandya, D., and E. H. Yeterian. 1985. Architecture and connections of cortical association areas. In *Cerebral Cortex: Association and Auditory Cortices,* ed. A. Peters and E. G. Jones, 3–61. New York: Plenum Press.

Paul, F. 1971. Biometrische Analyse der Frischvolumina des Grosshirnrinde und des Prosencephalon von 31 menschlichen, adulten Gehirnen. *Zeitschrift für Anatomie Entwicklunsgeschichte* 133:325–368.

Petersen, S. E., P. T. Fox, M. I. Posner, M. Mintum, and M. E. Raichle. 1988. Positron emission tomographic studies of the cortical anatomy of single-word processing. *Nature* 331:585–589.

Posner, M. I., S. E. Petersen, P. T. Fox, and M. E. Raichle. 1988. Localization of cognitive operations in the human brain. *Science* 240:1627–1631.

Riva, D., and I. Cazzaniga. 1986. Late effects of unilateral brain lesions on intelligence and early left versus right brain injury. *Neuropsychologia* 24:423–428.

Rosenberger, P. B., and D. B. Hier. 1980. Cerebral asymmetry and verbal intellectual deficits. *Annals of Neurology* 8:300–304.

Sacks, J., D. N. Kennedy, P. A. Filipek, and V. S. Caviness, Jr. 1990. MRI-based three-dimensional analysis of shape. *Magnetic Resonance in Medicine* 9:100.

Steinmetz, H., J. Rademacher, Y. Huang, H. Hefter, K. Zilles, A. Thron, and H. -J. Freund. 1989. Cerebral asymmetry: MR planimetry of the human planum temporale. *Journal of Compu constructive deficit. Brain and Cognition* 4:388–412.

Talairach, J., and P. Tournoux. 1988. *Co-planar stereotaxic atlas of the human brain.* New York: Thieme Medical Publishers, Inc.

Vargha-Khadem, F., A. M. O'Gorman, and G. V. Watters. 1985. Aphasia and handedness in relation to hemispheric side, age at injury and severity of cerebral lesion during childhood. *Brain* 108:677–696.

Visch-Brink, E. G., and M. Van de Sandt-Koenderman. 1984. The occurrence of paraphasias in the spontaneous speech of children with an acquired aphasia. *Brain* 23:258–271.

Voeller, K. K. S. 1986. Right-hemisphere deficit syndrome in children. *American Journal of Psychiatry* 143:1004–1009.

Wessely, W. 1970. Biometrische Analyse der Frischvolumina des Rhombencephalon, des Cerebellum und der Ventrikel von 31 adulten menschlichen Gehirnen. *Journal für Hirnforsch* 12:11–28.

Woods, B. T. 1980. The restricted effects of right-hemisphere lesions after age one: Wechsler test data. *Neuropsychologia* 18:65–70.

Woods, B. T., and S. Carey. 1979. Language deficits after apparent clinical recovery from childhood aphasia. *Annals of Neurology* 6:405–409.

Woods, B. T., and H. L. Teuber. 1978. Changing patterns of childhood aphasia. *Annals of Neurology* 3:272–280.

14. Studies of Handedness and Anomalous Dominance

Annett, M. 1970. Classification of hand preference by association analysis. *British Journal of Psychology* 61:303–321.

——— 1978. *Single Gene Explanation of Right and Left Handedness and Brainedness.* Coventry, England: Lanchester Polytechnic Press.

——— 1985. *Left, Right, Hand and Brain: The Right Shift Theory.* London: Lawrence Erlbaum.

Annett, M., and D. Kilshaw. 1982. Mathematical ability and lateral asymmetry. *Cortex* 18:547–568.

——— 1984. Lateral preference and skill in dyslexics: Implications of the right shift theory. *Journal of Child Psychology and Psychiatry* 25:357–377.

Annett, M., and M. Manning. 1990. Arithmetic and laterality. *Neuropsychologia* 28:61–69.

Bear, D., D. Schiff, J. Saver, M. Greenberg, and R. Freeman. 1986. Quantitative analysis of cerebral asymmetries: Fronto-occipital correlation, sexual dimorphism, and association with handedness. *Archives of Neurology* 43:598–603.

Beck, E. 1955. Typologie des Gehirns am Beispiel des dorsalen menschlichen Schlafenlappens nebst weiteren Beitragen zur Frage der Links-Rechtshirnigkeit. *Deutsche Zeitschrift für Nervenkrankheiten* 173:267–308.

Behan, P., and N. Geschwind. 1985. Dyslexia, congenital anomalies, and immune disorders: The role of the fetal environment. *Annals of the New York Academy of Science* 457:13–18.

Benbow, C. P. 1986. Physiological correlates of extreme intellectual precocity. *Neuropsychologia* 24:719–725.

Bender, B. G., M. H. Puck, J. A. Salbenblatt, and A. Robinson. 1983. Hemispheric organisation in 47,XXY boys [letter]. *Lancet* 1(8316):132.

Benton, A. L., R. Meyers, and G. J. Polder. 1962. Some aspects of handedness. *Psychiatria et Neurologia* (Basel) 144:321–337.

Betancur, C., A. Velez, G. Cabanieu, M. Le Moal, and P. J. Neveu. 1990. Association between left handedness and allergy: A reappraisal. *Neuropsychologia* 28:223–227.

Bishop, D. V. M. 1980. Measuring familial sinistrality. *Cortex* 16:311–313.

——— 1986. Is there a link between handedness and hypersensitivity? *Cortex* 22:289–296.

——— 1989. Does hand proficiency determine hand preference? *British Journal of Psychology* 80:191–199.

Briggs, G. G., and R. D. Nebes. 1975. Patterns of hand preference in a student population. *Cortex* 11:230–238.

Bryden, M. P. 1977. Measuring handedness with questionnaires. *Neuropsychologia* 15:617–624.

Bryson, S. E. 1990. Autism and anomalous handedness. In *Left Handedness, Behavioral Implications and Anomalies*, ed. S. Coren, 441–456. Dordrecht: Elsevier Science Publishers.

Calnan, M., and K. Richardson. 1976a. Developmental correlates of handedness in a national sample of 11-year-olds. *Annals of Human Biology* 3:329–342.

——— 1976b. Speech problems among children in a national survey: Associations with hearing, handedness and therapy. *Community Health* (Bristol) 8:101–106.

Campain, R., and J. Minckler. 1976. A note on the gross configurations of the human auditory cortex. *Brain Language* 3:318–323.

Cernacek, J. 1964. Handedness as a quantitative estimation. *Journal of Neurological Science* 1:152–159.

Chapman, L. J., and J. P. Chapman. 1987. The measurement of handedness. *Brain Cognition* 6:175–183.

Chavance, M., G. Dellatolas, M. G. Bousser, B. Amor, B. Grardel, A. Kahan,

M. F. Kahn, J. P. Le Floch, and G. Tchobroutsky. 1990. Handedness, immune disorders and information bias. *Neuropsychologia* 28:429–441.

Chi, J. G., E. C. Dooling, and F. H. Gilles. 1977. Gyral development of the human brain. *Annals of Neurology* 1:86–93.

Chui, H. C., and A. R. Damasio. 1980. Human cerebral asymmetries evaluated by computed tomography. *Journal of Neurology, Neurosurgery and Psychiatry* 43:873–878.

Colby, K. M., and C. Parkison. 1977. Handedness in autistic children. *Journal of Autism and Child Schizophrenia* 7:3–9.

Connolly, B. H. 1983. Lateral dominance in children with learning disabilities. *Physical Therapy* 63:183–187.

Cosi, V., A. Citterio, and C. Pasquino. 1988. A study of hand preference in myasthenia gravis. *Cortex* 24:573–577.

Dellatolas, G., I. Annesi, P. Jallon, M. Chavance, and J. Lellouch. 1990. An epidemiological reconsideration of the Geschwind-Galaburda theory of cerebral lateralization. *Archives of Neurology* 47:778–782.

Deuel, R. K., and C. C. Moran. 1980. Cerebral dominance and cerebral asymmetries on computed tomogram in childhood. *Neurology* 30:934–938.

Diehl, C. F. 1958. *A Compendium of Research and Theory on Stuttering.* Springfield, IL: Charles C. Thomas.

Drake, W. E. 1968. Clinical and pathological findings in a child with a developmental learning disability. *Journal of Learning Disabilities* 1:486–502.

Eberstaller, O. 1884. Zur oberflachen Anatomie der grosshirn Hemispharen. *Wien Med Blatter* 7:479, 642, 644.

Falzi, G., P. Perrone, and L. A. Vignolo. 1982. Right left asymmetry in anterior speech region. *Archives of Neurology* 39:239–240.

Fleiss, J. L. 1981. *Statistical Methods for Rates and Proportions.* New York: John Wiley and Sons.

Fleminger, J. J., R. Dalton, and K. F. Standage. 1977. Age as a factor in the handedness of adults. *Neuropsychologia* 15:471–473.

Galaburda, A. M. 1983. Developmental dyslexia: Current anatomical research. *Annals of Dyslexia* 33:41–53.

——— 1987. Letter. *Arthritis and Rheumatism* 30:355–356.

Galaburda, A. M., and T. L. Kemper. 1979. Cytoarchitectonic abnormalities in developmental dyslexia: A case study. *Annals of Neurology* 6:94–100.

Galaburda, A. M., M. LeMay, T. L. Kemper, and N. Geschwind. 1978. Right-left asymmetries in the brain. *Science* 199:852–856.

Galaburda, A. M., G. D. Rosen, G. F. Sherman, and F. Aboitiz. 1984. Developmental dyslexia: Fourth consecutive case with cortical anomalies. *Society of Neuroscience Abstract* 957.

Geschwind, N., and P. Behan. 1982. Left handedness: Association with immune disease, migraine, and developmental learning disorder. *Proceedings of the National Academy of Sciences* (USA) 79:5097–5100.

——— 1984. Laterality, hormones, and immunity. In *Cerebral Dominance: The Biological Foundations,* ed. N. Gerschwin and A. M. Galaburda, 211–224. Cambridge: Harvard University Press.

Geschwind, N., and A. M. Galaburda, eds. 1984. *Cerebral Dominance: The Biological Foundations.* Cambridge: Harvard University Press.

——— 1985a. Cerebral lateralization. Biological mechanisms, associations, and pathology: II. A hypothesis and a program for research. *Archives of Neurology* 42:521–552.

——— 1985b. Cerebral lateralization. Biological mechanisms, associations, and pathology: III. A hypothesis and a program for research. *Archives of Neurology* 42:634–654.

Geschwind, N., and W. Levitsky. 1968. Human brain: Left-right asymmetries in temporal speech region. *Science* 161:186–187.

Goldman, P. S., and T. W. Galkin. 1978. Prenatal removal of frontal association cortex in the fetal rhesus monkey: Anatomical and functional consequences in postnatal life. *Brain Research* 152:451–485.

Gotestam, K. O. 1990. Left handedness among students of architecture and music. *Perceptual and Motor Skills* 70:1323–1327.

Hauser, S. L., G. R. DeLong, and N. P. Rosman. 1975. Pneumographic findings in the infantile autism syndrome. A correlation with temporal lobe disease. *Brain* 98:667–688.

Hier, D. B., M. LeMay, P. B. Rosenberger, and V. P. Perlo. 1978. Developmental dyslexia: Evidence for a subgroup with a reversal of cerebral asymmetry. *Archives of Neurology* 35:90–92.

Hochberg, F. H., and M. LeMay. 1975. Arteriographic correlates of handedness. *Neurology* 25:218–222.

Hugdahl, K., B. Ellertsen, P. E. Waaler, and H. Klove. 1989. Left- and right-handed dyslexic boys: An empirical test of some assumptions of the Geschwind-Behan hypothesis. *Neuropsychologia* 27:223–231.

Humphrey, M. 1951. *Handedness and Cerebral Dominance.* Bachelor of Science thesis, Oxford University.

Humphreys, P., W. E. Kaufmann, and A. M. Galaburda. 1990. Developmental dyslexia in women: Neuropathological findings in three patients. *Annals of Neurology* 28:727–738.

Hynd, G. W., M. Semrud-Clikeman, A. R. Lorys, E. S. Novey, and D. Eliopulos. 1990. Brain morphology in developmental dyslexia and attention deficit disorder/hyperactivity. *Archives of Neurology* 47:919–926.

James, W. H. 1989. Foetal testosterone levels, homosexuality and handedness: A research proposal for jointly testing Geschwind's and Dorner's hypotheses. *Journal of Theoretical Biology* 136:177–180.

Johnstone, J., D. Galin, and J. Herron. 1979. Choice of handedness measures in studies of hemispheric specialization. *International Journal of Neuroscience* 9:71–80.

Jouandet, M. L., M. J. Tramo, D. M. Herron, A. Hermann, W. C. Loftus,

J. Bazell, and M. S. Gazzaniga. 1990. Brainprints: Computer-generated two-dimensional maps of the human cerebral cortex in vivo. *Journal of Cognitive Neuroscience* 1:88–117.

Kertesz, A., S. E. Black, M. Polk, J. Howell. 1986. Cerebral asymmetries on magnetic resonance imaging. *Cortex* 22:117–127.

Kaufmann, W. E., and A. M. Galaburda. 1989. Cerebrocortical microdysgenesis in neurologically normal subjects: A histopathologic study. *Neurology* 39:238–244.

Koff, E., M. A. Naeser, J. M. Pieniadz, A. L. Foundas, H. L. Levine. 1986. Computed tomographic scan hemispheric asymmetries in right- and left-handed male and female subjects. *Archives of Neurology* 43:487–491.

Lahita, R. G. 1988. Systemic lupus erythematosus: Learning disability in the male offspring of female patients and relationship to laterality. *Psychoneuroendocrinology* 13:385–396.

Lansky, L. M., H. Feinstein, and J. M. Peterson. 1988. Demography of handedness in two samples of randomly selected adults (N = 2083). *Neuropsychologia* 26:465–477.

LeMay, M. 1977. Asymmetries of the skull and handedness. *Journal of Neurological Sciences* 32:243–253.

LeMay, M., and A. Culebras. 1972. Human brain-morphologic differences in the hemispheres demonstrable by carotid arteriography. *New England Journal of Medicine* 287:168–170.

LeMay, M., and D. K. Kido. 1978. Asymmetries of the cerebral hemispheres on computed tomograms. *Journal of Computer Assisted Tomography* 2:471–476.

Lindesay, J. 1987. Laterality shift in homosexual men. *Neuropsychologia* 25:965–969.

Marx, J. L. 1982. Autoimmunity in left handers. Left handedness may be associated with an increased risk of autoimmune disease. Is testosterone the link between the two? *Science* 217:141–142, 144.

McCormick, C. M., S. F. Witelson, and E. Kingstone. 1990. Left handedness in homosexual men and women: Neuroendocrine implications. *Psychoneuroendocrinology* 15:69–76.

McGee, M. G., and T. Cozad. 1980. Population genetic analysis of human hand preference: Evidence for generation differences, familial resemblance, and maternal effects. *Behavior Genetics* 10:263–275.

McManus, I. C. 1984. Genetics of handedness in relation to language disorder. *Advances in Neurology* 42:125–138.

McMeekan, E. R. L., and W. A. Lishman. 1975. Retest reliabilities and interrelationship of the Annett hand preference questionnaire and the Edinburgh handedness inventory. *British Journal of Psychology* 66:53–59.

McRae, D. L., C. L. Branch, and B. Milner. 1968. The occipital horns and cerebral dominance. *Neurology* 18:95–98.

Messinger, H. B., M. I. Messinger, and J. R. Graham. 1988. Migraine and left-handedness: Is there a connection? *Cephalalgia* 8:237–244.

Meyers, S., and H. D. Janowitz. 1985. Left-handedness and inflammatory bowel disease. *Journal of Clinical Gastroenterology* 7:33–35.

Nass, R., S. Baker, P. Speiser, R. Virdis, A. Balsamo, E. Cacciari, A. Loche, and M. Dumic. 1987. Hormones and handedness: Left-hand bias in female congenital adrenal hyperplasia patients. *Neurology* 37:711–715.

Neils, J. R., and D. M. Aram. 1986. Handedness and sex of children with developmental language disorders. *Brain Language* 28:53–65.

Netley, C., and J. Rovet. 1982. Handedness in 47,XXY males [letter]. 2(8292):267.

——— 1984. Hemispheric lateralization in 47,XXY Klinefelter's syndrome boys. *Brain Cognition* 3:10–18.

Oldfield, R. C. 1971. The assessment and analysis of handedness: The Edinburgh inventory. *Neuropsychologia* 9:97–113.

Orton, S. T. 1925. "Word-blindness" in school children. *Archives of Neurology and Psychiatry* 14:581–615.

——— 1937. *Reading, Writing and Speech Problems in Children.* London: Chapman Hall.

Pennington, B. F., S. D. Smith, W. J. Kimberling, P. A. Green, and M. M. Haith. 1987. Left handedness and immune disorders in familiar dyslexics. *Archives of Neurology* 44:634–639.

Persson, P. G., and A. Ahlbom. 1988. Relative risk is a relevant measure of association of left handedness with inflammatory bowel disease. *Neuropsychologia* 26:737–738.

Peterson, J. M. 1979. Left handedness: Differences between student artists and scientists. *Perceptual Motor Skills* 48:961–962.

Peterson, J. M., and L. M. Lansky. 1974. Left handedness among architects: Some facts and speculation. *Perceptual Motor Skills* 38:547–550.

Plato, C. C., K. M. Fox, and R. M. Garruto. 1984. Measures of lateral functional dominance: Hand dominance. *Human Biology* 56:259–275.

Porac, C., and S. Coren. 1981. *Lateral Preferences and Human Behavior.* New York: Springer-Verlag.

Raczkowski, D., and J. W. Kalat. 1974. Reliability and validity of some handedness questionnaire items. *Neuropsychologia* 12:43–47.

Ratcliff, G., C. Dila, L. Taylor, and B. Milner. 1980. The morphological asymmetry of the hemispheres and cerebral dominance for speech: A possible relationship. *Brain Language* 11:87–98.

Rosenstein, L. D., and E. D. Bigler. 1987. No relationship between handedness and sexual preference. *Psychological Reports* 60:704–706.

Rubens, A. B., M. W. Mahowald, and J. T. Hutton. 1976. Asymmetry of the lateral (sylvian) fissures in man. *Neurology* 26:620–624.

Salcedo, J. R., B. J. Spiegler, E. Gibson, and D. B. Magilavy. 1985. The autoimmune disease systemic lupus erythematosus is not associated with left handedness. *Cortex* 21:645–647.

Satz, P., K. Achenbach, and E. Fennell. 1967. Correlations between assessed manual laterality and predicted speech laterality in a normal population. *Neuropsychologia* 5:295–310.

Satz, P., and H. V. Soper. 1987. Left handedness, dyslexia, and autoimmune disorder: A critique. *Journal of Clinical and Experimental Neuropsychology* 8:453–458.

Schachter, S. C., and A. M. Galaburda. 1986. Development and biological associations of cerebral dominance: Review and possible mechanisms. *Journal of the American Academy of Child Psychiatry* 25:741–750.

Schachter, S. C., B. J. Ransil, and N. Geschwind. 1987. Associations of handedness with hair color and learning disabilities. *Neuropsychologia* 25:269–276.

Schneider, G. E. 1981. Early lesions and abnormal neuronal connections. *Trends in Neuroscience* 4:187–192.

Schur, P. H. 1986. Handedness in systemic lupus erythematosus. *Arthritis and Rheumatism* 29:419–420.

Schwartz, S., and K. Kirsner. 1984. Can group differences in hemispheric asymmetry be inferred from behavioral laterality indices? *Brain Cognition* 3:57–70.

Searleman, A., and A. K. Fugagli. 1987. Suspected autoimmune disorders and left handedness: Evidence from individuals with diabetes, Crohn's disease and ulcerative colitis. *Neuropsychologia* 25:367–374.

Shaywitz, S. E., B. A. Shaywitz, J. M. Fletcher, and M. D. Escobar. 1990. Prevalence of reading disability in boys and girls. Results of the Connecticut longitudinal study. *Journal of the American Medical Association* 264:998–1002.

Smith, J. 1987. Left handedness: Its association with allergic disease. *Neuropsychologia* 25:665–674.

Tan, U. 1988. The distribution of hand preference in normal men and women. *International Journal of Neuroscience* 41:35–55.

Teszner, D., A. Tzavaras, J. Gruner, and H. Hecen. 1972. L'asymétrie droite-gauche du planum temporale: A propos de l'étude anatomique de 100 cerveaux. *Revue de Neurologie* (Paris) 146:444–449.

Thomson, M. 1975. Laterality and reading attainment. *British Journal of Educational Psychology* 45:317–321.

Treffert, D. A. 1988. The idiot savant: A review of the syndrome. *American Journal of Psychiatry* 145:563–572.

Tsai, L. Y. 1983. The relationship of handedness to the cognitive, language, and visuo-spatial skills of autistic patients. *British Journal of Psychiatry* 142:156–162.

Tsai, L. Y., C. G. Jacoby, and M. A. Stewart. 1983. Morphological cerebral asymmetries in autistic children. *Biological Psychiatry* 18:317–327.

Tsai, L. Y., and M. A. Stewart. 1982. Handedness and EEG correlation in autistic children. *Biological Psychiatry* 17:595–598.

Urion, D. K. 1988. Nondextrality and autoimmune disorders among relatives of language-disabled boys. *Annals of Neurology* 124:267–269.

van Strien, J. W., A. Bouma, and D. J. Bakker. 1987. Birth stress, autoimmune diseases, and handedness. *Journal of Clinical and Experimental Neuropsychology* 9:775–780.

Wada, J. A., R. Clarke, and A. Hamm. 1975. Cerebral hemispheric asymmetry in humans. *Archives of Neurology* 32:239–246.

Walzer, S. 1985. X chromosome abnormalities and cognitive development: Implications for understanding normal human development. *Journal of Child Psychology and Psychiatry* 26:177–184.

Weinberger, D. R., D. J. Luchins, J. Morihisa, and R. J. Wyatt. 1982. Asymmetrical volumes of the right and left frontal and occipital regions of the human brain. *Annals of Neurology* 11:97–100.

Weinstein, R. E., and D. R. Pieper. 1988. Altered cerebral dominance in an atopic population. *Brain, Behavior and Immunity* 2:235–241.

Weis, S., H. Haug, B. Holoubek, and H. Orun. 1989. The cerebral dominances: Quantitative morphology of the human cerebral cortex. *International Journal of Neuroscience* 47:165–168.

White, K., and R. Ashton. 1976. Handedness assessment inventory. *Neuropsychologia* 14:261–264.

Williams, S. M. 1986. Factor analysis of the Edinburgh handedness inventory. *Cortex* 22:325–326.

Witelson, S. F., and W. Pallie. 1973. Left hemisphere specialization for language in the newborn: Neuroanatomical evidence of asymmetry. *Brain* 96:641–646.

Wofsy, D. 1984. Hormones, handedness, and autoimmunity. *Immunology Today* 5:169–170.

Contributors

Matteo Adinolfi
Paediatric Research Unit
Division of Medical and Molecular Genetics
United Medical and Dental Schools of Guy's and St. Thomas's Hospitals
London Bridge
London SE1 9RT

Frederic Assal
Institut d'Anatomie
University of Lausanne
9 rue du Bugnon
1005 Lausanne, Switzerland

Lluis Barraquer-Bordas
Llúria 102
Barcelona 9, Spain

Pere Berbel
Department of Histology
Institute of Neuroscience
University of Alicante
03690 Alicante, Spain

Verne S. Caviness, Jr.
Division of Child Neurology
Departments of Neurology and Radiology
Massachusetts General Hospital
Harvard Medical School
Boston, Massachusetts 02114

Damaso Crespo
Department of Anatomy and Cell Biology
University of Cantabria
Santander, Spain

J. C. DeFries
Institute for Behavioral Genetics
University of Colorado
Boulder, Colorado 80309

Pauline A. Filipek
Division of Child Neurology
Departments of Neurology and Radiology
Massachusetts General Hospital
Harvard Medical School
Boston, Massachusetts 02114

Barbara L. Finlay
Department of Psychology
Cornell University
Ithaca, New York 14853

Roslyn Holly Fitch
Center for Molecular and Behavioral Neuroscience
Rutgers University
Newark, New Jersey 07102

Albert M. Galaburda
Neurological Unit
Beth Israel Hospital
Harvard Medical School
330 Brookline Avenue
Boston, Massachusetts 02215

Jacquelyn J. Gillis
Institute for Behavioral Genetics
University of Colorado
Boulder, Colorado 80309

Kenneth Hugdahl
Department of Biological and Medical Psychology
Division of Somatic Psychology
University of Bergen
Årstadveien 21, N-5009
Bergen, Norway

Giorgio M. Innocenti
Institut d'Anatomie
University of Lausanne
9 rue du Bugnon
1005 Lausanne, Switzerland

Darcy B. Kelley
Department of Biological Sciences
Columbia University
New York, New York 10027

David N. Kennedy
Division of Child Neurology
Departments of Neurology and Radiology
Massachusetts General Hospital
Harvard Medical School
Boston, Massachusetts 02114

Margaret Livingstone
Department of Neurobiology
Harvard Medical School
220 Longwood Avenue
Boston, Massachusetts 02115

Miguel Marín-Padilla
Professor of Pathology and Maternal and Child Health
Dartmouth Medical School
Hanover, New Hampshire 03756

Brad Miller
Department of Psychology
Cornell University
Ithaca, New York 14853

Charles Njiokiktjien
Division of Pediatric Neurology
Free University Hospital
Amsterdam, The Netherlands

Glenn D. Rosen
Dyslexia Research Laboratory and Charles A. Dana Research Institute
Beth Israel Hospital
330 Brookline Avenue
Boston, Massachusetts 02215

Steven C. Schachter
Department of Neurology
Division of Clinical Neurophysiology
Beth Israel Hospital
330 Brookline Avenue
Boston, Massachusetts 02215

Gordon F. Sherman
Department of Neurology
Harvard Medical School
Beth Israel Hospital
330 Brookline Avenue
Boston, Massachusetts 02215

Paula Tallal
Center for Molecular and Behavioral Neuroscience
Rutgers University
Newark, New Jersey 07102

Sally J. Wadsworth
Institute for Behavioral Genetics
University of Colorado
Boulder, Colorado 80309

Index